GREAT EXPERIMENTS IN PHYSICS

GREAT EXPERIMENTS IN PHYSICS

Firsthand Accounts from Galileo to Einstein

Edited by

MORRIS H. SHAMOS

Professor of Physics (Emeritus)
New York University

Dover Publications, Inc., New York

Published in Canada by General Publishing Company, Ltd., 30 Lesmill Road, Don Mills, Toronto, Ontario.
Published in the United Kingdom by Constable and Company, Ltd., 10 Orange Street, London WC2H 7EG.

This Dover edition, first published in 1987, is an unabridged and slightly corrected republication of the work first published by Holt, Rinehart and Winston, New York, in 1959.

Manufactured in the United States of America
Dover Publications, Inc., 31 East 2nd Street, Mineola, N.Y. 11501

Library of Congress Cataloging-in-Publication Data

Great experiments in physics.

Reprint. Originally published: New York : Holt, Rinehart and Winston, 1959.
Includes index.
1. Physics—History.   I. Shamos, Morris H. (Morris Herbert), 1917–
QC7.G74   1987   530′.09   86-30942
ISBN 0-486-25346-5 (pbk.)

Preface

One of the most refreshing aspects of current thought in science education is the realization that to appreciate fully our modern views of the physical world the student should explore the historical growth of ideas from which these views were fashioned. Unfortunately, the trend in recent decades has been quite the opposite. The conventional introductory course in physics has become highly specialized; new developments do not simply supplant the old but add to the already huge body of knowledge, forcing aside the more humanistic elements of the science. When James Conant, in the first of his 1946 Terry Lectures, suggested that science is "accumulated knowledge," he had in mind the great progress made in this area during the last few centuries. But the same term may be used to characterize the *encyclopedic* nature of most present-day introductory courses in physics, or for that matter, in almost any of the laboratory sciences.

Viewed in retrospect, the most significant ideas in physics stand out in simple elegance against a background shadowed by confusion, clouded at times by bigotry, yet always illumined by man's search for truth. Because so many of our key conceptual schemes seem almost "self-evident" in the light of experience, we frequently lose sight of the fact that there is a history to physics; that the ideas we now take for granted represent major intellectual achievements in the endless effort to understand nature, and that the origin of these ideas forms an important segment of man's cultural development. It is encouraging to note, therefore, a growing interest in the history of science both as a scholarly activity and as a pedagogic tool.

This book has a twofold purpose. It was developed mainly for use by liberal arts students in a new laboratory physics course at Washington Square College designed on the "great experiments" idea. The course traces the development of physical principles by a detailed study of a number of the most telling experiments in physics. There was the hope, moreover, that as supplementary reading material the book might help fill a gap in the more conventional introductory courses in physics. Other students of science, or of the history of science, and the interested layman may also find it profitable to examine the original accounts of some of these great experiments.

The experiments were selected on the basis of two criteria: first, that they be among those generally recognized as the most important in the evolution of physics, and second, that the concepts be meaningful to beginning physics students and lend themselves to laboratory experience at this level. Most of the experiments, in fact, are found in some form in the usual introductory course. A few conceptual schemes that satisfy only the first requirement, but which are essential for an understanding of contemporary ideas, are included in an appendix.

Except for the use of modern spellings the experiments are presented either in their original, or in translation. Where possible, the original illustrations have been retained; for greater clarity others had to be retouched or redrawn. Each chapter contains a biographical sketch of the individual responsible for the particular discovery, as well as occasional glimpses at the political and cultural atmosphere of the period in which he worked. This material, together with frequent marginal editorial notation, will make it easier, I believe, for the reader to follow the original accounts and to view them in historical perspective.

I wish to acknowledge my gratitude to the many individuals who had some part in the preparation of this work. I am especially grateful to Dean J. W. Buchta of the University of Minnesota for his critical reading of the entire manuscript, and to Professor Henry H. B. Noss of the History Department of New York University for his constructive comments on the first chapter. To my own colleagues in physics I am indebted for generous advice and suggestions, much of which found its way into the manuscript. I owe special thanks to Professor Edgar N. Grisewood, who tested portions of the manuscript in the classroom and was able thereby to offer helpful suggestions based upon student reaction, and to Miss Eleanor Karasak, who participated with Professor Grisewood and me in the initial planning of the course which prompted this book.

A book of this kind obviously could not be written without the cooperation of many publishing companies, whose permission to reproduce the original writings, either directly or in translation, is hereby gratefully acknowledged. My debt to the published literature is indicated only partly by the references and suggested supplementary reading; I wish particularly to single out W. F. Magie's, *A Source Book in Physics*, which proved invaluable to me both as inspiration and as an unlimited source of information. The task of preparing the manuscript was made easier by the tireless assistance of Miss Lillian Pollack, who was responsible for most of the typing, and Mrs. Freda Cahn, who helped with the proofreading. Finally, I am very grateful to two young graduate assistants, Ernesto Barreto and John F. Koch, for their help with several of the translations.

M. H. S.

New York, N. Y.
April 1959

Contents

CHAPTER	PAGE
1. Introduction	1
2. Accelerated Motion—GALILEO GALILEI	13
3. Boyle's Law: Pressure-volume Relations in a Gas—ROBERT BOYLE	36
4. The Laws of Motion—ISAAC NEWTON	42
5. The Laws of Electric and Magnetic Force—CHARLES COULOMB	59
6. The Law of Gravitation—HENRY CAVENDISH	75
7. The Interference of Light—THOMAS YOUNG	93
8. The Diffraction of Light—AUGUSTIN FRESNEL	108
9. Electromagnetism—HANS CHRISTIAN OERSTED	121
10. Electromagnetic Induction and Laws of Electrolysis—MICHAEL FARADAY	128
11. Lenz's Law—HEINRICH LENZ	159
12. The Mechanical Equivalent of Heat—JAMES JOULE	166
13. Electromagnetic Waves—HEINRICH HERTZ	184
14. X-Rays—WILHELM K. ROENTGEN	198
15. Natural Radioactivity—HENRI BECQUEREL	210
16. The Electron—J. J. THOMSON	216
17. The Photoelectric Effect—ALBERT EINSTEIN	232
18. The Elementary Electric Charge—ROBERT A. MILLIKAN	238
19. Induced Transmutation—ERNEST RUTHERFORD	250

20. The Neutron—JAMES CHADWICK   266

Appendix   281

    1. The Electromagnetic Field—JAMES CLERK
       MAXWELL   283

    2. The Quantum Hypothesis—MAX PLANCK   301

    3. The Theory of Relativity—ALBERT EINSTEIN   315

    4. The Hydrogen Atom—NIELS BOHR   329

    5. The Compton Effect—ARTHUR COMPTON   348

Index   359

GREAT
EXPERIMENTS
IN PHYSICS

Introduction

The Origin of Modern Science

IT IS always inviting, particularly in science, to assign definite causes to each new phenomenon. The temptation is strong, therefore, when seeking the origins of modern science, to set the specific time when it may be said to have started and to determine the sequence of events responsible for it. There was no abrupt starting point, however, and at best we can perceive but a few of the many factors that contributed to the accumulation of this form of man's intellectual wealth. The history of science cannot be divorced from the political and cultural history of civilization if we seek to account for its development. The arts and the sciences, cultural activities both, tend to flourish in similar social and political environments. The same subtle forces that shape the general cultural atmosphere of a period provide impetus as well to its scientific advancement. This does not mean that periods of vigorous intellectual activity necessarily saw major scientific accomplishment. Far from it! It means only that whenever man felt the urge to engage in cultural pursuits, he contributed—not always constructively, it is true—to the development of science.

Modern science, which is characterized by rational thought and by methods that have led successfully to an understanding of natural phenomena, is relatively recent, having its origin in the seventeenth century. But its roots may be traced, by a sometimes tortuous path, to ancient Greek culture. It is important to distinguish between the orderly structure we know as science and the empirical, technological development resulting from man's efforts to control his environment. The latter, sometimes called "practical" or "applied" science and involving for the most part trial-and-error methods, dates back virtually to the dawn of civilization. However much these discoveries—for example, in metallurgy, ceramics, irrigation, and the mechanical arts—may have contributed to the growth of civilization, the methods—and motives—that led to them must not be confused with the *planned experimentation* by which we seek to confirm our modern views of nature.

It would be a mistake to conclude that only in the last three hundred years has man been concerned with the search for truth. Earlier civilizations were no less interested in knowledge nor less curious about the nature of things. The essential difference lies in the fact that the earlier methods by which such knowledge was sought were not adequate to reveal the truth about the physical universe.

The Golden Age of Greece, the fifth and fourth centuries B.C., saw the beginnings of a system of natural philosophy that was to dominate man's scientific thinking for centuries to come. This was the period of the famous teacher-pupil sequence: Socrates, Plato, and Aristotle, probably three of the most remarkable individuals in the history of thought. All believed in the existence of universal or absolute truths, which man could discover if he would simply pursue the proper methods. They looked upon knowledge not as a means to utilitarian ends but as a means to satisfy human curiosity. And they held that *true knowledge,* as distinguished from knowledge derived via the senses, could be deduced by purely formal methods, i.e., by a system of logic. The use of some forms of logic apparently dates back to the pre-Socratic period of the sixth and fifth centuries B.C., when philosophers first showed concern over the internal consistency of their arguments. But it was Socrates and Plato who put deductive logic to such skillful use and Aristotle who invented the logical device known as the *syllogism.*

The Socratic method of reasoning found ready acceptance among the knowledge-loving Greeks, who became masters of deduction—and the victims of its weaknesses. Deduction is the process by which one proceeds to the solution of particular problems through the application of general principles. It involves drawing *logical* or *valid* conclusions from given premises, by assuming the ultimate truth of these premises. If we assert, for example, that *x* is greater than *y,* and *y* greater than *z,* we conclude that *x* is greater than *z.* The conclusion is a necessary consequence of the premises, and while the deduction is clearly correct the *truth* of the conclusion rests upon the reliability of the initial statements. It should be evident, then, that deductive reasoning does not necessarily lead to new knowledge, for it provides no recipe for testing the truth of the basic assertions.

The syllogism represents a particular form of deductive argument, having a structure frequently found in ordinary discourse. It consists of a general premise, assumed to be true, followed by a statement which makes use of the general premise in a specific case, and finally a logical conclusion. As an example, we might use the syllogism to derive a typical Aristotelianlike explanation for the acceleration of falling bodies:

The traveler hastens when approaching his destination.
A falling object may be likened to a traveler. . . .
Hence, falling objects accelerate as they approach the earth.

The argument is obviously defective in its premises. Even granting the (questionable) validity of the first premise, the analogy drawn in the second has no basis in fact. Yet the conclusion follows logically from these statements and illustrates how the followers of Aristotle employed his deductive system to "account" for natural phenomena. Such reasoning from false premises to a *correct* conclusion is unfortunately all too common, even today, in the area of explanation. The conclusion is correct, of course, because it stems from observation and the premises designed accordingly, not because it follows from true statements.

The example given represents but one of many variations of the syllogistic form of argument. It is actually more an example of *dialectic* than of *scientific* argument. In the final analysis the latter reasons from true premises while the former only from "probable," or even "plausible," statements. The formal logical methods by which such reasoning may be examined for self-consistency were already developed to a high degree by the start of the Hellenistic period (with the death in 323 B.C. of Alexander the Great, who had been tutored by Aristotle), during which they formed the foundation for the remarkable mathematical proofs of Euclid (c. 323–c. 285 B.C.) and Archimedes (287?–212 B.C.). However appropriate the deductive method may be in many branches of mathematics, it cannot serve alone as the means for understanding nature. The essence of physics, indeed of any natural science, is to account for nature in the simplest possible terms; that is, to reduce all that we observe to basic principles, or causes. This is what we mean by an explanation in science and is the way we discover new scientific knowledge. It is this economy of thought and of expression that characterizes explanation in modern physical science. But how do we find the basic causes of things? How do we establish the *truths* of our initial premises when arguing deductively? It is here that the Aristotelian procedure fails and we must seek other methods.

Aristotle differed from Plato and Socrates in part by his marked interest in natural phenomena and his higher regard for practical matters. He was not adverse to experimentation, although he could hardly be considered a thorough experimenter. Sometimes called the encyclopedist of ancient science because of his careful systematic observations in descriptive natural history, on which rest his chief qualifications as a "scientist," Aristotle nevertheless held the most naive and confused views regarding the nature of the physical world. Much as he contributed to the development of biology, it is because of his inferior physical reasoning that one finds so regrettable his great influence over succeeding centuries of scientific thought. There can be little doubt that it was largely his authority that served to delay so long the full evolution of such areas as dynamics, atomism, and astronomy. Some two thousand years later we are to find the founders of modern physics, such as Gilbert, Galileo, Boyle, and Newton, having to reject the prevailing doctrines of Aristotle before setting science on a firm foundation.

There are a number of probable causes for Aristotle's unsound doctrines in physical science. First of all, he failed to make effective use of the process of *induction*. This is the inverse of the deductive method and consists essentially in proceeding from the particular to the general; that is, in reasoning from a limited number of observations to a general conclusion that embraces all similar events. It is by this method that we establish the initial truths, from which we can then proceed, by deduction, to account for our particular experiences, and thus to test the reliability of the induction. Aristotle was familiar with the inductive method; in fact, he was the first to outline its principles. Perhaps he distrusted it as a means of gaining knowledge, or perhaps he failed to recognize its place in scientific reasoning. At any rate he did not make use of this highly significant form of argument. Instead, the generalizations that were to form the initial premises in his deductive arguments were derived either by pure intuition or by appealing to the ends or purposes (generally human) they served, that is, by *teleological* argument. This is not to say that intuition has no place in inductive reasoning; but it is intuition born of experience rather than the introspective inventions of Aristotle that distinguishes the modern use of induction.

All of our physical theories are generalizations from experience. When Newton declared that "Every body perseveres in its state of rest, or of uniform motion in a straight line . . . , " he clearly had proceeded by induction from a limited number of observations to a sweeping generalization that applies to the entire universe. How valid is such a procedure? By its nature it cannot yield absolute certainty. Its reliability, then, must be measured by its success in meeting the challenge of continued scientific testing. Each new observation that agrees with the conclusions implied by a hypothesis reached through induction adds both to the strength of the hypothesis and to the power of the inductive method generally.

Aristotle's physical arguments were largely nonmathematical; they were qualitative rather than quantitative and lacked the abstraction that is the power of modern physics. Nor did he resort to critical experiment to test the conclusions deducible from his premises. Leonardo da Vinci (1452–1519), the remarkable Florentine painter who was among the first to hammer at the crumbling walls of medieval science, held that *true science* began with observation; if mathematical reasoning were applied, greater certainty might be reached, but "those sciences are vain and full of errors which are not born from experiment, the mother of all certainty, and which do not end with one clear experiment."[1]

Consider, for example, Aristotle's views on the nature of matter, about which there had been considerable speculation even in pre-Socratic Greece. Is matter continuous and divisible indefinitely, or is it made up of basic units beyond which it cannot be further divided? This question had puzzled curious men for ages and was not finally resolved until the last century. An atomistic view of nature, which postulated a huge number of invisible particles in a sea of empty space, was invented by Leucippos about the middle of the fifth century B.C. and developed more fully some thirty years later by Democritus. "According to convention," said Democritus, "there is a sweet and a bitter, a hot and a cold, and there is color. In truth, there are atoms and there is a void." Certainly this was an oversimplification of nature, and a scheme based purely on philosophical speculation that could not be subject to test—but one which, in form, bears some resemblance to the modern atomic viewpoint.

At the other extreme were the continuists, among them Anaxagoras, the famous Athenian teacher of the fifth century B.C., and later, Aristotle. According to this school of thought all matter was composed of the same primordial stuff called *hyle,* and the differences among substances resulted from the presence in them of various amounts of certain properties given the hyle by four basic *elements:* fire, earth, air, and water. It was Aristotle's notion that these elements tend to arrange themselves in concentric fashion about the center of the world, with the earth at the center surrounded by successive shells of water, air, and fire. Coupled with this was the doctrine of *natural places:* Everything was assumed to have its "proper place" in nature, that of heavy things was below and that of light things, above. This "accounted" for the descent of stones, for example, and the fact that air and fire tend to rise.

Space will not permit a thorough discussion of Aristotle's scientific doctrines. His astronomical speculations and his views on motion were no less obscure than his conception of matter. It is perhaps inevitable that the science practiced by Aristotle should invite little more than ridicule from students of modern physics. One should bear in mind, however, that the Greeks generally did not have the experimental tradition in science that we now consider so essential, and that Aristotle's views would appeal to many of his contemporaries simply because of its speculative and seemingly ra-

---

[1] See Sir William Dampier, *A History of Science,* 4th ed. (New York: Cambridge University, 1949), p. 105. Also published by The Macmillan Company.

tional nature, and their agreement with what appeared to be "common sense."

It should not be concluded that all Greek physics was as barren as Aristotle's; on the contrary, there were occasional flashes of brilliance throughout ancient Greece. We have mentioned the mathematical accomplishments of Euclid and Archimedes. The latter solved physical problems in a completely modern fashion; he made full use of the mathematical tools available to him, and he exhibited a remarkable degree of abstraction for his period. His treatment of hydrostatic problems *(Archimedes' principle)* stands out as the most significant accomplishment in physics prior to the scientific revolution. In applied science Hero of Alexandria, who lived in the last century B.C., was responsible for an impressive array of practical inventions. By comparison with the Middle Ages, Greek science was truly prodigious.

Science in the Middle Ages

We have said nothing of Latin physics during the Hellenistic period for the reason that, while her engineering achievements were substantial, Rome had virtually no independent science. Instead, under the influence of Greek civilization the Romans became students of Greek science and were content to follow its dictates—even while constructing great systems of aqueducts, sewers, roads, harbors, and public buildings. They were too much concerned with practical problems to make original contributions in science. Of particular significance to scientists of the early modern period (sixteenth and seventeenth centuries) was a didactic piece written by the Roman poet Lucretius in 57 B.C. Entitled *Of the Nature of Things (De Rerum Natura),*[2] it popularized the atomistic view of matter in the most eloquent and lyrical fashion. While not a work of science, it was to have considerable impact centuries later upon scientists seeking to overthrow the prevailing Aristotelian doctrine of the four elements.

Throughout the period of the Roman Empire[3] and the Middle Ages that followed, physical science virtually stood still. Greek science was lost from view in the first few centuries of the Christian era; it was replaced largely by superstition and mysticism. Rational thought gave way to divine revelation as the test for truth, and the authority of the Scriptures dominated all philosophy. Interest in natural phenomena was replaced by an ethical and moral philosophy that reflected man's primary concern with religion. In such an atmosphere intellectual activity could hardly exist, let alone prosper.

---

[2] The translation by W. E. Leonard (New York: Dutton, 1957), is available in a paperback edition.

[3] Roughly to the end of the A. D. fifth century.

An important pseudoscientific product of the early Christian era was the practice of *alchemy,* which included the art of transmuting base metals into precious metals such as gold and silver. It was Aristotle's view of matter, his use of the "four elements," that provided the theoretical basis. If substances differed only in their relative proportions of the four elements, it followed that by appropriate changes in these amounts matter could be altered at will. Alchemy had its origin in Alexandria; it was practiced by the Saracens in Arabia during the early Middle Ages, and introduced into western Europe in the twelfth century, where it flourished until the latter part of the eighteenth.[4] The *philosopher's stone*[5] was never found, of course, and while a number of new substances were discovered and new chemical techniques developed, the art was clouded, on the whole, with mystic ritual and fraudulent practices. It must not be concluded that alchemy was entirely specious and that all alchemists were charlatans. On the contrary, true alchemy was a very broad system of natural philosophy, by which many genuine scholars sought an understanding not only of the physical world but of life itself. Their methods may seem strange by modern scientific standards, but one cannot doubt their sincerity. Unfortunately, certain aspects of their work, such as transmutation, and the secrecy with which they surrounded their techniques invited the participation of impostors and quacks who brought alchemy into disrepute.

Efforts to reconcile reason with theology led, in the late Middle Ages,[6] to the system of thought known as *Scholasticism.* This was an authoritarian, man-centered philosophy aimed primarily at discovering how to achieve salvation in the hereafter, and concerns us here chiefly because of its effect upon scientific thought. The Scholastics were given to the kind of explanation made popular centuries earlier by Socrates and Plato. We have seen that this consisted in discovering the ends or purposes which things served—their ultimate *good.* Everything was regarded as having been intended to serve some human need; indeed, the universe itself had been created for the benefit of man! There was little interest in observation and a distrust of man's senses as guides to the ultimate nature of things. Aristotle's writings were "rediscovered," having been translated to Arabic from the Greek, in which they had been preserved by Byzantine scholars, and then to Latin. Aristotelian philosophy suited the beliefs of the Church fathers; his work was established as authoritative in all matters not involving religious dogma and, as a result, opposition to his scientific doctrines ran the risk of censure

---

[4] It is interesting to note that even so distinguished a *modern* scientist as Newton practiced alchemy during most of his career.

[5] The transmuting agent, supposedly capable of transforming one substance into another.

[6] The twelfth and thirteenth centuries.

by the Church. It is against this sort of background that we find all the more impressive the remarkable accomplishments of late Renaissance science.

The Scientific Revolution

A false scientific hypothesis is best refuted by showing that the "facts" deduced from it are not true; that is, by scientific test or experiment. Thus, the obvious failure of alchemy contributed to the downfall of the four-element theory. Aristotle's views on motion could not be supported by experiment and were to fall before critical attacks by Galileo. In few areas of human activity is it possible to devise definitive tests for truth, but this is a characteristic feature of the physical sciences. By the beginning of the fourteenth century Scholastic philosophy was the subject of considerable dispute. Revived interest in classical learning, growing scepticism that *revealed truth* was the sole key to knowledge, greater concern with natural phenomena and human activities, were some of the many factors—social, political, and cultural—that led man from the semidarkness of the Middle Ages to the brighter intellectual atmosphere of the *Renaissance.* It was not a sudden transition, nor was it the result of a change in man's mentality. It was rather a change in *outlook,* a departure from classical traditions, and, fortunately for physics, a new approach to the question of what constituted a satisfactory explanation in science. A few men began to lean more heavily on experimentation than on intuition. They tended toward the method of explanation advocated centuries earlier by Democritus and other atomists: explanation in terms of the primary causes of things rather than in terms of the final causes or "goods" of Scholastic philosophers. While men generally were more concerned with books than with the study of nature, the few independent minds among them exerted a subtle influence. In the face of mounting criticism the existing scientific structure gradually weakened and crumbled. As pointed out by Francis Bacon[7] early in the seventeenth century:

> The subtlety of nature greatly exceeds that of sense and understanding; so that those fine meditations, speculations and fabrications of mankind are unsound, but there is no one to stand by and point it out. And just as the sciences we now have are useless for making discoveries of practical use, so the present logic is useless for the discovery of the sciences.

---

[7] Sir Francis Bacon (1561–1626), probably the greatest of English Renaissance philosophers, yet a statesman of questionable integrity. The quotation is from *Novum Organum,* Book I, pp. x-xi.

By this time the experimental approach, coupled with the use of mathematical abstraction, produced such impressive results that the entire transitional period of the late Renaissance has come to be known as the *scientific revolution*. It is in the sense of this *modern* approach that we set the beginning of modern physics in the seventeenth century. There were earlier significant developments, of course, such as the extraordinary grasp of mechanical principles exhibited by da Vinci early in the sixteenth century and the heliocentric theory published by Copernicus (1473–1543) in the year of his death. But these independent spirits, while contributing greatly to the development of scientific thought along correct lines, added little to the formal structure of physics, and it remained for Galileo, in the seventeenth century, to establish the first definitive concepts regarding the physics of motion. The notable achievements of Galileo stirred scientific activity throughout Europe; with every accomplishment the *new science* gathered momentum and surged irresistibly onward to its present highly developed state.

We should mention two important factors that sparked the growth of physics in the seventeenth century: the formation of scientific academies and the development of scientific instruments. Freedom of inquiry was not yet the hallmark of universities, largely because of their domination by the Church. In this void the scientific academies provided the needed encouragement and support of research that we now expect to find in institutions of higher learning.

The first organization that could be considered a scientific academy was the *Accademia dei Lincei,*[8] founded in Rome in 1603 through the patronage of Duke Federigo Cesi. This society, which elected Galileo its sixth member in 1611, held frequent meetings at which members discussed the results of their individual investigations. With the death of its patron in 1630 the Academy dispersed; evidently Cesi had provided the necessary organizational incentives as well as the financial support.

Some two decades later the *Accademia del Cimento*[9] was founded in Florence by two Medici brothers, Grand Dukes Ferdinand II and Leopold, and was supplied with a laboratory containing the finest instruments then available throughout Europe. It differed from the Lincean Academy, and from most modern societies, by encouraging its member to work jointly rather than individually on the most important problems of the time. In this respect it was more like a modern scientific institute, where groups of scientists frequently work in common on major problems. The academy

---

[8] Academy of the Lynx-eyed. The lynx was intended to symbolize the sharp eyes of science.
[9] Academy of Experiment. While organized formally in 1657, its members had met informally since 1651.

lasted but a short time—it was disbanded in 1667—but in this period it sponsored numerous investigations in physics, including observations on the pressure of the atmosphere and thermal properties of liquids and solids, and a measurement of the velocity of sound. In its final year the members published a joint account of their experiments and discoveries,[10] which had considerable impact on contemporary thought because of its purely scientific character and the diverse nature of the investigations covered by it.

In London, meanwhile, a group of men, meeting informally about once a week (beginning about 1645) to discuss natural philosophy, formed the nucleus of the Royal Society,[11] which was chartered in 1662. They established the custom of assigning specific problems or investigations to one or a group of colleagues, who were then required to report to the Society on their findings. In 1665 the Society began publication of its *Philosophical Transactions,* the first journal to contain original communications, including papers read before the Society, and one which the Society continues to publish to this day.

About the same time and in much the same way the French *Académie des Sciences* developed from informal gatherings in Paris of a group of interested philosophers and mathematicians. When it was established in 1666 as a regular Academy by Louis XIV, he provided its members with pensions and financial support for their researches, which they conducted jointly in the manner of the *Accademia del Cimento*. The Academy sponsored investigations on a variety of subjects covered by its two classes: mathematics (including astronomy and physics) and natural philosophy (including chemistry, medicine, anatomy, and so on). After its reorganization in 1699, it began publication of a regular series of original papers in its *Mémoires,* and following the Revolution, during which the old French academies were abolished, it was refounded in 1795 to form the basis for *L'Institut de France*.

As one after another scientific academy appeared on the scene the pace of physical inquiry quickened noticeably. While the available means of communication left much to be desired, men learned in the course of time what their colleagues were doing and could build on the discoveries of one another. The exchange of knowledge, facilitated by the publication of scientific journals, became—and remains—one of the most significant factors in the growth of physical science.

With increased sophistication in science the seventeenth-century philosopher needed new instruments with which to conduct his experiments. It is

---

[10] *Saggi di naturali esperienze fatte nell' Accademia del Cimento,* Florence, 1667.
[11] Chartered as the *Royal Society of London for the Promotion of Natural Knowledge.*

difficult to conceive of modern physics without precise instruments. Indeed, one of the chief characteristics of modern science is its reliance on measurement and the use of scientific instruments. They make it possible to study phenomena under controlled conditions, thereby justifying *reliable* conclusions about the nature of things. In rapid succession, during the seventeenth century, six important instruments were invented: the microscope, telescope, thermometer, barometer, air pump, and pendulum clock. It takes little imagination to realize the great effect these instruments, crude as they were by present standards, had on seventeenth-century physics.

These were the beginnings of physics. As men learned how to ask reasonable questions of nature, its secrets unfolded before them. Experiment followed experiment, each adding new evidence on the nature of the physical universe. But facts alone do not constitute a science. One must be able to link together the observed data; otherwise they would be simply a collection of meaningless details. One therefore searches for regularity in nature and, when found, expresses it in the form of broad unifying principles. The essence of physics is the success with which these basic ideas and abstract principles serve to coordinate the observed facts.

In the three centuries since Galileo discovered the first principles of dynamics a handful of experiments stand out sharply for the clarity of concept and great depth of understanding which they provided. In the following chapters we shall examine some of these *great experiments* and look briefly at the remarkable individuals responsible for them.

SUPPLEMENTARY READING

Burns, E. M., *Western Civilization, Their History and Their Cultures* (New York: Norton, 1949).

Butterfield, H., *The Origins of Modern Science* (New York: Macmillan, 1952).

Clagett, M., *Greek Science in Antiquity* (New York: Abelard-Schuman, 1955).

Cohen, M. and I. E. Drabkin, *A Source Book in Greek Science* (New York: McGraw-Hill, 1948).

Dampier, W. C., *A History of Science,* 4th ed. (New York: Cambridge University, 1949).

Hall, A. R., *The Scientific Revolution* (Boston: Beacon, 1954).

Ornstein, M., *The Role of the Scientific Societies in the Seventeenth Century* (Chicago: University Press, 1928).

Sarton, G., *A History of Science* (Cambridge, Mass.: Harvard University, 1952).

Taylor, A. E., *Aristotle* (New York: Dover, 1955).

Taylor, F. S., *The Alchemists* (New York: Henry Schuman, 1949).

Wiener, P. P., and A. Noland, *Roots of Scientific Thought* (New York: Basic Books, 1957).

Wightman, W. P. D., *The Growth of Scientific Ideas* (Edinburgh: Oliver and Boyd, 1951)

Wolf, A., *A History of Science, Technology, and Philosophy in the 16th and 17th Centuries* (London: Allen and Unwin, 1935), particularly Chs. 1, 4, and 5.

2

Galileo Galilei
1564-1642

Accelerated Motion

RICHLY ENDOWED with the memorable poetry of Dante and Petrarch, the satirical comedy of Machiavelli, and the incomparable art of da Vinci, Raphael, and Michelangelo, Renaissance Italy set the stage for the evolution of modern physics.[1] Prior to the seventeenth century, as we have seen, there was very little practice of physics in the modern sense.

Galileo was the first modern physicist. The brilliant work of Archimedes some 2000 years earlier involved a different level of abstraction and sophistication than we generally associate with the modern period. While in his methods of proof he must be regarded a modern scientist, his physical insight did not extend beyond objects at rest, whereas a clear understanding of motion was the essential passkey to the development of physics. No doubt Galileo profited greatly from Archimedes' works,[2] for he studied them avidly and was inspired by them to the combination of experiment and mathematical demonstration which he applied so successfully.

Galileo Galilei was born at Pisa on February 15, 1564, the year of Shakespeare's birth and the very day on which Michelangelo died. His father, Vincenzio Galilei, himself an accomplished scholar,[3] instilled in the son his own love for learning and intense distaste for reliance on classical authority. That Galileo adopted his father's intellectual attitude is evident

---

[1] The term is used here in the sense of the *modern period*, starting with the seventeenth century, rather than with reference to the most recent developments of the twentieth century.

[2] Some of Archimedes' works had been translated into Latin by the mathematician, Tartaglia (1500–1557).

[3] In the field of musical theory, in which he published his *Dialogue on Ancient and Modern Music* (1581).

from his remarkable career, characterized throughout by his independent spirit.

Almost a century earlier the universal talents of Leonardo da Vinci had illuminated the Italian scene. Truly modern in his approach to physical problems, he foreshadowed the principle of inertia long before Galileo demonstrated it experimentally and thereby laid the groundwork for the science of dynamics. It is difficult to assess the influence that da Vinci's scientific work may have had upon Galileo, for the famous painter never published his notes,[4] perhaps wisely in view of the political atmosphere. We may assume, however, that through friends and admirers some of his ideas were preserved and eventually brought to Galileo's attention.

When he was seventeen, following several years of classical training in a Benedictine abbey,[*] Galileo enrolled at the University of Pisa. It was his father's wish that he study medicine, but young Galileo soon tired of the doctrinaire texts of Aristotle and Galen, and became interested instead in the study of mathematics and physics. He showed little respect for books as authoritative sources of knowledge and quickly became known among his professors for his bold attacks on the Aristotelian views then widely held. He left without taking a degree, but with a reputation for extraordinary mathematical talents. With the aid of influential friends, this led to his appointment, at the age of twenty-five, to the poorly paid chair of mathematics at the university. There he spent the next three years on his first investigations of accelerated motion. It was here that he experimented with balls rolling down inclines; since he could not make measurements on freely falling objects with his water clock, he took this means of "slowing" the motion sufficiently to permit observation. During this period he arrived at a clear understanding of acceleration, as well as of the concept of inertia. In fact, it was at Pisa that Galileo constructed the framework for the physics of motion (dynamics). Tradition has it that he disproved Aristotle's doctrine of motion—that bodies fall with speeds proportional to their weights— by dropping two different objects simultaneously from the Leaning Tower of Pisa. However inviting to the imagination the account may seem, the evidence for it is not very strong. Galileo made no mention of it in his well-kept notes. Instead, the event was related briefly by one of his students, Viviani, some sixty years later. Such an experiment, in fact, had been performed by the Dutch physicist Simon Stevin (1548–1620), or Stevinus (the Latinized form of his name).

Galileo's predilection for argumentation made him unpopular with his

---

[4] The notes were first edited and published in 1906: the most recent edition is Edward MacCurdy (ed.), *The Notebooks of Leonardo da Vinci* (New York: Braziller, 1954).

[*] The Monastery of Santa Maria of Vallombrosa.

tradition-bound colleagues, and it was apparently with mutual feelings of relief that he left Pisa in 1592 for the chair of mathematics at the University of Padua. There he remained for eighteen years, the happiest period of his life, as he later recalled, and one of the most productive, for he completed the bulk of his work in mechanics during this period, although it was not published until much later.

It was at Padua that Galileo had his first experience with polemic composition, a style which evidently suited his temperament, for he used it effectively throughout his career. He had devised an improved form of calculating instrument, known as the *sector,* which he manufactured for sale throughout Europe, and he prepared an instruction manual, in Italian, for users of the instrument. Baldassar Capra, a student at Padua, preparing a similar work in Latin for the use of scholars outside Italy, essentially translated Galileo's book but claimed original authorship, thereby implying that Galileo had stolen the invention from him. Considerably annoyed, Galileo pressed charges against the student before the university administration and succeeded in having the book suppressed. Nevertheless, on the chance that some copies had escaped confiscation, Galileo attempted to balance the record by publishing a full account of the circumstances surrounding the affair.[5]

While his major contributions to physics were in mechanics, Galileo is perhaps best known for his accomplishments—and misfortunes—in the field of astronomy. He was a firm supporter of the Copernican theory, favoring it publicly following the appearance of a brilliant new star (a supernova) in 1604, which jarred the Aristotelian doctrine of the immutability of the heavens. In 1609 Johannes Kepler (1571–1630) published his famous work *(Astronomia Nova)* on astronomy, and Galileo learned of the discovery, in Holland, of the telescope. Kepler's work had little direct effect upon Galileo,[6] but he was greatly excited by the telescope and proceeded at once to make astronomical observations. The following year he published his findings in a pamphlet entitled *The Messenger of the Stars (Sidereus Nuncius).* Written in the language of scholars, the more quickly to reach the philosophers of Europe, it precipitated him headlong into a bitter feud with the followers of Aristotle. This was aggravated two years later by his *Discourse on Floating Bodies,* a treatise on hydrostatics (written in Italian) that attacked Aristotle's physical principles with a vengence.

Meanwhile, Galileo had returned to Florence as chief mathematician

---

[5] *Defense against the Calumnies and Impostures of Baldassar Capra* (1607).

[6] It remained for Newton to find in Kepler's work on planetary motion confirmation of his law of universal gravitation.

and philosopher[7] to the Grand Duke of Tuscany, and chief mathematician of the University of Pisa without teaching duties. Despite the fact that he had been awarded life tenure at Padua, and the political climate was considered healthier for him in the Venetian Republic,[8] Galileo longed to return to the province of his ancestors. In Florence he became the target of attacks by various theologians for his "anticlerical" views, to which he replied without compromise in typical spirited fashion, his arguments fairly bristling with logic. Ultimately, he was called to Rome to account for his views to the Inquisition, and in 1616 he was warned not to "teach, hold, or defend" the Copernican theory.

Galileo returned to Florence, disillusioned, and devoted himself to his work, completing his masterful *Dialogue concerning the two chief Systems of the World, the Ptolemaic and the Copernican*[9] in 1630. The book, published in 1632, was a thinly veiled attack on the Aristotelian view of the heavens and clearly supported the Copernican theory. His style of writing was the *dialogue,* to which he may have been partial because of his great admiration for Plato's literary accomplishments. Or a desire to be cautious, in view of the stormy political climate, may have influenced Galileo's use of the indirect style of the dialogue. In any event, of the three characters in the *Dialogue,* two represented Galileo's views by supporting the Copernican theory while the third, Simplicio,[10] was the awkward defender of Aristotelian tradition. Galileo used the same three interlocutors in his later work on mechanics. Needless to say, the book was not entirely impartial, Simplicio being made to "look the fool" in a number of arguments. As was his custom, Galileo wrote the *Dialogue* for the layman; he wrote in Italian in a style noted for its readability and literary elegance. He demolished opposing views with a mixture of biting sarcasm and convincing logic. It was a style quickly outmoded by the more formal scientific writing to which we are accustomed, wherein the discussion is confined to reports of experiences and inferences drawn therefrom.

With the publication of the *Dialogue,* which had been passed by the official censor, the full wrath of the Holy Office descended upon Galileo. He was accused of ridiculing the pope by placing the defense of his Aristotelian views into the mouth of the clumsy Simplicio and of violating the earlier injunction against supporting the Copernican theory. He was examined by the Inquisition, forced to recant his views, and sentenced to deten-

---

[7] The latter part of the title was specifically requested by Galileo, and he employed it to good advantage.

[8] The Venetians used their northern Protestant neighbors as a lever to preserve a measure of independence from the Vatican.

[9] English translation by T. Salusbury (1661).

[10] Literally, simpleton.

tion for an indefinite period. His *Dialogue* was placed on the Index of prohibited books, where it remained, together with other Copernican works, until 1822.

After several months of semidetention[11] Galileo was permitted to return to his villa at Arcetri, near Florence, to live out the remainder of his life in seclusion. Here he devoted himself to his final great work, *Dialogues concerning New Sciences (Dialoghi delle Nuove Scienze)*. The manuscript was completed in 1636, but because the Inquisition had banned Galileo's works, it was smuggled to Holland, where it was published by the Elzevirs in 1638. In that same year Galileo was visited by the famous poet, John Milton, who found him old and blind, but still working with his students Viviani and Torricelli. Milton wrote of his visit:

> . . . There it was that I found and visited the famous *Galileo* grown old, a prisoner to the Inquisition, for thinking in Astronomy otherwise than the Franciscan and Dominican Licensers thought.[12]

Galileo died on January 8, 1642. He had turned the science of physics to its proper course, and established for all time a monument to the power of the human intellect.

The *Dialogues concerning New Sciences,* generally known as *Discourses concerning Two New Sciences,* was Galileo's greatest scientific achievement. His contributions to astronomy, and to freedom of thought generally, were very important, of course, but judged purely from the point of view of physics, were not the equal of the *Discourses.* Writing to a friend concerning his plans for the *Discourses* Galileo spoke of them as being "superior to everything else of mine hitherto published" and said that "they contain results which I consider the most important of all my studies." He built up the science of moving bodies by reasoning from experience, using mathematics as the medium of deduction and the definition of acceleration as the starting point. He made use of—indeed, emphasized—the role of experiment in testing physical theory, although he was aware that experiment alone did not constitute science, that the concept of acceleration, for example, transcends laboratory experience. While his talents did not extend to the induction of all-inclusive first principles of dynamics (this remained for Newton to accomplish), his procedure or *scientific method* was entirely modern.

The two new sciences contained in the book were on the resistance of solid bodies to fracture[13] and motion *(movimenti locali)*. The book was

---

[11] He remained with the Archbishop of Siena, under conditions that might be considered house arrest.

[12] *Areopagitica*, ed. T. H. White (London, R. Hunter, 1819), pp. 116f.

[13] Essentially the strength of materials and structures.

divided into four parts or *days,* of which the third dealt with accelerated motion and the fourth with projectiles, or *violent motions.* The following extracts, relating to the laws of falling bodies, are taken from the third *day,* in the translation of Crew and de Salvio.[14] The interlocutors are Salviati, through whom Galileo presents his own views by having him read aloud from a manuscript of an otherwise unidentified author or academician, Sagredo;[15] another scholar skilled in mechanics, and Simplicio, the antagonist who advances the Aristotelian views.

[14] H. Crew and A. de Salvio (trans.), *Dialogues concerning Two New Sciences* (New York: Macmillan, 1914). Available in a paperbound edition from Dover Publications, New York.
[15] Named for a close friend of Galileo's.

Galileo's Experiment

CHANGE OF POSITION

My purpose is to set forth a very new science dealing with a very ancient subject. There is, in nature, perhaps nothing older than motion, concerning which the books written by philosophers are neither few nor small; nevertheless I have discovered by experiment some properties of it which are worth knowing and which have not hitherto been either observed or demonstrated. Some superficial observations have been made, as, for instance, that the free motion■ of a heavy falling body is continuously accelerated; but to just what extent this acceleration occurs has not yet been announced; for so far as I know, no one has yet pointed out that the distances traversed, during equal intervals of time, by a body falling from rest, stand to one another in the same ratio as the odd numbers beginning with unity.■

Originally, natural motion.

It has been observed that missiles and projectiles describe a curved path of some sort; however no one has pointed out the fact that this path is a parabola. But this and other facts, not few in number or less worth knowing, I have succeeded in proving; and what I consider more important, there have been opened up to this vast and most excellent science, of which my work is merely the beginning, ways and means by which other minds more acute than mine will explore its remote corners.

$s = \frac{1}{2}at^2$; hence the distances traversed in successive (equal) time intervals follow as $(t+1)^2 - t^2$ or $2t + 1$, where $t = 0,1,2,\ldots$ This theorem is demonstrated below.

This discussion is divided into three parts; the first part deals with motion which is steady or uniform; the second treats of motion as we find it accelerated in nature; the third deals with the so-called violent motions∎ and with projectiles.

UNIFORM MOTION

In dealing with steady or uniform motion, we need a single definition which I give as follows:

Definition∎

By steady or uniform motion, I mean one in which the distances traversed by the moving particle during any equal intervals of time, are themselves equal.

Caution

We must add to the old definition (which defined steady motion simply as one in which equal distances are traversed in equal times) the word "any," meaning by this, all equal intervals of time; for it may happen that the moving body will traverse equal distances during some equal intervals of time and yet the distances traversed during some small portion of these time intervals may not be equal, even though the time intervals be equal.∎

From the above definition, four axioms follow, namely:

Axiom I In the case of one and the same uniform motion, the distance traversed during a longer interval of time is greater than the distance traversed during a shorter interval of time.∎

Axiom II In the case of one and the same uniform motion, the time required to traverse a greater distance is longer than the time required for a less distance.

Axiom III In one and the same interval of time, the distance traversed at a greater speed is larger than the distance traversed at a less speed.

Axiom IV The speed required to traverse a longer distance is greater than that required to traverse a shorter distance during the same time interval.

Indicates omission of part of the original. Omitted here are several theorems leading to the definition of velocity; in modern notation $v = \frac{s}{t}$ or, for instantaneous velocity, $v = \frac{ds}{dt}$.

■..

SALV: The preceding is what our author has written concerning uniform motion. We pass now to a new and more discriminating consideration of naturally accelerated motion, such as that generally experienced by heavy falling bodies; following is the title and introduction.

NATURALLY ACCELERATED MOTION

The properties belonging to uniform motion have been discussed in the preceding section; but accelerated motion remains to be considered.

And first of all it seems desirable to find and explain a definition best-fitting natural phenomena. For anyone may invent an arbitrary type of motion and discuss its properties; thus, for instance, some have imagined helices and conchoids■ as described by certain motions which are not met with in nature, and have very commendably established the properties which these curves possess in virtue of their definitions; but we have decided to consider the phenomena of bodies falling with an acceleration such as actually occurs in nature and to make this definition of accelerated motion exhibit the essential features of observed accelerated motions.■ And this, at last, after repeated efforts we trust we have succeeded in doing. In this belief we are confirmed mainly by the consideration that experimental results are seen to agree with and exactly correspond with those properties which have been, one after another, demonstrated by us. Finally, in the investigation of naturally accelerated motion we were led, by hand as it were, in following the habit and custom of nature herself, in all her various other processes, to employ only those means which are most common, simple, and easy.

For I think no one believes that swimming or flying can be accomplished in a manner simpler or easier than that instinctively employed by fishes and birds.

When, therefore, I observe a stone initially at rest falling from an elevated position and continually acquiring new increments of speed, why should I not believe that such increases take place in a manner which is exceedingly simple and rather obvious to everybody?■ If now we examine the matter carefully we find no addition or increment more simple than that which repeats itself always in the same manner. This we readily understand when we consider the intimate relationship between time and motion; for just as uniformity of motion is defined by and conceived through equal times and equal spaces (thus we call a motion uniform when equal distances are traversed during equal time

Paths having spiral or shell-like forms.

Galileo's insistence that the description given the motion must agree with experience.

This is based upon the belief, held by most scientists, that nature behaves in an essentially simple fashion, and is best explained with economy of thought.

intervals), so also we may, in a similar manner, through equal time intervals, conceive additions of speed as taking place without complication; thus we may picture to our mind a motion as uniformly and continuously accelerated when, during any equal intervals of time whatever, equal increments of speed are given to it.■ Thus if any equal intervals of time whatever have elapsed, counting from the time at which the moving body left its position of rest and began to descend, the amount of speed acquired during the first two time intervals will be double that acquired during the first time interval alone; so the amount added during three of these time intervals will be treble; and that in four, quadruple that of the first time interval. To put the matter more clearly, if a body were to continue its motion with the same speed which it had acquired during the first time interval and were to retain this same uniform speed, then its motion would be twice as slow as that which it would have if its velocity had been acquired during *two* time intervals.

The definition of uniform acceleration, in the absence of experimental evidence, was a major accomplishment.

And thus, it seems, we shall not be far wrong if we put the increment of speed as proportional to the increment of time; hence the definition of motion which we are about to discuss may be stated as follows: A motion is said to be uniformly accelerated, when starting from rest, it acquires, during equal time intervals, equal increments of speed.

SAGR: ■Although I can offer no rational objection to this or indeed to any other definition, devised by any author whomsoever, since all definitions are arbitrary, I may nevertheless without offense be allowed to doubt whether such a definition as the above, established in an abstract manner, corresponds to and describes that kind of accelerated motion which we meet in nature in the case of freely falling bodies. And since the author apparently maintains that the motion described in his definition is that of freely falling bodies, I would like to clear my mind of certain difficulties in order that I may later apply myself more earnestly to the propositions and their demonstrations.

Here Galileo stresses the difficulty which the average individual would experience in visualizing this concept.

SALV: It is well that you and Simplicio raise these difficulties. They are, I imagine, the same which occurred to me when I first saw this treatise, and which were removed either by discussion with the author himself, or by turning the matter over in my own mind.

SAGR: When I think of a heavy body falling from rest, that is, starting with zero speed and gaining speed in proportion to the time from the beginning of the motion; such a motion as would, for instance, in eight beats of the pulse acquire eight degrees of speed;■ having at the end of the fourth beat acquired four degrees; at the end of the second, two; at the end of the first, one; and since time is divisible without limit, it follows from all these considerations that if the earlier speed of a body is

A common method of measuring short intervals of time prior to the invention of the pendulum clock.

less than its present speed in a constant ratio, then there is no degree of speed however small (or, one may say, no degree of slowness however great)■ with which we may not find this body traveling after starting from infinite slowness, i.e., from rest. So that if that speed which it had at the end of the fourth beat was such that, if kept uniform, the body would traverse two miles in an hour, and if keeping the speed which it had at the end of the second beat, it would traverse one mile an hour, we must infer that, as the instant of starting is more and more nearly approached, the body moves so slowly that, if it kept on moving at this rate, it would not traverse a mile in an hour, or in a day, or in a year or in a thousand years; indeed, it would not traverse a span■ in an even greater time; a phenomenon which baffles the imagination, while our senses show us that a heavy falling body suddenly acquires great speed.

SALV: This is one of the difficulties which I also at the beginning, experienced, but which I shortly afterwards removed; and the removal was effected by the very experiment which creates the difficulty for you. You say the experiment appears to show that immediately after a heavy body starts from rest it acquires a very considerable speed: and I say that the same experiment makes clear the fact that the initial motions of a falling body, no matter how heavy, are very slow and gentle. Place a heavy body upon a yielding material, and leave it there without any pressure except that owing to its own weight; it is clear that if one lifts this body a cubit■ or two and allows it to fall upon the same material, it will, with this impulse, exert a new and greater pressure than that caused by its mere weight; and this effect is brought about by the [weight of the]■ falling body together with the velocity acquired during the fall, an effect which will be greater and greater according to the height of the fall, that is according as the velocity of the falling body becomes greater. From the quality and intensity of the blow we are thus enabled to accurately estimate the speed of a falling body. But tell me, gentlemen, is it not true that if a block be allowed to fall upon a stake from a height of four cubits and drives it into the earth, say, four finger-breadths, that coming from a height of two cubits it will drive the stake a much less distance, and from the height of one cubit a still less distance; and finally if the block be lifted only one finger breadth how much more will it accomplish than if merely laid on top of the stake without percussion? Certainly very little. If it be lifted only the thickness of a leaf, the effect will be altogether imperceptible. And since the effect of the blow depends upon the velocity of this striking body, can any one doubt the motion is very slow and the speed more than small whenever the effect is imperceptible?■ See now the power of truth; the same experiment

---

That is, it does not change speed abruptly, but accelerates gradually.

The *span* of the arms; now the English *fathom*.

A *cubit* was the length of the forearm, taken as eighteen inches.

Words in brackets have been inserted by the translator.

Note that the concepts of momentum and energy had not yet been discovered. Galileo evidently had some intuitive notion of these.

which at first glance seemed to show one thing, when more carefully examined, assures us of the contrary.

But without depending upon the above experiment, which is doubtless very conclusive, it seems to me that it ought not to be difficult to establish such a fact by reasoning alone. Imagine a heavy stone■ held in the air at rest; the support is removed and the stone set free; then since it is heavier than the air it begins to fall, and not with uniform motion but slowly at the beginning and with a continuously accelerated motion. Now since velocity can be increased and diminished without limit, what reason is there to believe that such a moving body starting with infinite slowness, that is, from rest, immediately acquires a speed of ten degrees rather than one of four, or of two, or of one, or of a half, or of a hundredth; or, indeed, of any of the infinite number of small values?■ Pray listen. I hardly think you will refuse to grant that the gain of speed of the stone falling from rest follows the same sequence as the diminution and loss of this same speed when, by some impelling force, the stone is thrown to its former elevation: but even if you do not grant this, I do not see how you can doubt that the ascending stone, diminishing in speed must before coming to rest pass through every possible degree of slowness.

> *Heavy so as not to be influenced in its fall by the resistance of the air.*

> *The continuity of motion was not at all obvious in Galileo's time.*

SIMP: But if the number of degrees of greater and greater slowness is limitless, they will never be all exhausted, therefore such an ascending heavy body will never reach rest, but will continue to move without limit always at a slower rate; but this is not the observed fact.

SALV: This would happen, Simplicio, if the moving body were to maintain its speed for any length of time at each degree of velocity; but it merely passes each point without delaying more than an instant: and since each time interval however small may be divided into an infinite number of instants, these will always be sufficient to correspond to the infinite degrees of diminished velocity.

That such a heavy rising body does not remain for any length of time at any given degree of velocity is evident from the following: because if, some time interval having been assigned, the body moves with the same speed in the last as in the first instant of that time interval,■ it could from this second degree of elevation be in like manner raised through an equal height, just as it was transferred from the first elevation to the second, and by the same reasoning would pass from the second to the third and would finally continue in uniform motion forever.

> *Proof of the continuity of natural motion.*

SAGR: From these considerations it appears to me that we may obtain a proper solution of the problem discussed by philosophers, namely, what causes the acceleration in the natural motion of heavy bodies? Since, as it seems to me, the force impressed by the agent projecting the body up-

We would now
account for this
very simply in
terms of energy.

wards diminishes continuously, this force, so long as it was greater than the contrary force of gravitation, impelled the body upwards;■ when the two are in equilibrium the body ceases to rise and passes through the state of rest in which the impressed impetus is not destroyed, but only its excess over the weight of the body has been consumed—the excess which caused the body to rise. Then as the diminution of the outside impetus continues, and gravitation gains the upper hand, the fall begins, but slowly at first on account of the opposing impetus,■ a large portion of which still remains in the body; but as this continues to diminish it also continues to be more and more overcome by gravity, hence the continuous acceleration of motion.

The *opposing
impetus* is the
inertia of the
body.

SIMP: The idea is clever, yet more subtle than sound; for even if the argument were conclusive, it would explain only the case in which a natural motion is preceded by a violent motion, in which there still remains active a portion of the external force; but where there is no such remaining portion and the body starts from an antecedent state of rest, the cogency of the whole argument fails.

SAGR: I believe that you are mistaken and that this distinction between cases which you make is superfluous or rather nonexistent. But, tell me, cannot a projectile receive from the projector either a large or a small force such as will throw it to a height of a hundred cubits, and even twenty or four or one?

SIMP: Undoubtedly, yes.

SAGR: So therefore this impressed force may exceed the resistance of gravity so slightly as to raise it only a finger-breadth; and finally the force of the projector may be just large enough to exactly balance the resistance■ of gravity so that the body is not lifted at all but merely sustained. When one holds a stone in his hand does he do anything but give■ it a force impelling it upwards equal to the power of gravity drawing it downwards? And do you not continuously impress this force upon the stone as long as you hold it in the hand? Does it perhaps diminish with the time during which one holds the stone?

Opposing force.

And what does it matter whether this support which prevents the stone from falling is furnished by one's hand or by a table or by a rope from which it hangs? Certainly nothing at all. You must conclude, therefore, Simplicio, that it makes no difference whatever whether the fall of the stone is preceded by a period of rest which is long, short, or instantaneous provided only the fall does not take place so long as the stone is acted upon by a force opposed to its weight and sufficient to hold it at rest.

SALV: The present does not seem to be the proper time to investigate the cause of the acceleration of natural motion concerning which various opinions have been expressed by various philosophers, some explaining it by attraction to the center, others to repulsion between the very small parts of the body, while still others attribute it to a certain stress in the surrounding medium which closes in behind the falling body and drives it from one of its positions to another.■ Now, all these fantasies, and others too, ought to be examined; but it is not really worthwhile. At present it is the purpose of our author merely to investigate and to demonstrate some of the properties of accelerated motion (whatever the cause of this acceleration may be)—meaning thereby a motion, such that the momentum■ of its velocity goes on increasing after departure from rest, in simple proportionality to the time, which is the same as saying that in equal time intervals the body receives equal increments of velocity; and if we find the properties (of accelerated motion) which will be demonstrated later are realized in freely falling and accelerated bodies, we may conclude that the assumed definition includes such a motion of falling bodies and that their speed goes on increasing as the time and the duration of the motion.

SAGR: So far as I see at present, the definition might have been put a little more clearly perhaps without changing the fundamental idea, namely, uniformly accelerated motion is such that its speed increases in proportion to the space traversed; so that, for example, the speed acquired by a body in falling four cubits would be double that acquired in falling two cubits and this latter speed would be double that acquired in the first cubit.■ Because there is no doubt but that a heavy body falling from the height of six cubits has, and strikes with, a momentum double that it had at the end of three cubits, triple that which it had at the end of one.

SALV: It is very comforting to me to have had such a companion in error; and moreover let me tell you that your proposition seems so highly probable that our author himself admitted, when I advanced this opinion to him, that he had for some time shared the same fallacy. But what most surprised me was to see two propositions so inherently probable that they commanded the assent of everyone to whom they were presented, proven in a few simple words to be not only false, but impossible.

SIMP:■ I am one of those who accept the proposition, and believe that a falling body acquires force in its descent, its velocity increasing in proportion to the space, and that the momentum of the falling body is doubled when it falls from a doubled height; these propositions, it appears to me, ought to be conceded without hesitation or controversy.

Aristotle had accounted for *violent* motion by assuming that the air which was displaced by the moving object rushed around behind it to provide the motive force. That is, the medium aided the motion.

A strict definition of momentum had not yet been advanced, but the term was used in somewhat the same sense as the current usage; i.e. a quantity proportional both to the mass and velocity of a body. It was frequently confused, however, with energy.

Not so. The speed is proportional to the square root of the distance, as proved below ($v^2 = 2as$).

This was, after all, the simplest scheme imaginable.

SALV: And yet they are as false and impossible as that motion should be completed instantaneously; and here is a very clear demonstration of it. If the velocities are in proportion to the spaces traversed, or to be traversed, then these spaces are traversed in equal intervals of time; if, therefore, the velocity with which the falling body traverses a space of eight feet were double that with which it covered the first four feet (just as the one distance is double the other), then the time intervals required for these passages would be equal.■ But for one and the same body to fall eight feet and four feet in the same time is possible only in the case of instantaneous motion; but observation shows us that the motion of a falling body occupies time, and less of it in covering a distance of four feet than of eight feet; therefore it is not true that its velocity increases in proportion to the space.

The falsity of the other proposition■ may be shown with equal clearness. For if we consider a single striking body the difference of momentum in its blows can depend only upon difference of velocity; for if the striking body falling from a double height were to deliver a blow of double momentum, it would be necessary for this body to strike with a doubled velocity; but with this doubled speed it would traverse a doubled space in the same time interval; observation however shows that the time required for fall from the greater height is longer.

SAGR: You present these recondite matters with too much evidence and ease; this great facility makes them less appreciated than they would be had they been presented in a more abstruse manner. For, in my opinion, people esteem more lightly that knowledge which they acquire with so little labor than that acquired through long and obscure discussion.■

SALV: If those who demonstrate with brevity and clearness the fallacy of many popular beliefs were treated with contempt instead of gratitude the injury would be quite bearable; but on the other hand it is very unpleasant and annoying to see men, who claim to be peers of anyone in a certain field of study, take for granted certain conclusions which later are quickly and easily shown by another to be false.■ I do not describe such a feeling as one of envy, which usually degenerates into hatred and anger against those who discover such fallacies; I would call it a strong desire to maintain old errors, rather than accept newly discovered truths. This desire at times induces them to unite against these truths, although at heart believing in them, merely for the purpose of lowering the esteem in which certain others are held by the unthinking crowd. Indeed, I have heard from our academician many such fallacies held as true but easily refutable; some of these I have in mind.

Since the average velocity would be the same in each of the successive four-foot intervals, which was contrary to experience. This could have been Galileo's argument, rather than that given here, which seems to confuse average with instantaneous velocity.

That the momentum of a body doubles when it falls from a doubled height.

Perhaps so, yet simplicity is the key to good explanation in science.

One of Galileo's frequent references to the followers of classical authority.

SAGR: You must not withhold them from us, but, at the proper time, tell us about them even though an extra session be necessary. But now, continuing the thread of our talk, it would seem that up to the present we have established the definition of uniformly accelerated motion which is expressed as follows:

> A motion is said to be equally or uniformly accelerated when, starting from rest, its momentum■ receives equal increments in equal times.

SALV: This definition established, the author makes a single assumption, namely,

> The speeds acquired by one and the same body moving down planes of different inclinations are equal when the heights of these planes are equal.

By the height of an inclined plane we mean the perpendicular let fall from the upper end of the plane upon the horizontal line drawn through the lower end of the same plane. Thus, to illustrate, let the line *AB* be horizontal, and let the planes *CA* and *CD* be inclined to it; then the author calls the perpendicular *CB* the "height" of the planes *CA* and *CD;* he supposes that the speeds acquired by one and the same body, descending along the planes *CA* and *CD* to the terminal points *A* and *D* are equal since the heights of these planes are the same, *CB;* and also it must be understood that this speed is that which would be acquired by the same body falling from *C* to *B.*

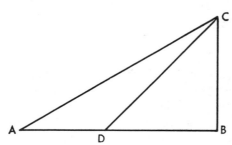

SAGR: Your assumption appears to me so reasonable that it ought to be conceded without question,■ provided of course there are no chance or outside resistances, and that the planes are hard and smooth, and that the figure of the moving body is perfectly round, so that neither plane nor moving body is rough. All resistance and opposition having been removed, my reason tells me at once that a heavy and perfectly round ball descending along the lines *CA, CD, CB* would reach the terminal points *A, D, B,* with equal momenta.

Or its velocity. In this case it is clear that momentum was intended to be directly proportional to velocity, as in current usage.

Actually, this probably would not be at all obvious to Galileo's contemporaries.

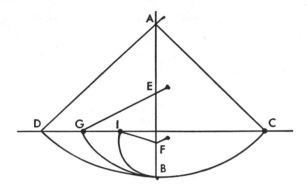

SALV: Your words are very plausible; but I hope by experiment to increase the probability to an extent which shall be little short of a rigid demonstration.

Imagine this page to represent a vertical wall, with a nail driven into it; and from the nail let there be suspended a lead bullet of one or two ounces by means of a fine vertical thread, *AB*, say from four to six feet long, on this wall draw a horizontal line *DC*, at right angles to the vertical thread *AB*, which hangs about two finger-breadths in front of the wall. Now bring the thread *AB* with the attached ball into the position *AC* and set it free; first it will be observed to descend along the arc *CBD*, to pass the point *B*, and to travel along the arc *BD*, till it almost reaches the horizontal *CD*, a slight shortage being caused by the resistance of the air and the string; from this we may rightly infer that the ball in its descent through the arc *CB* acquired a momentum on reaching *B*, which was just sufficient to carry it through a similar arc *BD* to the same height.■ Having repeated this experiment many times, let us now drive a nail into the wall close to the perpendicular *AB*, say at *E* or *F*, so that it projects out some five or six finger-breadths in order that the thread, again carrying the bullet through the arc *CB*, may strike upon the nail *E* when the bullet reaches *B*, and thus compel it to traverse the arc *BG*, described about *E* as center. From this we can see what can be done by the same momentum which previously starting at the same point *B* carried the same body through the arc *BD* to the horizontal *CD*. Now, gentlemen, you will observe with pleasure that the ball swings to the point *G* in the horizontal, and you would see the same thing happen if the obstacle were placed at some lower point, say at *F*, about which the ball would describe the arc *BI*, the rise of the ball always terminating exactly on the line *CD*. But when the nail is placed so low that the remainder of the thread below it will not reach to the height *CD* (which

This demonstration is still widely used to illustrate the conservation of mechanical energy. It is kinetic energy rather than momentum that suffices to carry it to the same height.

would happen if the nail were placed nearer *B* than to the intersection of *AB* with the horizontal *CD*) then the thread leaps over the nail and twists itself about it.■

This experiment leaves no room for doubt as to the truth of our supposition; for since the two arcs *CB* and *DB* are equal and similarly placed, the momentum acquired by the fall through the arc *CB* is the same as that gained by the fall through the arc *DB;* but the momentum acquired at *B*, owing to fall through *CB*, is able to lift the same body through the arc *BD;* therefore, the momentum acquired in the fall *BD* is equal to that which lifts the same body through the same arc from *B* to *D;* so, in general, every momentum acquired by fall through an arc is equal to that which can lift the same body through the same arc. But all these momenta which cause a rise through the arcs *BD, BG,* and *BI* are equal, since they are produced by the same momentum, gained by fall through *CB*, as experiment shows. Therefore all the momenta gained by fall through the arcs *DB, GB, IB* are equal.

SAGR: The argument seems to me so conclusive and the experiment so well adapted to establish the hypothesis that we may, indeed, consider it as demonstrated.

SALV: I do not wish, Sagredo, that we trouble ourselves too much about this matter, since we are going to apply this principle mainly in motions which occur on plane surfaces, and not upon curved, along which acceleration varies in a manner greatly different from that which we have assumed for planes.

So that, although the above experiment shows us that the descent of the moving body through the arc *CB* confers upon it momentum just sufficient to carry it to the same height through any of the arcs *BD, BG, BI*, we are not able, by similar means, to show that the event would be identical in the case of a perfectly round ball descending along planes whose inclinations are respectively the same as the chords of these arcs.■ It seems likely, on the other hand, that, since these planes form angles at the point *B*, they will present an obstacle to the ball which has descended along the chord *CB*, and starts to rise along the chord *BD, BG, BI*.

In striking these planes some of its momentum will be lost and it will not be able to rise to the height of the line CD; but this obstacle, which interferes with the experiment, once removed, it is clear that the momentum (which gains in strength with descent) will be able to carry the body to the same height. Let us then, for the present, take this as a postulate, the absolute truth of which will be established when we find that the inferences from it correspond to and agree perfectly with experiment.■ The author having assumed this single principle passes next to the propositions which he clearly demonstrates; the first of these is as follows:

As the length of thread decreases, the speed increases until the energy is dissipated with the ball striking the nail.

Evidently Galileo did not find it difficult to make the *mental* transition from the curved path to the planes.

Note the clear statement of *scientific method.*

Theorem I, Proposition I

The time in which any space is traversed by a body starting from rest and uniformly accelerated is equal to the time in which that same space would be traversed by the same body moving at a uniform speed whose value is the mean of the highest speed and the speed just before acceleration began.■

Let us represent by the line *AB* the time in which the space *CD* is traversed by a body which starts from rest at *C* and is uniformly accelerated; let the final and highest value of the speed gained during the interval *AB* be represented by the line *EB* drawn at right angles to *AB;* draw the line *AE,* then all lines drawn from equidistant points on *AB* and parallel to *BE* will represent the increasing values of the speed, beginning with the instant *A.* Let the point *F* bisect the line *EB;* draw *FG* parallel to *BA,* and *GA* parallel to *FB,* thus forming a parallelogram *AGFB* which will be equal in area to the triangle *AEB,* since the side *GF* bisects the side *AE* at the point *I;* for if the parallel lines in the triangle *AEB* are extended to *GI,* then the sum of all the parallels contained in the quadrilateral is equal to the sum of those contained in the triangle *AEB;* for those in the triangle *IEF* are equal to those contained in the triangle *GIA,* while those included in the trapezium *AIFB* are common. Since each and every instant of time in the time interval *AB* has its corresponding point on the line *AB,* from which points parallels drawn in and limited by the triangle *AEB* represent the increasing values of the growing velocity, and since parallels contained within the rectangle represent the values of a speed which is not increasing, but constant, it appears, in like manner, that the momenta assumed by the moving body may also be represented, in the case of the accelerated motion, by the increasing parallels of the triangle *AEB,* and, in the case of the uniform motion, by the parallels of the rectangle *GB.*■ For, what the momenta may lack in the first part of the accelerated motion (the deficiency of the momenta being represented by the parallels of the triangle *AGI*) is made up by the momenta represented by the parallels of the triangle *IEF.*

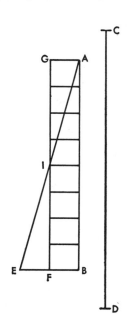

Hence it is clear that equal spaces will be traversed in equal times by two bodies, one of which, starting from rest, moves with a uniform acceleration, while the momentum of the other, moving with uniform speed, is one-half its maximum momentum under accelerated motion.

Q.E.D.

Theorem II, Proposition II

The spaces described by a body falling from rest with a uniformly accelerated motion are to each other as the squares of time intervals employed in traversing these distances.

Let the time beginning with any instant *A* be represented by the straight line *AB* in which are taken any two time intervals *AD* and *AE*. Let *HI* represent the distance through which the body, starting from rest at *H,* falls with uniform acceleration. If *HL* represents the space traversed during the time interval *AD,* and *HM* that covered during the interval *AE,* then the space *MH* stands to the space *LH* in a ratio which is the square of the ratio of the time *AE* to the time *AD;* or we may say simply that the distances *HM* and *HL* are related as the squares of *AE* and *AD.*

Draw the line *AC* making any angle whatever with the line *AB;* and from the points *D* and *E,* draw the parallel lines *DO* and *EP;* of these two lines, *DO* represents the greatest velocity attained during the interval *AD,* while *EP* represents the maximum velocity acquired during the interval *AE.* But it has just been proved that so far as distances traversed are concerned it is precisely the same whether a body falls from rest with a uniform acceleration or whether it falls during an equal time interval with a constant speed which is one half the maximum speed attained during the accelerated motion. It follows therefore that the distances *HM* and *HL* are the same as would be traversed, during the time intervals *AE* and *AD,* by uniform velocities equal to one half those represented by *DO* and *EP* respectively. If, therefore, one can show that the distances *HM* and *HL* are in the same ratio as the squares of the time intervals *AE* and *AD,* our proposition will be proven.

But in the fourth proposition of the first book it has been shown that the spaces traversed by two particles in uniform motion bear to one another a ratio which is equal to the product of the ratio of the velocities by the ratio of the times.■ But in this case the ratio of the velocities is the same as the ratio of the time intervals (for the ratio of *AE* to *AD* is the same as that of ½ *EP* to ½ *DO* or of *EP* to *DO*). Hence the

In modern notation
$s_1 = v_1 t_1; \; s_2 = v_2 t_2$
$\therefore s_1/s_2 = v_1 t_1/v_2 t_2$

ratio of the spaces traversed is the same as the squared ratio of the time intervals.

Q.E.D.

Falling from rest $v^2 = 2gs$, where g is the acceleration due to gravity. Thus, the distance is proportional to the square of the final velocity.

Evidently then the ratio of the distances is the square of the ratio of the final velocities,■ that is, of the lines *EP* and *DO*, since these are to each other as *AE* to *AD*.

COROLLARY I

Hence it is clear that if we take any equal intervals of time whatever, counting from the beginning of the motion, such as *AD, DE, EF, FG*, in which the spaces *HL, LM, MN, NI* are traversed, these spaces will bear to one another the same ratio as the series of odd numbers, 1, 3, 5, 7; for this is the ratio of the differences of the squares of the lines (which represent time), differences which exceed one another by equal amounts, this excess being equal to the smallest line (viz., the one representing a single time interval) : or we may say [that this is the ratio] of the differences of the squares of the natural numbers beginning with unity.

While, therefore, during equal intervals of time the velocities increase as the natural numbers, the increments in the distances traversed during these equal time intervals are to one another as the odd numbers beginning with unity.

SAGR: Please suspend the discussion for a moment since there just occurs to me an idea which I want to illustrate by means of a diagram in order that it may be clearer both to you and to me.

Let the line *AI* represent the lapse of time measured from the initial instant *A;* through *A* draw the straight line *AF* making any angle whatever; join the terminal points *I* and *F;* divide the time *AI* in half at *C;* draw *CB* parallel to *IF*. Let us consider *CB* as the maximum value of the velocity which increases from zero at the beginning, in simple proportionality to the intercepts on the triangle *ABC* of lines drawn parallel to *BC;* or what is the same thing, let us suppose the velocity to increase in proportion to the time; then I admit without question, in view of the preceding argument, that the space described by a body falling in the aforesaid manner will be equal to the space traversed by the same body during the same length of time traveling with a uniform speed equal to *EC,* the half of BC. Further let us imagine that the body has fallen with accelerated motion so that, at the instant *C*, it has the velocity BC. It is clear that if the body continued to descend with the same speed *BC,* without acceleration, it would in the next time interval *CI* traverse double the distance covered during the interval *AC,* with the uniform speed *EC* which is half of *BC;* but since the falling body acquires equal increments of speed during equal increments of time, it follows that the velocity *BC,* during the next time interval *CI* will be increased by an amount represented by the parallels of the triangle *BFG* which is equal to the triangle *ABC.* If, then, one adds to the velocity *GI* half of the velocity *FG,* the highest speed acquired by the accelerated motion and determined by the parallels of the triangle *BFG,* he will have the uniform velocity with which the same space would have been described in the time *CI;* and since this speed *IN* is three times as great as *EC* it follows that the space described during the interval *CI* is three times as great as that described during the interval *AC.* Let us imagine the motion extended over another equal time interval *IO,* and the triangle extended to *APO;* it is then evident that if the motion continues during the interval *IO,* at the constant rate *IF* acquired by acceleration during the time *AI,* the space traversed during the interval *IO* will be four times that traversed during the first interval *AC,* because the speed *IF* is four times the speed *EC.* But if we enlarge our triangle so as to include *FPQ* which is equal to *ABC,* still assuming the acceleration to be constant, we shall add to the uniform speed an increment *RQ,* equal to *EC;* then the value of the

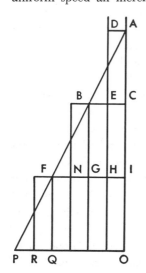

equivalent uniform speed during the time interval *IO* will be five times that during the first time interval *AC;* therefore the space traversed will be quintuple that during the first interval *AC.* It is thus evident by simple computation that a moving body starting from rest and acquiring velocity at a rate proportional to the time, will, during equal intervals of time, traverse distances which are related to each other as the odd numbers beginning with unity, 1, 3, 5;■ or considering the total space traversed, that covered in double time will be quadruple that covered during unit time; in triple time, the space is nine times as great as in unit time. And in general the spaces traversed are in the duplicate ratio of the times, i.e., in the ratio of the squares of the times.

Thus, $s = \frac{1}{2} at^2$, where a is replaced by g in the case of free fall, from which this follows immediately, as shown earlier.

Simp:■ In truth, I find more pleasure in this simple and clear argument of Sagredo than in the author's demonstration which to me appears rather obscure; so that I am convinced that matters are as described, once having accepted the definition of uniformly accelerated motion. But as to whether this acceleration is that which one meets in nature in the

Even Simplicio is convinced.

That is, whether the acceleration due to gravity is of this uniform type. One must resort to experiment for the answer to this question.

The premises from which deductions are to be drawn.

case of falling bodies, I am still doubtful;▪ and it seems to me, not only for my own sake but also for all those who think as I do, that this would be the proper moment to introduce one of those experiments—and there are many of them, I understand—which illustrate in several ways the conclusions reached.

SALV: The request which you, as a man of science, make is a very reasonable one; for this is the custom—and properly so—in those sciences where mathematical demonstrations are applied to natural phenomena, as is seen in the case of perspective, astronomy, mechanics, music, and others where the principles,▪ once established by well-chosen experiments, become the foundations of the entire superstructure. I hope therefore it will not appear to be a waste of time if we discuss at considerable length this first and most fundamental question upon which hinge numerous consequences of which we have in this book only a small number, placed there by the author, who has done so much to open a pathway hitherto closed to minds of speculative turn. So far as experiments go they have not been neglected by the author; and often, in his company, I have attempted in the following manner to assure myself that the acceleration actually experienced by falling bodies is that above described.

A piece of wooden moulding or scantling, about 12 cubits long, half a cubit wide, and three finger-breadths thick, was taken; on its edge was cut a channel a little more than one finger in breadth; having made this groove very straight, smooth, and polished, and having lined it with parchment, also as smooth and polished as possible, we rolled along it a hard, smooth, and very round bronze ball. Having placed this board in a sloping position, by lifting one end some one or two cubits above the other, we rolled the ball, as I was just saying, along the channel, noting, in a manner presently to be described, the time required to make the descent. We repeated this experiment more than once in order to measure the time with an accuracy such that the deviation between two observations never exceeded one tenth of a pulse beat.▪ Having performed this operation and having assured ourselves of its reliability, we now rolled the ball only one quarter the length of the channel; and having measured the time of its descent, we found it precisely one half of the former. Next we tried other distances, comparing the time for the whole length with that for the half, or with that for two thirds, or three fourths, or indeed for any fraction; in such experiments, repeated a full hundred times, we always found that the spaces traversed were to each other as the squares of the times, and this was true for all inclinations of the plane, i.e., of the channel, along which we rolled the ball. We also observed that the times of descent, for various inclinations of the plane,

Roughly 0.1 second, which could not be estimated accurately, particularly with the water clock used by Galileo. In fact, the experiment was extremely crude by modern standards. Others had difficulty repeating it.

bore to one another precisely that ratio which, as we shall see later, the author had predicted and demonstrated for them.

For the measurement of time, we employed a large vessel of water placed in an elevated position; to the bottom of this vessel was soldered a pipe of small diameter giving a thin jet of water, which we collected in a small glass during the time of each descent, whether for the whole length of the channel or for a part of its length; the water thus collected was weighed, after each descent, on a very accurate balance;■ the differences and ratios of these weights gave us the differences and ratios of the times, and this with such accuracy that although the operation was repeated many, many times, there was no appreciable discrepancy in the results.

SIMP: I would like to have been present at these experiments; but feeling confidence in the care with which you performed them, and in the fidelity with which you relate them, I am satisfied and accept them as true and valid.

■..

Apart from the reaction time of the observer in opening and closing the end of the pipe, such a timer would be quite accurate if the vessel were reasonably large; otherwise, the velocity of the jet would vary.

Omitted are a large number of special cases of descent along inclined planes.

SUPPLEMENTARY READING

Cohen, I. B., "Galileo," in *Lives in Science* by the Editors of Scientific American (New York: Simon and Schuster, 1957), pp. 3ff.

Cooper, L., *Aristotle, Galileo, and the Tower of Pisa* (Ithaca, N. Y.: Cornell University Press, 1935).

Crew, H., "Portraits of Famous Physicists," *Scripta Mathematica,* Pictorial Mathematics, Portfolio No. 4, 1942.

Crowther, J. G., *Six Great Scientists* (London: Hamilton, 1955), pp. 49ff.

Drake, S., *Discoveries and Opinions of Galileo* (New York: Doubleday, 1957).

Galilei, Galileo, *Dialogue on the Two Chief World Systems,* trans. by S. Drake (Berkeley: University of California, 1953).

Hall, A. R., *The Scientific Revolution* (Boston: Beacon, 1954).

Knedler, J. W. (ed.), *Masterworks of Science* (New York: Doubleday, 1947), pp. 75ff.

Magie, W. F., *A Source Book in Physics* (New York: McGraw-Hill, 1935), pp. 1ff.

Santillana, G. de, *The Crime of Galileo* (Chicago: University of Chicago, 1955).

Taylor, F. S., *Galileo and the Freedom of Thought* (London: 1938).

Wolf, A., *A History of Science, Technology, and Philosophy in the 16th and 17th Centuries* (London: Allen and Unwin, 1935), pp. 25ff. and 82ff.

3

Robert Boyle

1627-1691

Boyle's Law: Pressure-volume Relations in a Gas

GALILEO ILLUMINED the first part of the seventeenth century, Newton the last. The intervening period was notable chiefly for the insight it provided on the mechanics of fluids. To Galileo's dynamics of solids, his student Evangelista Torricelli (1608–1647) added the motion of liquids. Moreover, he suspected that air had weight and that the liquid in a barometer was supported by the pressure of the atmosphere rather than by nature's *horror vacui*.[1] His early death prevented him from establishing this hypothesis, which was proved later by Blaise Pascal (1623–1662). Thus, by the middle of the seventeenth century it was known that the height of the liquid in a barometer varied inversely with altitude, and that this probably was caused by variations in the pressure of the atmosphere.

Against this background Robert Boyle began his investigations on the pressure and elastic properties of the atmosphere. Boyle was born in Ireland on January 25, 1627, the seventh son of the Earl of Cork. Sometimes called the "father of chemistry," Boyle was a man of considerable wealth who devoted his entire life to science and religion. His scientific accomplishments were by no means considerable; he made no discovery of great conceptual significance, nor did he carry his work beyond the barest preliminary stages. Yet he achieved an immense reputation for the number of his contributions and, perhaps more important, for the general support of science which his stature provided.

Boyle was sent to Eton at the age of eight, where he early showed an aptitude for disciplined study and an insatiable craving for knowledge. Fol-

---

[1] "Nature abhors a vacuum," an Aristotelian precept which somehow has found its way into our modern speech.

lowing several years at Eton he continued his education briefly with tutors until he was nearly twelve, when he went on an extended tour of the Continent for the next six years. He was staying in Florence during the winter of 1641–1642 when Galileo died, and it was there that he first studied Galileo's works. He returned to England in 1644, a polished gentleman of seventeen, and applied himself vigorously to the practice of science. Despite his youth, he shortly became leader of the Philosophical College, which was later incorporated as the Royal Society following the Restoration of the Stuarts.

His earliest interests were in the field of chemistry. Like Galileo, he rejected alchemy, and with it the four-element theory of the Scholastics. While he did not carry his own ideas very far, others subsequently built upon them the foundations of modern analytic chemistry. He thereby contributed greatly toward establishing chemistry as a pure science. In 1661 he climaxed his chemical studies by publishing a book, written in dialogue style, entitled *The Sceptical Chymist: or Chymico-Physical Doubts and Paradoxes,* in which he defined a chemical element as a single, pure substance that could not be broken down into anything simpler, even by the most violent physical or chemical action. His views, founded largely on the old atomic hypothesis of Democritus, were little different, in principle, than those upon which we base our modern concepts of matter. He distinguished between mixtures and compounds, prepared phosphorus, and developed methods of analysis for various substances. Unfortunately, owing to the dominant position of the four-element theory, Boyle's chemical ideas had little immediate effect; it was not until Antoine Lavoisier (1743–1794) revived them about a century later that chemistry actually began to prosper.

In 1654, following a brief period at his ancestral estate, Boyle settled in Oxford, where he established a private laboratory[2] and held frequent meetings with others who, like him, were interested in *experimental philosophy.* Three years later, having learned of Otto von Guericke's (1602–1686) invention of the vacuum pump, he set about to build one, and thereby began his investigations of the properties of air. He showed its elastic properties, that it had weight; he studied its role in respiration and combustion and in the propagation of sound. His first results were published at Oxford in 1660, under the title *New Experiments Physico-Mechanical Touching the Spring of the Air and its Effects.* The second edition (1662) contained his experimental proof of the reciprocal relation between the volume and pressure of a gas (air), commonly known as *Boyle's law.* While the accuracy of his measurements left much to be desired, they were sufficiently convincing to invite prompt acceptance. Yet Boyle did not attach much significance to this dis-

---

[2] His assistant was Robert Hooke (1635–1703), who later investigated the elastic properties of solids.

covery, and five years later, Edmé Mariotte (?1620–1684) discovered the same relationship independently.

Boyle moved to London in 1668, perhaps because the Royal Society, of which he had been a founding member, held its meetings there. He took a very active part in its proceedings, presented his own results before its members, including the distinguished Newton, and was elected its president in 1680. He felt compelled to decline this honor because he would not subscribe to oaths, but he otherwise contributed greatly to the growth of the Society.

Early in his life Boyle exhibited strong leanings toward theology. By the time he was twenty-one he already had written several moral and religious essays, as well as a treatise on ethics. Throughout his life he continued to support various movements aimed at spreading the Gospel in different quarters, and his will endowed the annual *Boyle Lectures* for the defense of Christianity against unbelievers. In 1690, the year before his death, he published *The Christian Virtuoso,* an effort to show how science and religion afford mutual support to one another.

He was in constant touch with the most fertile minds of his time. So great was his prestige that until his death on December 30, 1691, he exerted considerable influence over the development of physical science in England. He carried on a voluminous correspondence with others of scientific bent, constantly encouraging them in scholarly pursuits.

The extract that follows is taken from *The Works of the Honorable Robert Boyle,* in an English translation by R. Boulton (Boyle wrote in Latin), vol. 1 (1699), pages 404ff. The spellings have been modernized, but the contents are otherwise in the original form.

Boyle's Experiment

By *spring,* Boyle meant *elasticity.*

To make it evident that the spring■ of the air is able to do much more than we have attributed to it, on account of its spring and weight, we tried the following experiments.

THE AIR'S CONDENSATION MEASURED

In the form of a
J.

Having poured mercury into a glass tube, which consisted of a long leg and a short one, and which were so bent as to lie almost parallel to each other;■ we pasted papers upon each, which were divided by marks into inches, and each of these inches into eight parts; and upon pouring mercury into the longer tube, we observed that the air contained in the shorter, which was hermetically sealed at the top, was condensed by 29 inches of mercury, into half the space it possessed before; from whence it appears, that if it were able in so compressed a state, by virtue of its spring, to resist a cylinder of mercury of 29 inches, besides the atmospheric cylinder■ incumbent upon that, it follows that its compression in the open air, being but half as much, it must have but half that weight from the atmosphere that lies upon it in that compressed state.■

But to be more exact, we took a tube represented by the figure, pasting upon the shorter leg a paper, divided into twelve inches, and each of those into quarters; and another upon the longer leg, which made up several feet, which were likewise divided into inches, and those subdivided again into quarters. The tube being thus marked, the lower end was placed in a wooden box, that the mercury might run into it if the pipe chanced to break: And one being assigned to pour in mercury at the top of the tube, another was stationed to observe when the mercury in the small tube rose up to each of the divisions above mentioned; and to take notice, likewise, how high it stood in the long tube at the same time; wherefore the several observations were set down, and are contained in the following table.■

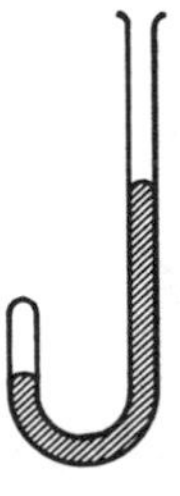

The atmospheric pressure.

Thus, doubling the pressure decreased the volume to one half.

Boyle assumed that the bore of the tube was sufficiently uniform so that the heights of the mercury columns were accurate measures of the volumes.

The first A column gives the length in quarter-inch divisions; the second gives the same length expressed in inches. Column E assumes the well-known relation PV = constant. The reasonably close agreement with experiment (column D) indicates that other effects, such as temperature, were not very important in these measurements.

A TABLE OF THE CONDENSATION OF THE AIR

A	A	B	C	D	E	
48	12	00		29 2/16	29 2/16	AA. The number of equal spaces in the shorter leg that contained the same parcel of air diversely extended.
46	11½	01 7/16		30 9/16	30 6/16	
44	11	02 13/16		31 15/16	31 12/16	
42	10½	04 6/16		33 8/16	33½	
40	10	06 3/16		35 5/16	35	
38	9½	07 14/16		37	36 15/19	B. The height of the mercurial cylinder in the longer leg, that compressed the air into those dimensions.
36	9	10 2/16		39 5/16	38 6/8	
34	8½	12 8/16		41 10/16	41 2/17	
32	8	15 1/16		44 3/16	43 11/16	
30	7½	17 15/16	Added to 29⅛ makes	47 1/16	46 6/8	
28	7	21 3/16		50 5/16	50	C. The height of the mercurial cylinder, that counterbalanced the pressure of the atmosphere.
26	6½	25 3/16		54 5/16	53 10/13	
24	6	29 11/16		58 13/16	58 6/8	
23	5¾	32 3/16		61 5/16	60 18/23	
22	5½	34 15/16		64 1/16	63 6/11	
21	5¼	37 15/16		67 1/16	66 4/7	D. The aggregate of the two last columns, B and C, exhibiting the pressure sustained by the included air.
20	5	41 9/16		70 1/16	70	
19	4¾	45		74 2/16	73 11/19	
18	4½	48 12/16		77 1/16	77 6/8	
17	4¼	53 11/16		82 12/16	82 4/17	
16	4	58 2/16		87 14/16	87 6/8	E. What that pressure should be according to the hypothesis, that supposes the pressures and expansions to be in reciprocal proportion.
15	3¾	63 15/16		93 1/16	93½	
14	3½	71 5/16		100 7/16	99 6/8	
13	3¼	78 11/16		107 13/16	107 7/13	
12	3	88 7/16		117 9/16	116 6/8	

A precautionary note to future experimenters. The mercury must be poured in slowly.

But in trying this experiment, whoever pours in the mercury must do it by degrees,■ and according to the directions of the other, who takes notice of the ascent of the mercury below; for if it be poured in without caution, it may rise up above the marks placed on the tubes, before due observations can be made.

From which it appeared that the temperature should be kept constant.

Having, by the weight of so vast a cylinder of mercury, compressed the air into a quarter of the space it possessed before, we observed, though it could not be sensibly condensed further by cold, yet the flame of a candle brought near it gave us reason to think that a greater degree of heat would have expanded it;■ but fearing the cracking of the tube, we dared not try it.

From the experiment it appears, that as the air is more or less compressed, so it is able to counterbalance a heavier or lighter cylinder of mercury. And that the mercury was supported by the spring of that condensed air appeared by sucking up the air out of the tube when the mercury was 100 inches high in the pipe; for the pressure of the incumbent pillar of the atmosphere, being by that means removed, the mercury was raised in the long tube by the expansion of the air in the short leg: And not by any *funiculus,* since, as the *Objector* confesses, that cannot raise more than a cylinder of 30 inches.■

Franciscus Linus (1595-1675), a supporter of Aristotelian doctrines, had suggested that the Torricelli vacuum contained an invisible stretched membrane, called a *funiculus,* which could draw up a column of mercury to a maximum height of about 30 inches. The *Objector* refers to Linus.

THE AIR'S RAREFACTION CONSIDERED

But, together with what has been said, it may illustrate not a little our doctrine of the spring of air to observe how much its spring is weakened, as it is variously expanded and rarefied.

Note the lack of good agreement between columns D and E for small values of pressure. Boyle attributed this to occluded air in the mercury, which came out as the pressure was decreased.

A TABLE OF THE RAREFACTION OF THE AIR

A. The number of equal spaces at the top of the tube, that contained the same parcel of air.

B. The height of the mercurial cylinder, that together with the spring of the included air counterbalanced the pressure of the atmosphere.

C. The pressure of the atmosphere.

D. The complement of *B* to *C*, exhibiting the pressure sustained by the included air.

E. What that pressure should be according to the hypothesis.

A	*B*	*C*	*D*	*E*
1	$00\frac{0}{0}$	Subtracted from $29\frac{3}{4}$ leaves	$29\frac{3}{4}$	$29\frac{3}{4}$
$1\frac{1}{2}$	$10\frac{5}{8}$		$19\frac{1}{8}$	$19\frac{5}{6}$
2	$15\frac{3}{8}$		$14\frac{3}{8}$	$14\frac{7}{8}$
3	$20\frac{2}{8}$		$9\frac{4}{8}$	$9\frac{15}{12}$
4	$22\frac{5}{8}$		$7\frac{1}{8}$	$7\frac{7}{16}$
5	$24\frac{1}{8}$		$5\frac{5}{8}$	$5\frac{19}{25}$
6	$24\frac{7}{8}$		$4\frac{7}{8}$	$4\frac{23}{24}$
7	$25\frac{4}{8}$		$4\frac{2}{8}$	$4\frac{1}{4}$
8	$26\frac{0}{0}$		$3\frac{6}{8}$	$3\frac{23}{32}$
9	$26\frac{3}{8}$		$3\frac{3}{8}$	$3\frac{11}{36}$
10	$26\frac{6}{8}$		$3\frac{0}{0}$	$2\frac{39}{40}$
12	$27\frac{1}{8}$		$2\frac{5}{8}$	$2\frac{23}{48}$
14	$27\frac{4}{8}$		$2\frac{2}{8}$	$2\frac{1}{8}$
16	$27\frac{6}{8}$		$2\frac{0}{0}$	$1\frac{55}{64}$
18	$27\frac{7}{8}$		$1\frac{7}{8}$	$1\frac{47}{72}$
20	$28\frac{0}{0}$		$1\frac{6}{8}$	$1\frac{9}{80}$
24	$28\frac{2}{8}$		$1\frac{4}{8}$	$1\frac{23}{96}$
28	$28\frac{3}{8}$		$1\frac{3}{8}$	$1\frac{1}{16}$
32	$28\frac{4}{8}$		$1\frac{2}{8}$	$0\frac{119}{128}$

In which experiment it is to be noted, first, that we made use of a glass tube, about 6 feet long, sealed at one end.

Secondly, we had in readiness a glass pipe, about the diameter of a swan's quill,■ which was marked with a paper stuck upon it, divided into inches, and half quarters; which, being immersed in the other cylinder of mercury, and open at both ends, that the mercury might rise in it, it helped to fill the other up. And with about an inch of it standing above the mercury, the orifice was sealed up; so that an inch of air was contained in the tube, which, by lifting up the tube, was gradually expanded to several inches. It being noted, in the meantime, in several steps, how much the mercury in the small tube was, by the expanded air, permitted to rise above the surface of the mercury in the other tube: By which method the former observations being made, we inverted the large tube, and found by trying the Torricellian experiment, that that day the air sustained the mercury at 29¾ inches; wherefore it was observed in making the foregoing observations that the difference between the account which answers our hypothesis and the other, probably resulted from a new access of air to that included inch; and indeed, by removing the tube when the observations were completed, we found that it had gained about half an eighth, which we judged might arise from some bubbles lodged in the pores of the mercury. From which experiment it appeared that the inch of air when expanded to double its dimensions, was able, with a cylinder of mercury about 15 inches, to counterbalance the pressure of the atmosphere, which would raise the mercury eight and twenty inches, when the spring of that air was lost by a further expansion.■ So that the atmosphere here below must consequently be as much compressed, as if twenty-eight inches of mercury gravitated upon it.

> Approximately ¼ inch.

> Since the doubled volume of air required an additional 15 inches of mercury to support the atmosphere, it followed that its pressure had been roughly halved.

SUPPLEMENTARY READING

Conant, J. B. (ed.), *Robert Boyles' Experiments in Pneumatics,* Harvard Case Histories in Experimental Science (Cambridge: Harvard University, 1957), pp. 3ff.

Dampier, W. C., *A History of Science* (New York: Cambridge University, 1946).

Lenard, P., *Great Men of Science* (New York: British Book Centre, 1934), pp. 62ff.

Magie, W. F., *A Source Book in Physics* (New York: McGraw-Hill, 1935), pp. 84ff.

Wolf, A., *A History of Science, Technology, and Philosophy in the 16th and 17th Centuries* (London: Allen and Unwin, 1935), pp. 102ff. and 235ff.

Woodruff, L. L. (ed.), *The Development of the Sciences* (New Haven, Conn.: Yale University, 1923).

4

Isaac Newton

1642 - 1727

The Laws of Motion

THE ESSENTIAL STRENGTH of the science of physics lies in the great depth of its conceptual schemes, in the relatively few principles that serve to unify the entire range of man's knowledge of the physical universe. The first and most remarkable *synthesizer* in the history of science was Isaac Newton, who discovered the law of universal gravitation, completed the formal development of the mechanical principles outlined by Galileo, made important contributions in other branches of physics, and inspired, in part, the *Enlightenment* which climaxed the intellectual revolution in philosophy toward the end of the seventeenth century. His was a modern approach to science in every major respect, yet he practiced alchemy. By this paradox one can visualize how little scientific understanding prevailed even in Newton's time. Physics was still a tottering infant in the last half of the seventeenth century.

Newton was born at Woolsthorpe, in Lincolnshire, England, on December 25, 1642, the year of Galileo's death. His childhood was uneventful, aside from the fact that it was shadowed by the civil war brought on by the excesses of Charles I. He attended village schools until he was twelve, when he was sent to the King's School at Grantham, several miles from his home. Here, though not a precocious youngster, he eventually reached top rank by applying himself diligently, perhaps, as the story goes, because of his desire to best another young man in his class. After attending school for four years he was withdrawn by his mother, recently widowed for the second time, to help with the management of their farm. He was singularly unsuccessful in this enterprise, having no interest whatever in farming; thus, after a short time, he returned to school to prepare for admission to the university.

Newton entered Trinity College, Cambridge, in 1661, the year after the Restoration of Charles II. Probably the effects of the Restoration were still noticeable at the university, judging from reports that the students, because of the unsettled atmosphere, were more than ever boisterous and rowdy. Being very much an introvert, and not given to frequent merrymaking, Newton tended to withdraw from his fellow students and to devote himself to scholarly pursuits. He did not distinguish himself in his early years at college until he came under the tutorship of Isaac Barrow (1630-1677), a distinguished mathematician and Greek scholar, who had transferred from a professorship of Greek to the Lucasian chair of mathematics in 1663. From that point on it seemed that Newton had found himself, for his extraordinary talents soon became evident. Several months after receiving his B.A. degree in 1665 the university was closed because of the prevalence of bubonic plague. Newton returned to his home in Woolsthorpe and spent this period of enforced "idleness" in contemplation of nature's ways. During the next two years, most of which were spent at home, he made his greatest discoveries.

Shortly before leaving Cambridge he had begun an experimental investigation of the nature of white light,[1] and, with the aid of a glass prism, showed it to be a mixture of colors. Apparently he was led to this study by his interest in optics and particularly in the lenses employed in telescopes. These lenses were not color corrected, of course, and the chromatic aberration had the effect of fringing the images with color. He concluded, incorrectly, that a refracting telescope could not be made free of chromatic effects,[2] and consequently constructed the first reflecting telescope. It was for this accomplishment, by the way, that he was elected a fellow of the Royal Society in 1672. He completed his theory of colors during his absence from Cambridge, discovered the binomial theorem, the direct method of *fluxions* (differential calculus), and the inverse method of fluxions (integral calculus), all within a period of several months. His experiments on light were the subject of Newton's first scientific paper, published in the *Philosophical Transactions* (1672).

It was during this period also that he conceived the idea of universal gravitation, the first great inductive generalization. In Newton's own words,

> . . . and in the same year I began to think of gravity extending to the orb of the Moon . . . and having thereby compared the force requisite to keep the Moon in her orb with the force of gravity at the surface of the earth, and found them to answer pretty nearly . . . All this was in the two plague years of 1665 and 1666, for in those days I was in the prime

---

[1] Using sunlight.
[2] The conclusion was correct for a single lens, but chromatic aberration can be reduced to a negligible degree with combinations of lenses.

of my age for invention, and minded mathematics and philosophy more
than at any time since. . . .[3]

The story that he was led to his discovery of universal gravitation by observing an apple fall, even if true, is obviously an exaggeration. Perhaps it was one link in his chain of reasoning, but much more was needed before he could take the flight of imagination that gave him the final *law* of gravitation. His boldest step was in concluding that the same force responsible for attracting an apple to the earth also served to hold the moon in its orbit; that the same laws of motion apply to the planets as pertain to objects falling to the earth. From Kepler's empirical Third Law he was able to deduce the actual law of force governing the planets in their motions about the sun, which he found to be the well-known inverse square attraction. It appears that Robert Hooke (1635–1703), who was curator of experiments to the Royal Society at the time, may have arrived independently at the same law. A controversy developed over priority from which Newton recoiled in typical fashion. He was not one to advance his own cause, and unlike his predecessor, Galileo, had little taste for public debate. There was also a quarrel between Newton's supporters and Gottfried Wilhelm von Leibnitz (1646–1716) concerning the invention of the calculus, which was probably discovered independently by the two men. The question of priority plagued much of the early development of physics. Then, as now, men were anxious to establish priority of discovery, but communication was much slower, which increased the likelihood of simultaneous, though independent, discoveries. In virtually all his major achievements, Newton became involved in arguments regarding priority.

Newton returned to Cambridge in 1667, in which year he was elected a fellow of Trinity College. He received his M.A. degree the following year, and in 1669 he succeeded Dr. Barrow as Lucasian professor of mathematics. He devoted much of his time thereafter to chemical experiments, including alchemy, until he was persuaded by Edmund Halley (1656–1742) in 1684 to occupy himself with research in theoretical mechanics. So intensively did he apply himself to this task that in 1687 he published, at Halley's expense, his crowning achievement, the *Principia*.[4]

The same year saw Newton's first brush with politics. He was a member of a delegation representing the university in its dispute with James II, who had attempted to use his royal powers to grant a degree by mandate. It was feared that King James, an avowed Catholic, had determined to make that

---

[3] Quoted from the *Catalogue of Portsmouth Papers* (Cambridge, Eng.: University Press, 1888), p. xviii. The reason that Newton found the two to agree only "pretty nearly" is attributed to the fact that he used a not too accurate estimate of the size of the earth.

[4] *Philosophiae Naturalis Principia Mathematica.*

faith the established religion of England. The opposition of the university probably was founded not so much on the king's attempt to pre-empt its authority in granting degrees as on the fact that the recipient of this honor was to be a Catholic monk. Shortly afterward James II was deposed in the revolution of 1688–1689 and the English throne presented by Parliament to William and Mary. In 1689 Newton was elected a member of the Convention Parliament for the university; thus he was engaged in political activity at a crucial moment in England's history—a time when Parliament passed a number of sweeping laws designed to safeguard the rights of men and affirmed its own authority over that of the monarchy. Thereafter, Newton remained more or less in public life. He was appointed Warden of the Mint in 1695 with the responsibility for completely recoining the silver currency, which had been severely debased as a result of the civil war. Upon completing this task in 1699, Newton was appointed Master of the Mint, which office he retained until his death in 1727.[5] He was the recipient of many honors and awards; following the death of Robert Hooke in 1703 he became president of the Royal Society, a post to which he was re-elected annually for the remainder of his life. In the same year he published his *Opticks,* a classic work which originated the science of physical optics. Now the elder statesman of science, Newton was knighted by Queen Anne in 1705, the first scientist to receive this distinction. He accomplished little more of scientific note in his remaining years, but devoted much of his time to alchemical studies, religious mysticism, and interpretation of the Scripture.[6] He sought the same things that men before and after him desired, the transmutation of base metals into gold and the elixir of life. His approach to these problems was that of the scientist rather than the mystic, except that his theological interests no doubt influenced his thoughts on alchemy and science generally. While he published no chemical treatises his notes indicate a grasp of chemistry far beyond that of his contemporaries, although far short of his physical insight. How difficult it is to reconcile these pursuits with his clear, incisive views of the physical universe. Yet we must bear in mind that the atomic theory of matter was not established until a century later. Newton recognized his good fortune in being able to build upon the accomplishments of others when he remarked, "If I have seen farther than others, it is because I have stood on the shoulders of giants."

The *Principia* has been termed the greatest of all scientific books. It was written in Latin, not for the educated layman as Galileo wrote, but for other physicists. Its cold, formal style, suggesting an aloof, almost majestic person-

---

[5] In his later years this was largely a sinecure.
[6] He wrote such theological works as *Observations on the Prophecies of Daniel,* and a *Church History.*

ality, resembles more nearly the present-day scientific writing than the fascinating polemics of Galileo. The two were as far apart in their personal habits and their views on science as one could imagine. Galileo was very much the extrovert, Newton cared little for society and companionship. Galileo was the pioneer who had to fashion his science out of the emptiness of scholastic philosophy and in the face of classical authority; Newton's work was in keeping with the intellectual atmosphere of his time. Where Newton devoted much time to his alchemical and theological studies, Galileo ridiculed alchemy and, like da Vinci before him, put theology aside until it conflicted with his science. To Galileo the study of nature was a source of great enjoyment; to Newton, with his remarkable grasp of the ways of nature, more a source of satisfaction.

The *Principia* consists of three "books" plus an introduction. The introduction and first book contain his famous *laws of motion,* which form the starting point of every dynamical proposition, and treat the motion of bodies, without resistance, under various laws of force. The second explores motion in resisting media, i.e., in fluids, an area in which Newton was not entirely successful, while the third book discusses universal gravitation with several astronomical applications.

The following excerpts include first, the definitions that form the bases for his work, followed by the laws of motion. The second and third extracts are taken from Book III of the *Principia,* and consist of his Rules of Reasoning in Philosophy and a part of the General Scholium, which contains his familiar statement that he frames no hypotheses[7]; yet his works are notable for the occasional speculations which they contain.

The translation is by Andrew Motte, taken from the first American edition, 1848.

---

[7] *Hypotheses non fingo.* By which he meant that supposition—that which is not deduced from observation—has no place in science.

Newton's "Experiment"

DEFINITIONS

Definition I

Here mass (quantity of matter) is defined as the product of the volume and the density, and density is not further defined. It would seem,

The quantity of matter is the measure of the same, arising from its density and bulk conjunctly.∎

Thus air of a double density, in a double space, is quadruple in quantity; in a triple space, sextuple in quantity. The same thing is to be understood of snow, and fine dust or powders, that are condensed by compression or liquefaction; and of all bodies that are by any causes whatever differently condensed. I have no regard in this place to a medium, if any such there is, that freely pervades the interstices between the parts of bodies. It is this quantity that I mean hereafter everywhere under the name of body or mass. And the same is known by the weight of each body; for it is proportional to the weight, as I have found by experiments on pendulums, very accurately made, which shall be shown hereafter.

therefore, that Newton took density as a fundamental unit, instead of mass. He later defined *equal densities* in terms of inertia, thereby equating mass with inertia, which is the modern view. To describe mass simply as the quantity of matter in a body, without reference to its inertial properties, would be incorrect.

Definition II

The quantity of motion∎ *is the measure of the same, arising from the velocity and quantity of matter conjunctly.*

By *quantity of motion* was meant momentum, mv.

The motion of the whole is the sum of the motions of all the parts; and therefore in a body double in quantity, with equal velocity, the motion is double; with twice the velocity, it is quadruple.

Definition III

The vis insita, or innate force of matter, is a power of resisting, by which every body, as much as in it lies, endeavors to persevere in its present state, whether it be of rest, or of moving uniformly forward in a right line.

∎This force is ever proportional to the body whose force it is; and differs nothing from the inactivity of the mass, but in our manner of conceiving it. A body, from the inactivity of matter, is not without difficulty put out of its state of rest or motion. Upon which account, this *vis insita*, may, by a most significant name, be called *vis inertiae*, or force of inactivity. But a body exerts this force only, when another force, impressed upon it, endeavors to change its condition; and the exercise of this force may be considered both as resistance and impulse; it is resistance, in so far as the body, for maintaining its present state, withstands the force impressed; it is impulse, in so far as the body, by not easily giving way to the impressed force of another, endeavors to change the state of that other. Resistance is usually ascribed to bodies at rest, and impulse to those in motion; but motion and rest, as commonly conceived, are only relatively distinguished; nor are these bodies always truly at rest, which commonly are taken to be so.∎

The description of inertia.

For example, a body at rest on the surface of the earth is nevertheless moving through space, etc. Newton was quite clear in his understanding of *relative motion.*

Definition IV

An impressed force is an action exerted upon a body, in order to change its state, either of rest, or of moving uniformly forward in a right line.

This force consists in the action only; and remains no longer in the body, when the action is over. For a body maintains every new state it acquires, by its *vis inertiae* only. Impressed forces are of different origins; as from percussion, from pressure, from centripetal force.

Definition V

Newton appears to have intended that all action-at-a-distance forces be included in this definition. Note that he had not yet equated gravity on the earth with the force holding the planets in their orbits, although he suggests (below) that gravity may be the agent.

Note that Newton did not fall into the common error of attributing the tendency to move outward to a *centrifugal force.*

Straight lines.

A centripetal force is that by which bodies are drawn or impelled, or any way tend, towards a point as to a center.■

Of this sort is gravity, by which bodies tend to the center of the earth; magnetism, by which iron tends to the loadstone; and that force, whatever it is, by which the planets are perpetually drawn aside from the rectilinear motions, which otherwise they would pursue, and made to revolve in curvilinear orbits. A stone, whirled about in a sling, endeavors to recede from the hand that turns it; and by that endeavor, distends the sling, and that with so much the greater force, as it is revolved with the greater velocity, and as soon as ever it is let go, flies away. That force which opposes itself to this endeavor, and by which the sling perpetually draws back the stone towards the hand, and retains it in its orbit, because it is directed to the hand as the center of the orbit, I call the centripetal force.■ And the same thing is to be understood of all bodies, revolved in any orbits. They all endeavor to recede from the center of their orbits; and were it not for the opposition of a contrary force which restrains them to, and detains them in their orbits, which I therefore call centripetal, would fly off in right lines,■ with an uniform motion. A projectile, if it was not for the force of gravity, would not deviate towards the earth, but would go off from it in a right line, and that with an uniform motion, if the resistance of the air was taken away. It is by its gravity that it is drawn aside perpetually from its rectilinear course, and made to deviate towards the earth, more or less, according to the force of its gravity, and the velocity of its motion. The less its gravity is, for the quantity of its matter, or the greater the velocity with which it is projected, the less will it deviate from a rectilinear course, and the farther it will go. If a leaden ball, projected from the top of a mountain by force of gunpowder with a given velocity, and in a direction parallel to the horizon, is carried in a curved line to the distance of two miles before it falls to the ground; the same, if the resistance of the air were taken away, with a double or decuple■ velocity, would fly twice or ten times as far. And by increasing the velocity, we may at pleasure increase the distance to which it might be projected, and diminish the curvature of the line, which it might describe, till at last it should fall at the distance of 10, 30, or 90 degrees, or even might go quite round the whole earth before it falls; or lastly, so that it might never fall to the earth, but go forward into the celestial spaces, and proceed in its motion *in infinitum.*■ And after the same manner that a projectile, by the force of gravity, may be made to revolve in an orbit, and go round the whole earth, the moon also, either by the force of gravity, if it is endued■ with gravity, or by any other force, that impels it towards the earth may be perpetually drawn aside towards the earth, out of the rectilinear way, which by its innate force it would pursue; and would be made to revolve in the orbit which it now describes; nor could the moon without some such force, be retained in its orbit. If this force was too small, it would not sufficiently turn the moon out of a rectilinear course: if it was too great, it would turn it too much, and draw down the moon from its orbit towards the earth. It is necessary, that the force be of a just quantity, and it belongs to the mathematicians to find the force, that may serve exactly to retain a body in a given orbit, with a given velocity; and *vice versa,* to determine the curvilinear way, into which a body projected from a given place, with a given velocity, may be made to deviate from its natural rectilinear way, by means of a given force.

■...

SCHOLIUM■

Hitherto I have laid down the definitions of such words as are less known, and explained the sense in which I would have them to be understood in the following discourse. I do not define time, space, place, and motion, as being well known to all. Only I must observe, that the vulgar conceive those quantities under no other notions but from the relation they bear to sensible objects. And thence arise certain prejudices, for the removing of which, it will be convenient to distinguish them into absolute and relative, true and apparent, mathematical and common.

I. Absolute, true, and mathematical time, of itself, and from its own nature flows equably without regard to anything external, and by another name is called duration: relative, apparent, and common time, is some sensible and external (whether accurate or unequable) measure of duration by the means of motion, which is commonly used instead of true time; such as an hour, a day, a month, a year.

Decuple = 10 times.

Thereby becoming a satellite of the earth or perhaps of the sun.

Invested with.

Omitted are several further definitions pertaining to centripetal forces.

An explanatory note.

As Newton himself pointed out (see below), there is no way to distinguish absolute from apparent or relative rectilinear motion. Hence, *absolute motion* or *absolute space* have little meaning. Modern relativity theory denies the existence of absolute space and time, emphasizing the relative nature of these concepts. Since Newtonian dynamics was formulated for inertial systems (i.e., non-accelerated systems) the question of absolute space and time was purely academic, inasmuch as the motion of the system did not affect the dynamical laws. It was not until much later, in connection with electrodynamics, that the problem of relative vs. absolute motion required serious attention.

Superfices: surfaces. That is, two solids may have equal volumes but different surface areas.

As can be seen, relative motion is a concept easily grasped. Not so the concept of absolute motion.

II. Absolute space,■ in its own nature, without regard to anything external, remains always similar and immovable. Relative space is some movable dimension or measure of the absolute spaces; which our senses determine by its position to bodies; and which is vulgarly taken for immovable space; such is the dimension of a subterraneous, an aereal, or celestial space, determined by its position in respect of the earth. Absolute and relative space, are the same in figure and magnitude; but they do not remain always numerically the same. For if the earth, for instance, moves, a space of our air, which relatively and in respect of the earth remains always the same, will at one time be one part of the absolute space into which the air passes; at another time it will be another part of the same, and so, absolutely understood, it will be perpetually mutable.

III. Place is a part of space which a body takes up, and is according to the space, either absolute or relative. I say, a part of space; not the situation, nor the external surface of the body. For the places of equal solids are always equal; but their superfices,■ by reason of their dissimilar figures, are often unequal. Positions properly have no quantity, nor are they so much the places themselves, as the properties of places. The motion of the whole is the same thing with the sum of the motions of the parts; that is, that translation of the whole, out of its place, is the same thing with the sum of the translations of the parts out of their places; and therefore the place of the whole is the same thing with the sum of the places of the parts, and for that reason, it is internal, and in the whole body.

IV. Absolute motion is the translation of a body from one absolute place into another; and relative motion, the translation from one relative place into another. Thus in a ship under sail, the relative place of a body is that part of the ship which the body possesses;■ or that part of its cavity which the body fills, and which therefore moves together with the ship; and relative rest is the continuance of the body in the same part of the ship, or of its cavity. But real, absolute rest, is the continuance of the body in the same part of that immovable space, in which the ship itself, its cavity, and all that it contains, is moved. Wherefore, if the earth is really at rest, the body, which relatively rests in the ship, will really and absolutely move with the same velocity which the ship has on the earth. But if the earth also moves, the true and absolute motion of the body will arise, partly from the true motion of the earth, in immovable space; partly from the relative motion of the ship on the earth; and if the body moves also relatively in the ship; its true motion will arise, partly from the true motion of the earth, in immovable space, and partly from the relative motions as well of the ship on the earth, as of the body in the ship; and from these relative motions will arise the relative motion of the body on the earth.■ As if that part of the earth, where the ship is, was truly moved toward the east, with a velocity of 10010 parts; while the ship itself, with a fresh gale, and full sails, is carried towards the west, with a velocity expressed by 10 of those parts; but a sailor walks in the ship towards the east, with 1 part of the said velocity; then the sailor will be moved truly in immovable space towards the east, with a velocity of 10001 parts, and relatively on the earth towards the west, with a velocity of 9 of those parts.

■..

It is indeed a matter of great difficulty to discover, and effectually to distinguish, the true motions of particular bodies from the apparent; because the parts of that immovable space, in which those motions are performed, do by no means come under the observation of our senses. Yet the thing is not altogether desperate; for we have some arguments to guide us, partly from the apparent motions, which are the differences of the true motions; partly from the forces, which are the causes and effects of the true motions. For instance, if two globes, kept at a given distance one from the other by means of a cord that connects them, were revolved about their common center of gravity, we might, from the tension of the cord, discover the endeavor of the globes to recede from the axis of their motion, and from thence we might compute the quantity of their circular motions.■ And then if any equal forces should be impressed at once on the alternate faces of the globes to augment or diminish their circular motions, from the increase or decrease of the tension of the cord, we might infer the increment or decrement of their motions; and thence would be found on what faces those forces ought to be impressed, that the motions of the globes might be most augmented; that is, we might discover their hindermost faces, or those which, in the circular motion, do follow. But the faces which follow being known, and consequently the opposite ones that precede, we should likewise know the determination of their motions. And thus we might find both the quantity and the determination of this circular motion, even in an immense vacuum, where there was nothing external or sensible with which the globes could be compared. But now, if in that space some remote bodies were placed that kept always a given position one to another, as the fixed stars do in our regions, we could not indeed determine from the relative translation of the globes among those bodies, whether the motion did belong to the globes or to the bodies. But if we

The *true* or absolute motion of the earth could be determined only by reference to *fixed* points in space. The so-called *fixed stars* actually are in motion, but are so distant as to appear relatively fixed.

A section dealing further with efforts to distinguish absolute from relative quantities has been omitted.

The experiment here described points to a basic distinction between linear and angular motion. The idea of absolute angular motion, without reference to the earth or any other system, is more meaningful. Nevertheless, the *absolute* motion of the individual globes could not be determined; the experiment gives only their relative motion with respect to each other.

observed the cord, and found that its tension was that very tension which the motions of the globes required, we might conclude the motion to be in the globes, and the bodies to be at rest; and then, lastly, from the translation of the globes among the bodies, we should find the determination of their motions. But how we are to collect the true motions from their causes, effects, and apparent differences; and, *vice versa,* how from the motions, either true or apparent, we may come to the knowledge of their causes and effects, shall be explained more at large in the following tract. For to this end it was that I composed it.

AXIOMS, OR LAWS OF MOTION■

These are the well-known laws of motion, which form the starting point of every argument in classical dynamics. The first two laws, which relate to inertia, were generalizations from Galileo's observations. The first law, usually known as the *law of inertia,* is clearly a special case of the second.

Law I

Every body perseveres in its state of rest, or of uniform motion in a right line, unless it is compelled to change that state by forces impressed thereon.

Projectiles persevere in their motions, so far as they are not retarded by the resistance of the air, or impelled downwards by the force of gravity. A top, whose parts by their cohesion are perpetually drawn aside from rectilinear motions, does not cease its rotation, otherwise than as it is retarded by the air. The greater bodies of the planets and comets, meeting with less resistance in more free spaces, preserve their motions both progressive and circular for a much longer time.

Law II

By *alteration of motion,* Newton had in mind rate of change of momentum. For the case of constant mass, this becomes $F = ma$. Thus, the second law provides a definition of force in terms of the acceleration given to a mass.

The alteration of motion■ is ever proportional to the motive force impressed; and is made in the direction of the right line in which that force is impressed.

If any force generates a motion, a double force will generate double the motion, a triple force triple the motion, whether that force be impressed altogether and at once, or gradually and successively, And this motion (being always directed the same way with the generating force), if the body moved before, is added to or subducted from the former motion, according as they directly conspire with or are directly contrary to each other; or obliquely joined, when they are oblique, so as to produce a new motion compounded from the determination of both.

Law III▪

To every action there is always opposed an equal reaction: or the mutual actions of two bodies upon each other are always equal, and directed to contrary parts.

Whatever draws or presses another is as much drawn or pressed by that other. If you press a stone with your finger, the finger is also pressed by the stone. If a horse draws a stone tied to a rope, the horse (if I may so say) will be equally drawn back towards the stone: for the distended rope, by the same endeavor to relax or unbend itself, will draw the horse as much towards the stone, as it does the stone towards the horse, and will obstruct the progress of the one as much as it advances that of the other. If a body impinge upon another, and by its force change the motion of the other, that body also (because of the equality of the mutual pressure) will undergo an equal change, in its own motion, towards the contrary part. The changes made by these actions are equal, not in the velocities but in the motions of bodies; that is to say, if the bodies are not hindered by any other impediments. For, because the motions are equally changed, the changes of the velocities made towards contrary parts are reciprocally proportional to the bodies. This law takes place also in attractions, as will be proved in the next scholium.

Corollary I

A body by two forces conjoined will describe the diagonal of a parallelogram, in the same time that it would describe the sides, by those forces apart.▪

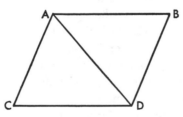

If a body in a given time, by the force *M* impressed apart in the place *A*, should with an uniform motion be carried from *A* to *B*; and by the force *N* impressed apart in the same place, should be carried from *A* to *C*; complete the parallelogram *ABCD*, and, by both forces acting together, it will in the same time be carried in the diagonal from *A* to *D*. For since the force *N* acts in the direction of the line *AC*, parallel to *BD*, this force (by the second law) will not at all alter the velocity generated by the other force *M,* by which the body is carried towards the line *BD.* The body therefore will arrive at the line *BD* in the same time, whether the force *N* be impressed or not; and therefore at the end of that time it will be found somewhere in the line *BD.* By the same argument, at the end of the same time it will be found somewhere in the line *CD.* Therefore it will be found in the point *D,* where both lines meet. But it will move in a right line from *A* to *D,* by Law I.

■.....................................

SCHOLIUM

Hitherto I have laid down such principles as have been received by mathematicians, and are confirmed by abundance of experiments. By the first two laws and the first two corollaries, *Galileo* discovered that the descent of bodies varied as the square of the time *(in duplicata ratione temporis)* and that the motion of projectiles was in the curve of a parabola; experience agreeing with both, unless so far as these motions are a little retarded by the resistance of the air. When a body is falling, the uniform force of its gravity acting equally, impresses, in equal intervals of time, equal forces upon that body, and therefore generates equal velocities; and in the whole time impresses a whole force, and generates a whole velocity proportional to the time. And the spaces described in proportional times are as the product of the velocities and the times; that is, as the squares of the times.■ And when a body is thrown upwards, its uniform gravity impresses forces and reduces velocities proportional to the times; and the times of ascending to the greatest heights are as the velocities to be taken away, and those heights are as the product of the velocities and the times, or as the squares of the velocities.■ And if a body be projected in any direction, the motion arising from its projection is compounded with the motion arising from its gravity. Thus, if the body *A* by its motion of projection alone could describe in a given time the right line *AB,* and with its motion of falling alone could describe in the same time the altitude *AC;* complete the parallelogram *ABCD,* and the body by that compounded motion will at the end of the time be found in the place *D;* and the curved line *AED,* which that body describes, will be a parabola, to which the right line *AB* will be a tangent at *A;* and whose ordinate *BD* will be as the square of the line *AB.*■ On the same laws and corollaries depend those

Omitted are a corollary regarding the composition of forces and two concerning the relative motion of bodies in systems which are themselves in motion.

$s = \frac{1}{2} gt^2$

$v^2 = 2gs$

Since the horizontal motion is uniform, *AB* is proportional to the time, while *BD* depends upon the square of the time. Hence *BD* is as the square of *AB.*

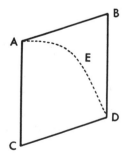

things which have been demonstrated concerning the times of the vibration of pendulums, and are confirmed by the daily experiments of pendulum clocks. By the same, together with Law III, Sir *Christopher Wren,*■ Dr. *Wallis,* and Mr. *Huygens,* the greatest geometers of our times, did severally determine the rules of the impact and reflection of hard bodies, and about the same time communicated their discoveries to the *Royal Society,* exactly agreeing among themselves as to those rules. Dr. *Wallis,* indeed, was somewhat earlier in the publication; then followed Sir *Christopher Wren,* and, lastly, Mr. *Huygens.* But Sir *Christopher Wren* confirmed the truth of the thing before the *Royal Society* by the experiments on pendulums, which M. *Mariotte* soon after thought fit to explain in a treatise entirely upon that subject.■ But to bring this experiment to an accurate agreement with the theory, we are to have due regard as well to the resistance of the air as to the elastic force of the concurring bodies.

Known also as an architect, Wren prepared official reports on the condition of public buildings. Among others, he criticized the design and construction of Old St. Paul's and Westminster Abbey in London and strengthened these structures with iron, rebuilding St. Paul's in the process.

Mariotte also discovered Boyle's law independently.

RULES OF REASONING IN PHILOSOPHY

Rule I

We are to admit no more causes of natural things than such as are both true and sufficient to explain their appearances.■

What might be termed the economy of scientific explanation.

To this purpose the philosophers say that Nature does nothing in vain, and more is in vain when less will serve; for Nature is pleased with simplicity, and affects not the pomp of superfluous causes.

Rule II

Therefore to the same natural effects we must, as far as possible, assign the same causes.■

These rules Newton evidently applied in reaching his law of universal gravitation.

As to respiration in a man and in a beast; the descent of stones in *Europe* and in *America;* the light of our culinary fire and of the sun; the reflection of light in the earth, and in the planets.

Rule III

The qualities of bodies, which admit neither intension nor remission of degrees, and which are found to belong to all bodies within the reach of our experiments, are to be esteemed the universal qualities of all bodies whatsoever.

For since the qualities of bodies are only known to us by experiments, we are to hold for universal all such as universally agree with experiments; and such as are not liable to diminution can never be quite taken away. We are certainly not to relinquish the evidence of experiments for the sake of dreams and vain fictions of our own devising; nor are we to recede from the analogy of Nature, which uses to be■ simple, and always consonant to itself. We no other way know the extension of bodies than by our senses, nor do these reach it in all bodies; but because we perceive extension in all that are sensible, therefore we ascribe it universally to all others also. That abundance of bodies are hard, we learn by experience; and because the hardness of the whole arises from the hardness of the parts, we therefore justly infer the hardness of the undivided particles not only of the bodies we feel but of all others.■ That all bodies are impenetrable, we gather not from reason, but from sensation. The bodies which we handle we find impenetrable, and thence conclude impenetrability to be an universal property of all bodies whatsoever. That all bodies are movable, and endowed with certain powers (which we call the *vires inertiae*■) of persevering in their motion, or in their rest, we only infer from the like properties observed in the bodies which we have seen. The extension, hardness, impenetrability, mobility, and *vis inertiae* of the whole, result from the extension, hardness, impenetrability, mobility, and *vires inertiae* of the parts; and thence we conclude the least particles of all bodies to be also all extended, and hard, and impenetrable, and movable, and endowed with their proper *vires inertiae*. And this is the foundation of all philosophy. Moreover, that the divided but contiguous particles of bodies may be separated from one another, is a matter of observation; and, in the particles that remain undivided, our minds are able to distinguish yet lesser parts, as is mathematically demonstrated. But whether the parts so distinguished, and not yet divided, may, by the powers of Nature, be actually divided and separated from one another, we cannot certainly determine. ■Yet, had we the proof of but one experiment that any undivided particle, in breaking a hard and solid body, suffered a division, we might by virtue of this rule conclude that the undivided as well as the divided particles may be divided and actually separated to infinity.

Lastly, if it universally appears, by experiments and astronomical observations, that all bodies about the earth gravitate towards the earth, and that in proportion to the quantity of matter which they severally contain; that the moon likewise, according to the quantity of its matter, gravitates towards the earth; that, on the other hand, our sea gravitates toward the moon; and all the planets mutually one towards another; and

Which uses to be: which is inclined to be.

Not on the modern view of the structure of matter, for if we change the state of a substance, say from solid to liquid, we do not alter its *parts* (molecules), but rather the forces acting between them.

Inertia.

Newton points out that one exception may destroy an hypothesis.

the comets in like manner towards the sun;■ we must, in consequence of this rule, universally allow that all bodies whatsoever are endowed with a principle of mutual gravitation. For the argument from the appearances concludes with more force for the universal gravitation of all bodies than for their impenetrability; of which, among those in the celestial regions, we have no experiments, nor any manner of observation. Not that I affirm gravity to be essential to bodies: by their *vis insita* I mean nothing but their *vis inertiae*. This is immutable. Their gravity is diminished as they recede from the earth.

> Induction of the general features of universal gravitation.

Rule IV

In experimental philosophy we are to look upon propositions collected by general induction from phenomena as accurately or very nearly true, notwithstanding any contrary hypotheses that may be imagined, till such time as other phenomena occur, by which they may either be made more accurate, or liable to exceptions.■

This rule must follow, that the argument of induction may not be evaded by hypotheses.

> That is, induction from observation is to be more highly regarded than pure imagination.

Hitherto we have explained the phenomena of the heavens and of our sea by the power of gravity, but have not yet assigned the cause of this power.■ This is certain, that it must proceed from a cause that penetrates to the very centers of the sun and planets, without suffering the least diminution of its force; that operates not according to the quantity of the surfaces of the particles upon which it acts (as mechanical causes used to do), but according to the quantity of the solid matter which they contain, and propagates its virtue on all sides to immense distances, decreasing always as the inverse square of the distances.■ Gravitation towards the sun is made up out of the gravitations towards the several particles of which the body of the sun is composed; and in receding from the sun decreases accurately as the inverse square of the distances as far as the orbit of Saturn, as evidently appears from the quiescence of the aphelion■ of the planets; nay, and even to the remotest aphelion of the comets, if those aphelions are also quiescent. But hitherto I have not been able to discover the cause of those properties of gravity from phenomena, and I frame no hypotheses; for whatever is not deduced from the phenomena is to be called an hypothesis; and hypotheses, whether metaphysical or physical, whether of occult qualities or mechanical, have no place in experimental philosophy.■ In this philosophy particular propositions are inferred from the phenomena, and afterwards rendered general by induction. Thus it was that the impenetrability, the mobility,

> The law of universal gravitation was included at the very end of the *Principia*.

> The modern statement relates the force to the product of the masses and the inverse square of the distance between *centers of gravity,* which may not coincide with the geometrical centers.

> Aphelion: The point in the orbit farthest from the sun.

> The modern view does not place such a restriction upon hypothesis. In fact, Newton

himself resorted many times to speculation. Presumably he meant only that experiment was to take precedence, in case of a conflict between the two.

Here Newton speculates on the nature of "action-at-a-distance" forces.

and the impulsive force of bodies, and the laws of motion and of gravitation, were discovered. And to us it is enough that gravity does really exist, and act according to the laws which we have explained, and abundantly serves to account for all the motions of the celestial bodies, and of our sea.

And now we might add something concerning a certain most subtle spirit which pervades and lies hid in all gross bodies;■ by the force and action of which spirit the particles of bodies attract one another at near distances, and cohere, if contiguous; and electric bodies operate to greater distances, as well repelling as attracting the neighboring corpuscles; and light is emitted, reflected, refracted, inflected, and heats bodies; and all sensation is excited, and the members of animal bodies move at the command of the will, namely, by the vibrations of this spirit, mutually propagated along the solid filaments of the nerves, from the outward organs of sense to the brain, and from the brain into the muscles. But these are things that cannot be explained in few words, nor are we furnished with that sufficiency of experiments which is required to an accurate determination and demonstration of the laws by which this electric and elastic spirit operates.

SUPPLEMENTARY READING

Andrade, E. N. da C., *Sir Isaac Newton* (New York: Chanticleer Press, 1950).

Cohen, I. B., "Isaac Newton" in *Lives in Science* by the Editors of Scientific American (New York: Simon and Schuster, 1957), pp. 21ff.

———— (ed.), *Isaac Newton's Papers and Letters on Natural Philosophy* (Cambridge, Mass.: Harvard University Press, 1958).

Crew, H., "Portraits of Famous Physicists," *Scripta Mathematica*, Pictorial Mathematics, Portfolio No. 4, 1942.

Crowther, J. G., *Six Great Scientists* (New York: Hamilton, 1955), pp. 89ff.

Knedler, J. W. (ed.), *Masterworks of Science* (New York: Doubleday, 1947), pp. 172ff.

Magie, W. F., *A Source Book in Physics* (New York: McGraw-Hill, 1935), pp. 30ff.

More, L. T., *Isaac Newton, a Biography* (New York: Scribner's, 1934).

Newton, I., *Mathematical Principles of Natural Philosophy and His System of the World,* trans. by A. Motte, ed. by F. Cajorie (Berkeley: University of California, 1947).

The Royal Society, *Newton Tercentenary Celebrations* (Cambridge, Eng.: Cambridge University, 1947).

Wolf, A., *A History of Science, Technology, and Philosophy in the 16th and 17th Centuries* (London: Allen and Unwin, 1935), pp. 145ff.

5

Charles Coulomb
·1736-1806

The Laws of Electric and Magnetic Force

WHILE various electric and magnetic phenomena had long been recognized, it was not until the late sixteenth century that William Gilbert (1540–1603) set the stage for the formal development of these branches of science, and not until the late eighteenth century that the first major advances were recorded.

Gilbert, court physician to Queen Elizabeth I and president of the College of Physicians, devoted much of his life to the study of magnetism. In 1600 he published his great work, *De Magnete,*[1] the result of some seventeen years of research during which he collected all that was then known about electricity and magnetism and performed many experiments of his own. He investigated the attraction between magnets and the electrostatic effects observed when certain bodies were rubbed.[2] His observations, important as they were to navigation and to experimental science generally, were otherwise largely qualitative and hence provided little understanding of the nature of the phenomena proper. He wrote, for the most part, within the framework of Aristotelian thought. Nevertheless, his work became a significant landmark because of its thoroughness and occasional departures from classical tradition. His book was the first important scientific work to be published in England.

During the seventeenth century much attention was given to terrestrial magnetism but very little to electricity. The *Accademia del Cimento*[3]

---

[1] *On the Magnet, Magnetic Bodies Also, and on the Great Magnet the Earth, a New Physiology* (Translated by S. P. Thompson, London: Chiswick Press, 1900). See also the English version edited by D. J. Price (New York: Basic Books, 1958).

[2] He coined the name *electricity* from the Greek work for amber.

[3] Academy of Experiment.

experimented with the electrification of various substances by rubbing, Otto von Guericke (1602–1686) developed a machine which produced *frictional* electricity,[4] and Newton speculated that the force of a magnet (the magnetic field) varied inversely as the cube of the distance.[5] Beyond these few investigations it would appear that seventeenth-century scientists were too much preoccupied with researches in other branches of physics (such as mechanics and optics) to be concerned with electrical experiments. In the eighteenth century, however, quite the opposite occurred. The earlier experiments of Gilbert and von Guericke on frictional electricity led the way to systematic investigations of this phenomenon, including attempts to account for its behavior. Again, the observations were almost wholly qualitative, consisting mainly in the development of new frictional machines and in numerous experiments conducted chiefly for the amusement they afforded. The famous American statesman, Benjamin Franklin (1706–1790), although not trained in science, made a notable contribution when he proposed his *one-fluid* theory of electricity to account for the observed phenomena.[6] The *art* of performing electrical demonstrations developed rapidly during this period, and no doubt contributed greatly to the over-all progress in this field. What was lacking, however, was some quantitative knowledge of the forces acting between charged bodies. It was this that Coulomb provided late in the eighteenth century.

Charles Augustin Coulomb was born on June 14, 1736, into a family of high social position in Angoulême, in the south of France. He grew up in an age of political unrest, when much of France felt the influence of the liberal theories of Voltaire (1694–1778) and the democratic ideals of Jean Jacques Rousseau (1712–1778). He was educated for a military career, mostly in Paris, where he studied science and mathematics. Suiting his capabilities he began his career as a military engineer, spending several years as an officer in Martinique, where he supervised the construction of fortifications.

Probably his scientific interests grew out of this work, and upon returning to France in 1776, he settled in Paris, where he began to devote full time to these interests. The next thirteen years, until the Revolution broke out in 1789, were Coulomb's most productive. He first attracted attention by winning a prize offered by the Academy of Sciences for the best method of constructing a ship's compass. His memoir, entitled *Théorie des Machines*

---

[4] It consisted of a sphere of sulphur, arranged so as to rotate while one held his hand on it, thereby producing a charge.

[5] *Principia*, Book III (Correct for distances large compared with the length of the magnet).

[6] Our present view of electricity is essentially a one-fluid theory, the electrons constituting the fluid. While we have two kinds of electricity, positive and negative, only the negative moves (in solid conductors).

*Simples,** won for him both the prize and membership in the Academy. It was while investigating this problem that Coulomb invented his torsion balance (about 1784). The Rev. John Michell, in England, also invented a torsion balance, used later by Henry Cavendish (1731–1810) to measure the density of the earth, but it appears that while Michell's discovery probably preceded Coulomb's, the two were arrived at independently.

Coulomb carried out intensive investigations with his torsion balances, described in considerable detail in papers published in the *Mémoires de l'Académie Royale des Sciences* beginning in 1784. In 1789, with the storming of the Bastille, Coulomb thought it prudent to leave Paris for the relative obscurity of his small estate near Blois. Accordingly, he resigned all his official positions, military and academic, and went into semiretirement. He continued his investigations, however, and when Napoleon established the Consulate in 1799, Coulomb returned to Paris, where he lived until his death in 1806.

Apart from his determination of the laws of electric and magnetic force, Coulomb was responsible for a number of other significant contributions to electricity. He showed how the charge was distributed on the surface of a conductor, and recognized this as a consequence of the mutual repulsion of like charges according to an inverse square law.[7] His accomplishments advanced electrostatics enormously, and with him ended the first major period in the development of this branch of physics.

The following extracts, which describe Coulomb's measurements relating to the laws of electric and magnetic force, are taken from his first and second memoirs on electricity and magnetism, published in the *Institut de France, Mémoires de l'Académie des Sciences* (1785), pages 569ff. and 578ff.

---

* The Theory of Simple Machines.

[7] In this he was anticipated by Henry Cavendish (1731–1810), who unfortunately failed to publish his conclusions, found much later among his unpublished papers.

Coulomb's Experiment

Experimental determination of the law according to which bodies charged with the same type of electricity repel each other.

In a memoir presented to the Academy in 1784,■ I determined by experiment the laws governing the torsion in a metal thread. I found that this force is proportional to the angle of twist, to the fourth power of the diameter, and inversely proportional to the thread's length. The

The prize memoir.

constant of proportionality depends upon the metal used, and can be determined experimentally.

In the same memoir I showed that by using this force of torsion it is possible to measure accurately very small forces; for instance one ten-thousandth of a grain. In the same memoir I gave the first application of this theory, an attempt to evaluate the constant force attributed to adhesion in the formula that expresses the friction on the surface of a solid body moving through a fluid.

I submit to the Academy today an electric balance built according to the same principles; it measures very exactly the state and electric force of a body, however small its charge.

CONSTRUCTION OF THE BALANCE

Figures 1-5 appear on page 72.

Experience has shown me that to perform several electric experiments in a convenient way I should correct some mistakes made when building the first balance of this type. However, since the balance is the first of this type, I shall describe it, keeping in mind that the accuracy and precision of the experiments I am going to perform may vary. Figure 1■ shows a detailed diagram of the complete balance.

1 line probably = $\frac{1}{12}$ inch

Over a glass cylinder 12 inches in diameter and 12 inches high, a flat piece of glass 13 inches in diameter was placed, which covers the entire structure. This plate has two holes of about 20 lines■ in diameter, one of them at its center, f, on which a glass tube 24 inches high is placed. This tube is cemented in place with the cement currently used in electrical apparatus. On top of the tube at h is placed a torsion micrometer, shown in detail in Figure 2. The top part of this micrometer, No. 1, has a knob b, index io, and a suspension clamp q, which fits into the hole G of part No. 2. Part No. 2 is made up of a circle ab, which has a 360° scale on its edge, and a copper tube Φ that fits into the hole H of part No. 3, which is attached to the top of the glass tube fh of Figure 1.

The clamp q (Fig. 2, No. 1) has nearly the form of the tip of a crayon holder, which can be narrowed by means of the ring q. It is in this clamp that one end of a very fine silver wire is placed. The other end of this wire is attached at P (Fig. 3) by means of a clamp on the rod Po. This rod is made of copper or iron and its diameter is barely one line. The upper end P is split, making a clamp that is closed by means of the sliding ring ϕ. This cylinder is enlarged and pierced at C by a sliding needle ag. The whole weight of the cylinder must be such that it can keep the silver wire stretched without breaking it. The needle ag, as can be seen in Figure 1, is suspended horizontally at its center and at about half the height of the glass container. It is made either of a silk thread coated with Spanish wax or of a straw similarly coated. It is about 18 lines long and terminates in a cylindrical thread of shellac; at one end of this needle is placed a small ball which is made of pith and is two or three lines in diameter. At the end *g* is a small piece of paper soaked in turpentine;■ this paper counterbalances the ball *a* and slows down the oscillations.

We have said that the glass cover *AC* has a second hole *m;* it is through this hole that a small rod *mϕb* is introduced. The lower part of this rod (*ϕb*) is made of shellac, and at *b* terminates in another small pith ball. Around the glass container, at the height of the needle, is a scale *ZQ* divided into 360 degrees. This scale was made for simplicity out of paper fastened around the container at about the height of the needle.

To start the operation of this instrument the flat glass cover is placed more or less in position by lining up the hole *m* with the first division of the scale *ZOQ* drawn on the glass container. The index *io* of the micrometer is placed at the zero mark of the micrometer scale. Then the entire micrometer is turned on the glass tube *fh* until, by looking past the vertical thread and the center of the ball, the needle *ag* is seen at the first division of the scale *ZOQ* (that is, when the center of the ball *a* is lined up with the zero mark). The ball *b,* suspended by the thread *mϕb,* is then introduced through the hole *m*, in such a way that it touches the ball *a,* and that the center of the ball (*b*) and the suspension thread are lined up with the zero mark of the scale *ZOQ.* The balance is now ready to operate; we shall, for example, determine the fundamental law by which electrified bodies repel each other.

FUNDAMENTAL LAW OF ELECTRICITY

The repulsive force between two small spheres charged with the same type of electricity is inversely proportional to the square of the distance between the centers of the two spheres.■

Experiment

A small conductor is charged,■ this is nothing but a large-headed pin insulated by inserting it at the end of a rod of Spanish wax (Fig. 4). This pin is introduced through the hole *m* and touches the ball *b*, which is in contact with ball *a*. Upon removing the pin the two balls are charged with the same nature of charge,■ and separate from each other by a distance that can be measured by lining up the suspension thread

The purpose of the turpentine is not apparent, unless it was to stiffen the paper somewhat.

Coulomb's law for like charges.

Probably by touching it to another charged object.

If the two balls were of the same diameter, they would assume equal charges.

and the center of the ball *a* with the corresponding division on the $\mathcal{Z}OQ$ scale. The index of the micrometer is now turned in the sense *pno;* the suspension thread *lP* is thus twisted and a torsional force proportional to the angle of twist is produced which tends to pull the ball *a* toward the ball *b*. Comparing the torsional force with the distance between the two balls, the law of repulsion is determinated. Here, I intend only to carry out some trials which are easily reproduced, and which will make evident the law of repulsion.

First Trial. Having charged the two balls with the head of the pin with the micrometer index set at *O,* the ball *a* of the needle is separated from the ball *t* by 36 degrees.

Second Trial. Turning the suspension thread through 126 degrees by means of the knob *O* of the micrometer, the two balls are found separated and at rest at 18 degress from one another.■

Third Trial. After turning the suspension thread through 567 degrees, the two balls are separated by 8 degrees and a half.

That is, turning the suspension so as to counteract the repulsion of the charged balls.

Explanation and Result of This Experiment

When the balls are not charged they touch each other and the center of ball *a,* held in place by the needle, is not displaced by more than half the diameters of the two balls from the point where the torque due to the suspension thread is negligible. It is worthwhile to note here that the silver thread *lP* which provides suspension is 28 inches long, and so thin that a foot of it is not heavier than $\frac{1}{16}$ grain.■ To calculate the force necessary to twist this thread, by acting on the point *a* which is four inches away from the wire (*lP*) or from the center of suspension, I have found by using the formulas explained in a paper on the laws of torsion in metal threads, published in the volume of the Academy for 1784, that to twist this thread through 360 degrees it is necessary to apply at point *a,* a force of $\frac{1}{340}$ grain to operate the lever *aP* which is four inches; since the torsional forces are proportional to the angle of twist, the least repulsive force between the two balls will cause a marked separation one from the other, as proved in the aforementioned paper.

In our first trial we found that where the index of the micrometer is over the point *O,* the balls are separated 36 degrees, which therefore produces a torsional force of $36° = \frac{1}{3400}$ grain; in the second trial the distance between the balls is 18 degrees, but since the micrometer has been turned through 126 degrees, it follows that at a distance of 18 degrees the repulsive force is equivalent to 144 degrees. That is, at half the first separation the repulsion of the balls is four times as great.

437.5 grains = 1 ounce (av.)

In the third trial the suspension thread was turned 567 degrees, and the two balls were only 8½ degrees apart, the total torsion is therefore 576 degrees; four times that of the second trial, and the distance between the balls lacked only half a degree of being decreased to half of that at which it stood in the second trial.

It results, then, from these three trials, that two balls charged with the same type of electricity exert a repulsive force on each other, inversely proportional to the square of the distance between them.

■...

The electric balance which I presented to the Academy in June 1784, measured accurately and in a simple and direct way, the repulsion of two balls carrying electricity of the same nature. By using this balance it has been proved that the repulsive action of the two balls, charged with the same type of electricity and placed at different distances, is very exactly inversely proportional to the square of the distance between them.

I wished to use the same method to determine the attractive force between two balls charged with a different nature of electricity but by using this same balance to measure the attractive force, I found an experimental difficulty that did not occur when measuring the repulsive force. The experimental difficulty arises when the two balls are drawn near to each other. The attractive force which increases, as we have clearly seen, according to the inverse square law of distances, frequently increases at a greater rate than the torsional force, which increases only directly as the angle of twist;■ consequently, if several observations are desired, the balls must be prevented from touching each other by means of an insulating stop in the path of the needle. But since the balance is often required to measure forces of less than ¹⁄₁₀₀₀ grain, the collision of the needle and the obstacle influences the results and causes part of the electric charge to be lost.

Figure 5, and the calculations that I shall make, consequently will show the difficulties of operation and, at the same time, the way in which it is necessary to reform the experiments to assure success.

Let *aca'* be the natural position of the needle before the suspension thread is twisted; *a* represents the pith ball attached to the needle *aa'* of an insulating material; *b* is the ball suspended through the hole in the balance. Suppose the two bodies are charged; one with electricity that will be called positive and the other with electricity that will be called negative;■ they will attract each other; the ball at *a*, in the needle, tends to approach the ball *b*, taking the position *φcφ'*. At this position we will show that the angle by which the suspension is twisted is equal to the attractive force between the two balls; and that this force is inversely

Omitted are some remarks concerning various technical problems and the interpretation of the results. The second memoir, dealing with the attractive force between unlike charges, begins here.

This applies when the distance between the balls is less than some critical value, as shown below.

Benjamin Franklin had by this time assigned such polarities on the basis of his *one-fluid* theory of electricity.

proportional to the square of the distance, as we have shown the repulsive force to be. Setting $ab = a$, $a\phi = x$, $D =$ the product of the electric charges of the two balls. The arcs a and x are small enough to be taken as the distance between the two balls (otherwise it is necessary to make the required corrections). It will then be found with these assumptions that equilibrium between the attraction of the two balls and the repulsive

Thus, n must be the torsion constant of the system, measured in terms of the force (at a) required to move the ball through unit distance.

torque, is given by the formula $nx = \dfrac{D}{(a - x)^2}$ ■ or, $D = nx(a - x)^2$, and from this, when $x = a$ or O the value of D is negligible, and there is a point ϕ between a and b where the quantity D is a maximum. Calculating this point it is found that $x = \frac{1}{3} a$.■ Substituting this value of x in the formula that represents D in the case of equilibrium, it is found that $D = \frac{4}{27} na^3$, and consequently every time that D is greater than $\frac{4}{27} na^3$, there will not be a point ϕ between a and b where the needle can remain in equilibrium and the balls will necessarily touch. It is worthwhile to mention that in practice, even if D is smaller than $\frac{4}{27} na^3$, the balls often join, because the flexibility of the suspension permits it to oscillate somewhat, and in passing $\frac{1}{3} a$ the attractive force increases at a greater rate than the force of torsion; therefore, when the ball ϕ,■ because of its amplitude of oscillation, arrives at a distance x where D is greater than $nx(a - x)^2$, the balls will continue to approach each other until they touch.

Found by setting
$\dfrac{dD}{dx} = 0$

That is, when the ball a reaches the position ϕ.

Guided by this theory I have placed in equilibrium, at different distances, the attractive force of the two charged balls, and the torsional force of my micrometer. Comparing then the different results I have concluded that the attractive force between the charged balls, one charged positively and the other negatively, is inversely proportional to the square of the distance between the centers of the balls, the same result that I have shown for repulsive forces.

■...

An alternate method, involving the attractive force between a large and a small object, is omitted.

Magnetic bodies attract or repel each other at finite distances in the same way as do charged bodies. The magnetic fluid seems to have, if not by its nature, at least by its properties an analogy with the electric fluid. Based on this analogy it can be assumed that the two fluids obey the same laws. In all other phenomena of attractions or repulsion that nature presents to us, for instance elasticity and chemical affinity, the forces seem to be exerted only at very small distances, and it seems, therefore, that they are nothing but the same laws of electricity and magnetism. As a matter of fact, by calculating by theoretical means the attraction and repulsion of the elements of a body, we know that the molecules repel or attract each other always by forces that are inversely proportional to the cube of the distances (or to a smaller power).■ For instance, bodies can act one upon the other, at finite distances; if the action of the molecules depended upon an inverse proportionality as the cube of the distance (or higher power),■ their bodies could not act one upon the other in general but only at infinitely small distances.

In this research we have used two methods to determine experimentally the law according to which the magnetic fluid acts. The first of these methods consists in suspending a magnetic needle in the magnetic meridian, approaching this needle with another one situated conveniently, and determining by calculations and observation, at different distances, the force with which the magnetic fluid of one of the needles acts on the magnetic fluid of the other. In the second method, use is made of a magnetic balance, more or less similar to the electric balance described previously. Having to report in detail our observation, it is necessary to review the known properties of magnetic needles that will be useful to us in this regard.

A needle 24 inches long or more, of good steel strongly tempered and magnetized by the double-touch method■ that Mr. Aepinus describes and practices as an application of his excellent theory of electricity and magnetism, has a pole at each of its ends, its magnetic center being more or less at its center.

With two magnetic needles, poles of the same kind repel each other while poles of different kind attract each other. This attraction increases as the distance between the ends of the needles decreases.

If a magnetic needle is suspended horizontally in such a way that it is able to turn freely about its own center, it will always place itself in the same direction, which is called the magnetic meridian. This meridian makes an angle with the earth's meridian which varies in the course of a day from hour to hour by a sort of periodic motion;■ it also varies, probably periodically, during the course of a year. How long it remains at any point on the earth is still unknown to us. If a horizontally suspended needle is made to oscillate it will displace itself equally from the two sides of the magnetic meridian, being attracted to its original position by a force which is easily determined by observing the time required for the oscillations and knowing the form and weight of the needle. (Refer to the 7th volume of Savant Etrangers, *Mémoirs de l'Académie*.)

DESCRIPTION OF THE EXPERIMENT

From a drawing plate■ I took a wire of excellent steel, 25 inches long and 1½ lines in diameter, which was magnetized by the method of

---

Margin notes:

Coulomb could not have had a very clear concept of the nature of matter, for Dalton's work came some twenty years later. But he evidently held some form of molecular view.

That is, the force would then have a shorter range.

By the use of two magnets stroked outward from the center of the needle.

The *magnetic declination*. Daily variations of 4-5 minutes on either side of the mean are typical.

A die for drawing wire to a given diameter.

double touch, with its magnetic center about at its middle. I then suspended a magnetic needle by means of a silk thread drawn from a cocoon and 3 inches long; when this needle came to rest I traced its magnetic meridian for a distance of two feet from the center of suspension. Then, I traced lines perpendicular to this magnetic meridian, and the steel wire, being placed along these perpendiculars, was moved until the needle aligned itself in the magnetic meridian; the latter was traced, of course, before the magnetic needle was brought nearby.

I then observed, by placing my magnetized wire closer or farther from the suspended needle, how far beyond the meridian the end of my wire is when the needle returns to its original position.

First Experiment

The wire was placed along that perpendicular to the meridian which was at a distance of 1 inch from the near end of the needle. The wire was then moved back and forth along this perpendicular until the needle was restored to the meridian, at which time the end of the wire was found to be displaced by +10 lines.

Trial 1   The wire is placed at a distance of 1 inch (from the end of the needle). The end displaced from the magnetic meridian + 10 lines.

Trial 2   The wire is placed at a distance of 2 inches (from the end of the needle). The end displaced from the magnetic meridian + 9 lines.

Trial 3   The wire is placed at a distance of 4 inches (from the end of the needle). The end displaced from the magnetic meridian + 8 lines.

Trial 4   The wire is placed at a distance of 8 inches (from the end of the needle). The end displaced from the magnetic meridian − 4 lines.

Trial 5   The wire is placed at a distance 16 inches (from the end of the needle). The end displaced from the magnetic meridian − 42 lines.

Second Experiment

Essentially a *magnetometer*.

A magnetic needle 2 inches in length was suspended horizontally by its center; it was free, aided only by the magnetic force of the earth, and made 34 oscillations in 60 seconds. Now we took the magnetic wire of the preceding experiment, which was 25 inches long, but instead of placing it horizontal and perpendicular to the magnetic meridian it was placed vertically in the meridian at 2 inches beyond the end of the suspended needle with its south pole down. ∎ The south pole of the vertical wire will thus attract the north pole of the needle and will make it dip vertically, the distance from the tip of the needle being kept at 2 inches, the number of oscillations that the needle makes was determined. All through these measurements the bottom of the steel wire was at or below the former level of the needle. Following are the results of this experiment:

Trial 1   The end of the wire at the same level as the needle; 120 osc in 60 sec.

Trial 2 The end is lowered 6 lines; 122 osc in 60 sec.
Trial 3 The end is lowered 1 inch; 122 osc in 60 sec.
Trial 4 The end is lowered 2 inches; 115 osc in 60 sec.
Trial 5 The end is lowered 3 inches; 112 osc in 60 sec.
Trial 6 The end is lowered 4 inches; 98 osc in 60 sec.
Trial 7 The end is lowered 8 inches; 39 osc in 60 sec.

Third Experiment

A needle 4 inches long was suspended at the position of the first; the steel wire was placed vertically 3 inches from the end of this needle, as in the preceding experiment where the procedure is given. The needle being free and acted upon only by the force of the earth's flux made 53 osc in 60 sec.

Trial 1 The end of the wire at the same level as the needle; 152 osc in 60 sec.
Trial 2 The end is lowered 1 inch; 152 osc in 60 sec.
Trial 3 The end is lowered 2 inches; 148 osc in 60 sec.
Trial 4 The end is lowered 4 inches; 120 osc in 60 sec.
Trial 5 The end is lowered 8 inches; 58 osc in 60 sec.

EXPLANATION AND RESULT OF THE THREE EXPERIMENTS

The preceding experiments demonstrate that the center of action of each half of our wire is located more or less at the end points of the wire in such a way that since our steel wire is 25 inches long, it can be assumed, without much error, that the magnetic fluid is concentrated within two or three inches from each end.

As a matter of fact, during the first experiment, where the steel wire was placed horizontally and perpendicular to the magnetic meridian in which the needle was suspended, there were two forces acting on the needle: the magnetic force that holds it in position, and the magnetic force from the different points of the magnetized wire, but since in our first experiment the needle remained over its magnetic meridian at all times, it follows that all the magnetic forces due to the steel wire, 25 inches long, acting on the needle were balanced among themselves. Therefore, in the first three trials, where the distances were 1, 2, and 4 inches, the magnetic forces acting 8 to 10 lines from the end of the needle are in equilibrium with all the rest of the needle. It seems, therefore, that it can be assumed that half of the magnetic fluid is concentrated within the last 10 lines of the end of the needle.

The second and third experiments yield the same results. In these experiments the steel wire was placed vertically in the magnetic meridian of the needle; consequently, the part at the top of the wire is very oblique to the suspended needle and acts at a greater distance, being able to influence the oscillation but a slight amount. It can be seen that in these two experiments the greater number of oscillations of the suspended needle resulted when the end of the wire was lowered somewhat less than 1 inch below the level of the suspended needle. It is seen, therefore, that the average force due to the lower half of the steel wire is concentrated at some 8 to 10 lines from the end, as we have shown in the first experiment. From this we conclude that a steel wire 25 inches long, magnetized by the double touch method, has its magnetic fluid concentrated about 10 lines from its extremes.■ This information is necessary in order to find the law of attraction and repulsion relative to distance.

The magnetic fluid acts by attraction or repulsion in proportion to the density of fluid and inversely proportional to the square of the distance of its molecules.

The first part of this statement does not require proof;■ hence, let us proceed to the second.

We have seen that the magnetic fluid in our steel wire 25 inches long was concentrated at the ends over a length of 2 to 3 inches; that the center of action of each half of the steel needle was about 10 lines from its ends. Therefore, by setting up a few inches away from our steel wire a very small needle which, as we shall see later, has its magnetic fluid concentrated 1 or 2 lines from its end points, we can calculate the mutual action of the wire on the needle, and of the needle on the wire by supposing that the magnetic fluid in the steel wire is concentrated at points 10 lines from its ends, and in a needle 1 inch in length, at points 1 to 2 lines from the ends. These observations will guide us in the following experiment:

The poles of a magnet are not located at the very ends of the magnet but at some small distance from the ends.

While true, it is somewhat surprising that Coulomb offered no proof of it.

Fourth Experiment

A steel wire weighing 70 grains and one inch long, magnetized by the method of double touch, was suspended by means of a silk thread 3 lines long, made from a single fiber drawn from a cocoon. After it was lined up with the magnetic meridian, a steel wire 25 inches long was placed vertically in the meridian at different distances in such a way that its lower end was always 10 lines below the level of the suspended needle. By varying the distance and causing the needle to oscillate the number of oscillations found in equal times was recorded, with the following results:

Trial 1   The needle free to oscillate due to the action of the earth, makes 15 osc in 60 sec.

Trial 2   The wire placed at 4 inches from the center of the needle makes 41 osc in 60 sec.

Trial 3   The wire placed 8 inches from the center of the needle makes 24 osc in 60 sec.

Trial 4   The wire placed 16 inches from the center of the needle makes 17 osc in 60 sec.

EXPLANATION AND RESULT OF
THIS EXPERIMENT

When a freely suspended pendulum is set into oscillation by forces acting in a given direction, the forces are proportional to the inverse square of the time required to make a given number of oscillations;■ or what is the same thing, to the square of the number of oscillations made in the same time.

For any simple harmonic motion, the period, T, is given by:
$$T = 2\pi \sqrt{m/k}$$
where m is the mass and k the force constant of the system. In the rotational case
$$T = 2\pi \sqrt{I/c_0}$$
where I is the moment of inertia and c_0 the restoring torque per unit twist.

In the preceding experiment the needle oscillates by virtue of two different forces; one is the magnetic force of the earth, the other the action of all points of the wire on points of the needle. In our experiment all forces are in the plane of the magnetic meridian; and the needle being suspended horizontally, can oscillate only because of the horizontal components of the forces.

Since we have seen, in the three preceding experiments, that the magnetic fluid is concentrated at the ends of the wire, perhaps at 10 lines from the end of the wire, and since the suspended needle is one inch long, the north end is attracted from a distance of $3\frac{1}{2}$ inches and the south end repelled by the lower pole of the needle, which is $4\frac{1}{2}$ inches distant from it; therefore, it may be supposed without serious error that the mean distance at which the lower pole of the steel wire exerts its action on the two poles of the needle is 4 inches. Consequently, if the action of the magnetic fluid is proportional to the inverse square of the distance, the action of the lower pole of the steel wire on the needle should be proportional to $\frac{1}{4}^2$, $\frac{1}{8}^2$, $\frac{1}{16}^2$; or to 1, $\frac{1}{4}$, $\frac{1}{16}$.■

Referring the others to $\frac{1}{4}^2$.

But, since the horizontal forces that make the needle oscillate are proportional to the square of the number of oscillations made in equal turns, and by virtue of the magnetic force of the earth alone the free needle makes 15 osc in 60 sec., this last force will be measured by the square of the number of oscillations, that is, by 15^2. In the second trial, the two forces due to the steel wire and the earth cause the needle to make 41 osc in 60 sec.; therefore these two forces together are measured

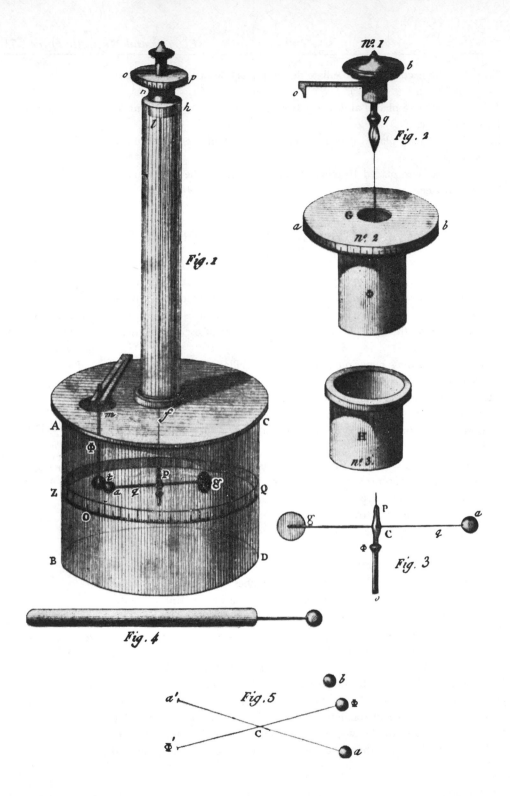

by 41^2. The force due to the action of the magnetized wire alone is thus measured by the difference between these two squares; it is therefore proportional to $41^2 - 15^2$. We have then, for the action of the wire on the needle:

	Distance	*Force due to the magnetic action of the steel wire*
For Trial 2	4 inches	$41^2 - 15^2 = 1456$
For Trial 3	8 inches	$24^2 - 15^2 = 351$
For Trial 4	16 inches	$17^2 - 15^2 = 64$

The second and third trials, in which the distances are as $1:2$, give very nearly the inverse square ratio of the distances for the forces. The fourth trial gives a number slightly smaller, but we note that in the fourth trial the distance of the lower pole from the center of the needle is 16 inches; and that the distance of the upper pole to this same center of the needle is about $\sqrt{16^2 + 23^2}$; therefore, if the action of the lower pole is represented by $\frac{1}{16^2}$, the horizontal action of the upper will be $\frac{16}{(16^2 + 23^2)^{3/2}}$ and this means that the action of the lower pole is to the action of the upper pole as about $100:19$;■ from which it follows that the oscillations of the needle are produced by the action of these two poles, and since the upper pole acts in a sense opposed to the lower pole, the square of the oscillations produced by the action of the lower pole of the magnetized wire alone is diminished by $\frac{19}{100}$ by the opposite action of the upper part of the same wire. Therefore, to find the action of the lower pole alone it is necessary, letting x be the true value of this force, to make $(x - \frac{19}{100}x) = 64$, from which $x = 79$. Substituting this quantity in the result of the 4th trial, we find that

> *Trial 2* For a distance of 4 inches the force is 1456
> *Trial 3* For a distance of 8 inches the force is 331
> *Trial 4* For a distance of 16 inches the force is   79

and these forces are very nearly proportional to the numbers 16, 4, 1, or to the inverse square of the distances.

I have repeated this experiment many times, suspending needles of 2 and 3 inches long, and I have always found, after making the necessary corrections just explained, that the action of the magnetic fluid, either repulsive or attractive, varies inversely as the square of the distances.

■....................................

The quantity $\frac{16}{(16^2 + 23^2)^{3/2}}$ arises from the factor $\frac{1}{16^2 + 23^2}$ for the inverse square effect and the factor $\frac{16}{(16^2 + 23^2)^{1/2}}$ which is the cosine of the angle that resolves the total action of the upper pole into the horizontal direction.

Note that better agreement could not be expected unless the time were measured to a fraction of a second.

Omitted are some details of the experimental technique and a discussion of incidental observations.

SUPPLEMENTARY READING

Conant, J. C. (ed.), *The Development of the Concept of Electric Charge,* Harvard Case Histories in Experimental Science (Cambridge: Harvard University, 1957), pp. 543ff.

Dampier, W. C., *A History of Science* (New York: Cambridge University, 1946), p. 206.

Lenard, P., *Great Men of Science* (New York: British Book Centre, 1934), pp. 149ff. Lenard's treatment of his contemporaries (the last decade of the nineteenth and the early twentieth century) is marked by several omissions which seriously distort the scientific history of the period. He makes no mention, for example, of Becquerel, Roentgen, and J. J. Thomson, although he refers to the areas of physics in which they achieved fame. These omissions appear to have been intentional, perhaps the result of some personal prejudice, since Lenard chose to eliminate from his book those scientists who had survived World War I.

Magie, W. F., *A Source Book in Physics* (New York: McGraw-Hill, 1935), pp. 97ff.

Wolf, A., *A History of Science, Technology, and Philosophy in the 18th Century,* 2d ed. (London: Allen and Unwin, 1952), pp. 245ff. and 268ff.

6

Henry Cavendish

1731 - 1810

The Law of Gravitation

MORE THAN a century elapsed between Newton's publication of his law of universal gravitation and its experimental proof. Not that physicists held serious doubts about the truth of his hypothesis; its agreement with astronomical data left little question of its validity. Nevertheless, as in all other new concepts in physics, *direct* experimental evidence was required: a measurement, in the laboratory, of the force of attraction between two masses. It is not surprising that the experiment was performed only after Newton's time; the technical difficulties were much too formidable. The force between any two masses of convenient size, such as could be employed in a laboratory investigation, is extremely minute, demanding great skill in its measurement. There was another reason why such an experiment was considered important: if the gravitational force between two known masses[1] were measured, the mass of the earth, and hence its density, would follow directly from such data.

The individual responsible for the first successful measurement of the gravitational force pioneered as well in other areas, notably chemistry and electricity, although he left unpublished most of his investigations in electricity. One of the wealthiest men of his day, he lived in virtual seclusion, devoting his entire life to science. Henry Cavendish was born on October 10, 1731, apparently at Nice (where his mother had gone for her health), the first son of Lord Charles Cavendish, himself an experimenter of some note. Details of his earliest education are lacking, but one would assume that he was tutored privately. When he was eleven he became a pupil of

---

[1] As determined by weighing.

the Rev. Dr. Newcombe, master of Hackney Seminary, and in 1749 entered Peterhouse College, Cambridge. At that time, Cambridge was not yet the center of scientific inquiry for which it later became so well known. In England science was, for the most part, privately supported, either by individuals or through the Royal Society. The great universities, Cambridge and Oxford, were noted for the liberal education they provide but lacked the tradition of scholarly research. Not until the middle of the nineteenth century did these universities become the focal points of scientific research in England. It is not surprising, therefore, to find science in the eighteenth century practiced in many cases by nonacademic persons, and even by those with little or no formal preparation.

Cavendish left Cambridge after three years without completing work for his degree and took up residence in London. His secluded life and the obscurity which surrounds his activities make it difficult to determine what led him into science. At any rate he appears to have taken a great interest in mathematics and experimental science. His earliest investigations were in chemistry and heat, but he published nothing until 1766, when he sent to the Royal Society a paper on *Factitious Airs*.[2] There followed many other papers on chemical investigations, climaxed in 1681 by his discovery that hydrogen (which he obtained by dissolving metals in dilute sulphuric acid) and oxygen, when burned together, formed water. He showed, furthermore, that the weight of water produced was equal to that of the gases that disappeared. Despite these advances, the combustion process was poorly understood; the *phlogiston theory*[3] was widely held, and even Cavendish referred to hydrogen as *phlogiston* or *inflammable air*, and oxygen as *dephlogisticated air*. He is generally regarded also as the discoverer of nitric acid, produced in his combustion chamber.

His investigation of the composition of water led him into a dispute over prior discovery. Cavendish did not publish his *Experiments on Air* until 1783. In the meanwhile Joseph Priestly (1733–1804) had made a somewhat similar observation in 1781, unknown to Cavendish in as much as his retiring nature kept him out of touch with his scientific colleagues. James Watt (1736–1819), claiming ignorance of Priestly's experiments, proposed in 1783 that water was composed of dephlogisticated and inflammable air. For a time there was lively debate on the subject; the final conclusion was that Cavendish and Watt had conducted much the same investigations at about the same time, and came to similar conclusions. Had Cavendish been more communicative, probably the dispute would not have developed.

---

[2] Evidently he meant gases in the form of chemical compounds. These gases could be separated from alkaline substances by solution in acids.

[3] Based on the theory that air was the only gaseous element; *phlogiston* was that which escaped during combustion.

Prior to his researches on the composition of water, Cavendish spent several years investigating electrical phenomena, but he published only two papers on the subject, in 1772 and 1776. Much later, when Maxwell edited a volume[4] of the unpublished papers of Cavendish, it appeared that the most significant of his results had not been put into print, but that he had anticipated many of the phenomena later discovered by Michael Faraday (1791–1867) and others.[5] Evidently Cavendish carried out these investigations mainly to satisfy his own curiousity and saw no compelling reason to publish his results.

Regarding his measurement of the density of the earth, there is nothing to indicate how Cavendish became interested in the problem, except that he had some interest in the torsion balance, which was the instrument employed in the experiment, and had discussed the problem with the Reverend Michell. He conducted a remarkably detailed series of measurements and obtained a result within one percent of that presently accepted.

Cavendish continued his investigations in various fields until his death in 1810, when he left behind a considerable fortune, allowed to accumulate during his lifetime, and a large stack of manuscripts which attested to his diverse interests and capabilities. The latter were suitably recognized when Cambridge University, late in the nineteenth century, named its new Cavendish Laboratory for him.

The following extract, published in the *Philosophical Transactions,* vol. 17 (1798), page 469, describes his measurement of the density of the earth.

---

[4] Maxwell, J. C., *The Electrical Researches of the Hon. Henry Cavendish* (Cambridge, Eng.: Cambridge University, 1879).
[5] He recognized that the inverse square law of electric force could be deduced from the fact that charge resides on the surface of a body, a principle later discovered by Coulomb.

Cavendish's Experiment

Many years ago, the late Rev. John Michell, of this Society,■ contrived a method of determining the density of the earth, by rendering sensible the attraction of small quantities of matter; but, as he was engaged in other pursuits, he did not complete the apparatus till a short time before his death, and did not live to make any experiments with it. After his death, the apparatus came to the Rev. Francis John Hyde Wollaston, Jacksonian professor at Cambridge, who, not having conveniences

The Royal Society.

for making experiments with it, in the manner he could wish, was so good as to give it to me.

The apparatus is very simple; it consists of a wooden arm, 6 feet long, made so as to unite great strength with little weight.■ This arm is suspended in an horizontal position, by a slender wire 40 inches long, and to each extremity is hung a leaden ball, about 2 inches in diameter; and the whole is inclosed in a narrow wooden case, to defend it from the wind.

As no more force is required to make this arm turn round on its center, than what is necessary to twist the suspending wire, it is plain, that if the wire is sufficiently slender, the most minute force, such as the attraction of a leaden weight a few inches in diameter, will be sufficient to draw the arm sensibly aside. The weights which Mr. Michell intended to use were 8 inches in diameter. One of these was to be placed on one side of the case, opposite to one of the balls, and as near it as could conveniently be done, and the other on the other side, opposite to the other ball, so that the attraction of both these weights would conspire in drawing the arm aside; and, when its position, as affected by these weights, was ascertained, the weights were to be removed to the other side of the case, so as to draw the arm the contrary way, and the position of the arm was to be again determined; and, consequently, half the difference of these positions would show how much the arm was drawn aside by the attraction of the weights.

In order to determine from hence the density of the earth, it is necessary to ascertain what force is required to draw the arm aside through a given space.■ This Mr. Michell intended to do, by putting the arm in motion, and observing the time of its vibrations, from which it may easily be computed.*

Mr. Michell had prepared two wooden stands, on which the leaden weights were to be supported, and pushed forwards, till they came almost in contact with the case; but he seems to have intended to move them by hand.

As the force with which the balls are attracted by these weights is excessively minute, not more than $\frac{1}{50,000,000}$■ of their weight, it is plain, that a very minute disturbing force will be sufficient to destroy the success of the experiment; and, from the following experiments it will

The arm was constructed in the form of a truss.

That is, the torsion constant of the apparatus is required.

Note the absence of scientific notation, which did not come into common usage until much later.

In his experiments on the laws of electric and magnetic force.

---

* Mr. Coulomb has, in a variety of cases, used a contrivance of this kind for trying small attractions;■ but Mr. Michell informed me of his intention of making this experiment, and of the method he intended to use, before the publication of any of Mr. Coulomb's experiments.

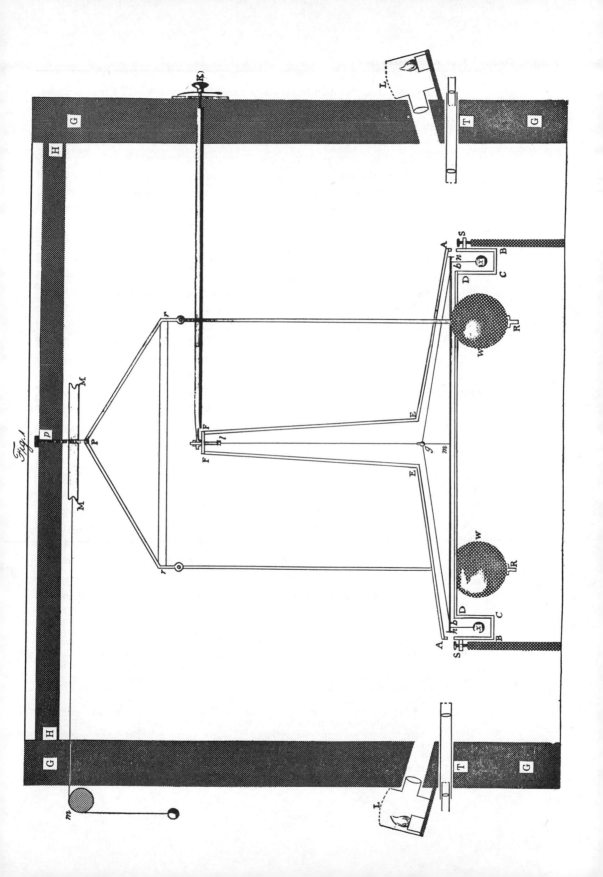

Fig. 1.

appear, that the disturbing force most difficult to guard against, is that arising from the variations, of heat and cold; for, if one side of the case is warmer than the other, the air in contact with it will be rarefied, and, in consequence, will ascend, while that on the other side will descend, and produce a current which will draw the arm sensibly aside.†

As I was convinced of the necessity of guarding against this source of error, I resolved to place the apparatus in a room which should remain constantly shut, and to observe the motion of the arm from without, by means of a telescope; and to suspend the leaden weights in such manner, that I could move them without entering into the room. This difference in the manner of observing, rendered it necessary to make some alteration in Mr. Michell's apparatus; and, as there were some parts of it which I thought not so convenient as could be wished, I chose to make the greatest part of it afresh.

Figure 1 is a longitudinal vertical section through the instrument, and the building in which it is placed. *ABCDDCBAEFFE* is the case; *x* and *x* are the two balls, which are suspended by the wires *bx* from the arm *gbmb,* which is itself suspended by the slender wire *gl*. This arm consists of a slender deal■ rod *bmb,* strengthened by a silver wire *bgb;* by which means it is made strong enough to support the balls, though very light.*

Deal: fir or pine.

The case is supported, and set horizontal, by four screws, resting on posts fixed firmly into the ground: two of them are represented in the figure, by *S* and *S;* the two others are not represented, to avoid confusion. *GG* and *GG* are the end walls of the building. *W* and *W* are the leaden weights; which are suspended by the copper rods *RrPrR,* and the wooden bar *rr,* from the center pin *Pp*. This pin passes through a hole in the beam *HH,* perpendicularly over the center of the instrument, and turns round in it, being prevented from falling by the plate *p*. *MM* is a pulley, fastened to this pin; and *Mm,* a cord wound round the pulley,

---

† M. Cassini, in observing the variation compass placed by him in the observatory (which was constructed so as to make very minute changes of position visible, and in which the needle was suspended by a silk thread), found that standing near the box, in order to observe, drew the needle sensibly aside; which I have no doubt was caused by this current of air. It must be observed, that his compass box was of metal, which transmits heat faster than wood, and also was many inches deep; both which causes served to increase the current of air. To diminish the effect of this current, it is by all means advisable to make the box, in which the needle plays, not much deeper than is necessary to prevent the needle from striking against the top and bottom.

* Mr. Michell's rod was entirely of wood, and was much stronger and stiffer than this, though not much heavier; but, as it had warped when it came to me, I chose to make another, and preferred this form, partly as being easier to construct and meeting with less resistance from the air, and partly because, from its being of a less complicated form, I could more easily compute how much it was attracted by the weights.

and passing through the end wall; by which the observer may turn it round, and thereby move the weights from one situation to the other.

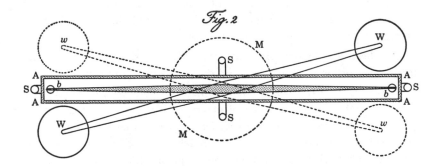

Figure 2 is a plan of the instrument. *AAAA* is the case. *SSSS*, the four screws for supporting it; *bb*, the arm and balls. *W* and *W*, the weights. *MM*, the pulley for moving them. When the weights are in this position, both conspire in drawing the arm in the direction *bW*; but, when they are removed to the situation *w* and *w*, represented by the dotted lines, both conspire in drawing the arm in the contrary direction *bw*. These weights are prevented from striking the instrument, by pieces of wood, which stop them as soon as they come within ⅕ of an inch of the case. The pieces of wood are fastened to the wall of the building; and I find that the weights may strike them with considerable force, without sensibly shaking the instrument.

In order to determine the situation of the arm, slips of ivory are placed within the case, as near to each end of the arm as can be done without danger of touching it, and are divided to 20ths of an inch. Another small slip of ivory is placed at each end of the arm, serving as a vernier, and subdividing these divisions into five parts; so that the position of the arm may be observed with ease to 100ths of an inch, and may be estimated to less. These divisions are viewed, by means of the short telescopes *T* and *T* (Fig. 1.) through slits cut in the end of the case, and stopped with glass; they are enlightened by the lamps *L* and *L*,■ with convex glasses, placed so as to throw the light on the divisions; no other light being admitted into the room.

Which burned lamp oil. The diagram shows the tops of the lamp housings to be ventilated.

The divisions on the slips of ivory run in the direction *Ww* (Fig. 2.) so that, when the weights are placed in the positions *w* and *w*, represented by the dotted circles, the arm is drawn aside, in such direction as to make the index point to a higher number on the slips of ivory; for which reason, I call this the positive position of the weights.

FK (Fig. 1.) is a wooden rod, which, by means of an endless screw, turns round the support to which the wire *gl* is fastened, and thereby enables the observer to turn round the wire, till the arm settles in the middle of the case, without danger of touching either side. The wire *gl* is fastened to its support at top and to the center of the arm at bottom, by brass clips, in which it is pinched by screws.

In these two figures, the different parts are drawn nearly in the proper proportion to each other, and on a scale of one to thirteen.

Before I proceed to the account of the experiments, it will be proper to say something of the manner of observing. Suppose the arm to be at rest, and its position to be observed, let the weights be then moved, the arm will not only be drawn aside thereby, but it will be made to vibrate, and its vibrations will continue a great while; so that, in order to determine how much the arm is drawn aside, it is necessary to observe the extreme points of the vibrations, and from thence to determine the point which it would rest at if its motion was destroyed, or the point of rest, as I shall call it.■ To do this, I observe three successive extreme points of a vibration, and take the mean between the first and third of these points, as the extreme point of vibration in one direction, and then assume the mean between this and the second extreme, as the point of rest; for, as the vibrations are continually diminishing, it is evident, that the mean between two extreme points will not give the true point of rest.

It may be thought more exact, to observe many extreme points of vibration, so as to find the point of rest by different sets of three extremes, and to take the mean result; but it must be observed, that notwithstanding the pains taken to prevent any disturbing force, the arm will seldom remain perfectly at rest for an hour together;■ for which reason, it is best to determine the point of rest, from observations made as soon after the motion of the weights as possible.

The next thing to be determined is the time of vibration, which I find in this manner;■ I observe the two extreme points of a vibration, and also the times at which the arm arrives at two given divisions between these extremes, taking care, as well as I can guess, that these divisions shall be on different sides of the middle point, and not very far from it. I then compute the middle point of the vibration, and, by proportion, find the time at which the arm comes to this middle point. I then, after a number of vibrations, repeat this operation, and divide the interval of time, between the coming of the arm to these two middle points, by the number of vibrations, which gives the time of one vibration. The following example will explain what is here said more clearly.

In the same way that the resting point of a chemical balance, for example, is determined by the *method of swings.*

Apparently what was meant here is that the disturbing forces would also influence the motion of the arm.

Instruments for the measurement of time were by then rather highly developed. Not only the pendulum clock, but also the lever escapement were used routinely.

Extreme points	Division	Time			Point of rest	Time of middle of vibration		
27.2		h.	′	″		h.	′	″
	25	10	23	4⎫	—	10	23	23
	24			57⎭				
22.1	—	—	—		24.6			
27	—	—	—		24.7			
22.6	—	—	—		24.75			
26.8	—	—	—		24.8			
23	—	—	—		24.85			
26.6	—	—	—		24.9			
	25	11	5	22⎫	—	11	5	22
	24		6	48⎭				
23.4								

h. ′ ″ denote hours, minutes, and seconds.

The first column contains the extreme points of the vibrations. The second, the intermediate divisions. The third, the time at which the arm came to these divisions; and the fourth, the point of rest, which is thus found: the mean between the first and third extreme points is 27.1, and the mean between this and the second extreme point is 24.6, which is the point of rest, as found by the three first extremes. In like manner, the point of rest found by the second, third, and fourth extremes is 24.7, and so on. The fifth column is the time at which the arm came to the middle point of the vibration, which is thus found: the mean between 27.2 and 22.1 is 24.65, and is the middle point of the first vibration; and, as the arm came to 25 at 10ʰ 23′ 4″, and to 24 at 10ʰ 23′ 57″, we find, by proportion, that it came to 24.65 at 10ʰ 23′ 23″. In like manner, the arm came to the middle of the seventh vibration at 11ʰ 5′ 22″; and, therefore, six vibrations were performed in 41′ 59″, or one vibration in 7′ 0″. ∎

To judge of the propriety of this method, we must consider in what manner the vibration is affected by the resistance of the air, and by the motion of the point of rest.

Note the long period, 7 minutes, the result of the large moment of inertia of the arm

∎ ....

An analysis of the errors contributed by these sources is omitted. Cavendish concluded that they were negligible.

ACCOUNT OF THE EXPERIMENTS

In my first experiments, the wire by which the arm was suspended was 39¼ inches long, and was of copper silvered, one foot of which weighed 2⁴⁄₁₀ grains: its stiffness was such, as to make the arm perform a vibration in about 15 minutes. I immediately found, indeed, that it was not stiff enough, as the attraction of the weights drew the balls so much aside, as to make them touch the sides of the case; I, however, chose to make some experiments with it, before I changed it.

In this trial, the rods by which the leaden weights were suspended were of iron; for, as I had taken care that there should be nothing magnetical in the arm, it seemed of no signification whether the rods were magnetical or not; but, for greater security, I took off the leaden weights, and tried what effect the rods would have by themselves. Now I find, by computation, that the attraction of gravity of these rods on the balls, is to that of the weights, nearly as 17 to 2500;■ so that, as the attraction of the weights appeared, by the foregoing trial, to be sufficient to draw the arm aside by about 15 divisions, the attraction of the rods alone should draw it aside about $\frac{1}{10}$ of a division; and, therefore, the motion of the rods from one near position to the other, should move it about $\frac{1}{5}$ of a division.

The result of the experiment was, that for the first 15 minutes after the rods were removed from one near position to the other, very little motion was produced in the arm, and hardly more than ought to be produced by the action of gravity; but the motion then increased, so that, in about a quarter or half an hour more, it was found to have moved $\frac{1}{2}$ or $1\frac{1}{2}$ division, in the same direction that it ought to have done by the action of gravity. On returning the irons back to their former position, the arm moved backward, in the same manner that it before moved forward.

It must be observed, that the motion of the arm, in these experiments, was hardly more than would sometimes take place without any apparent cause; but yet, as in three experiments which were made with these rods, the motion was constantly of the same kind, though differing in quantity from $\frac{1}{2}$ to $1\frac{1}{2}$ division, there seems great reason to think that it was produced by the rods.

As this effect seemed to me to be owing to magnetism, though it was not such as I should have expected from that cause, I changed the iron rods for copper, and tried them as before; the result was, that there still seemed to be some effect of the same kind, but more irregular, so that I attributed it to some accidental cause, and therefore hung on the leaden weights, and proceeded with the experiment.

It must be observed, that the effect which seemed to be produced by moving the iron rods from one near position to the other, was, at a medium, not more than one division; whereas the effect produced by moving the weight from the midway to the near position, was about 15 divisions; so that, if I had continued to use the iron rods, the error in the result caused thereby, could hardly have exceeded $\frac{1}{30}$ of the whole.

It must be observed, that in this experiment the attraction of the weights drew the arm from 11.5 to 25.8, so that, if no contrivance had

Simply by the ratio of their masses, and taking into account the inverse square law.

EXPERIMENT I. AUG. 5. Weights in midway position

Extreme points	Division	Time			Point of rest	Time of middle vibration			Difference	
		h.	′	″		h.	′	″	′	″
	11.4	9	42	0						
	11.5		55	0						
	11.5	10	5	0	11.5					

At 10ʰ 5′, weights moved to positive position.

Extreme points	Division	Time			Point of rest	Time of middle vibration			Difference	
23.4										
27.6	—	—	—		25.82					
24.7	—	—	—		26.07					
27.3	—	—	—		26.1					
25.1	—	—	—							

At 11ʰ 6′, weights returned back to midway position.

Extreme points	Division	Time			Point of rest	Time of middle vibration			Difference	
5.										
	11	0	0	48 ⎫	—	0	1	13		
	12		1	30 ⎭						
18.2	—	—	—		12	—	—		14	56
	12		16	29 ⎫	—		16	9		
	11		17	20 ⎭						
6.6	—	—	—		11.92	—	—		14	36
	11		30	24 ⎫	—		30	45		
	12		31	11 ⎭						
16.3	—	—	—		11.72	—	—		15	13
	12		45	58 ⎫	—		45	58		
	11		47	4 ⎭						
7.7										

Motion on moving from midway to pos. = 14.32
pos. to midway = 14.1
Time of one vibration = 14′ 55″

been used to prevent it, the momentum acquired thereby would have carried it to near 40, and would, therefore, have made the balls to strike against the case. To prevent this, after the arm had moved near 15 divisions, I returned the weights to the midway position, and let them remain there, till the arm came nearly to the extent of its vibration, and then again moved them to the positive position, whereby the vibrations were so much diminished, that the balls did not touch the sides; and it was this which prevented my observing the first extremity of the vibration. A like method was used, when the weights were returned to the midway position, and in the two following experiments.

The vibrations, in moving the weights from the midway to the positive position, were so small,■ that it was thought not worth while to observe the time of the vibration. When the weights were returned to the midway position, I determined the time of the arm's coming to the middle point of each vibration, in order to see how nearly the times of the different vibrations agreed together. In great part of the following

Because of the technique employed, which had the effect of damping the vibrations.

experiments, I contented my self with observing the time of its coming to the middle point of only the first and last vibration.

■....................................

Omitted are the data from two additional experiments, which yielded substantially the same results.

These experiments are sufficient to show, that the attraction of the weights on the balls is very sensible, and are also sufficiently regular to determine the quantity of this attraction pretty nearly, as the extreme results do not differ from each other by more than $\frac{1}{10}$ part. But there is a circumstance in them, the reason of which does not readily appear, namely, that the effect of the attraction seems to increase, for half an hour, or an hour, after the motion of the weights; as it may be observed, that in all three experiments, the mean position kept increasing for that time, after moving the weights to the positive position; and kept decreasing, after moving them from the positive to the midway position.

The first cause which occurred to me was, that possibly there might be a want of elasticity, either in the suspending wire, or something it was fastened to, which might make it yield more to a given pressure, after a long continuance of that pressure, than it did at first.

To put this to the trial, I moved the index so much, that the arm, if not prevented by the sides of the case, would have stood at about 50 divisions, so that, as it could not move farther than to 35 divisions, it was kept in a position 15 divisions distant from that which it would naturally have assumed from the stiffness of the wire; or, in other words, the wire was twisted 15 divisions. After having remained two or three hours in this position, the index was moved back, so as to leave the arm at liberty to assume its natural position.

That is, exceeds the elastic limit.

It must be observed, that if a wire is twisted only a little more than its elasticity admits of,■ then instead of setting as it is called, or acquiring a permanent twist all at once, it sets gradually, and, when it is left at liberty, it gradually loses part of that set which it acquired; so that if, in this experiment, the wire, by having been kept twisted for two or three hours, had gradually yielded to this pressure, or had begun to set, it would gradually restore itself, when left at liberty, and the point of rest would gradually move backwards; but, though the experiment was twice repeated, I could not perceive any such effect.

A discussion of the results obtained with a stiffer suspension (having a period of about 7 minutes) is omitted. The same long-term drift was found.

■....................................

By induced magnetism.

My next trials were, to see whether this effect was owing to magnetism. Now, as it happened, the case in which the arm was inclosed, was placed nearly parallel to the magnetic east and west, and therefore, if there was any thing magnetic in the balls and weights, the balls would acquire polarity from the earth;■ and the weights also, after having remained some time, either in the positive or negative position, would acquire polarity in the same direction, and would attract the balls; but, when the weights were moved to the contrary position, that pole which before pointed to the north, would point to the south, and would repel the ball it was approached to; but yet, as repelling one ball towards the south has the same effect on the arm as attracting the other towards the north, this would have no effect on the position of the arm. After some time, however, the poles of the weight would be reversed, and would begin to attract the balls, and would therefore produce the same kind of effect as was actually observed.

To try whether this was the case, I detached the weights from the upper part of the copper rods by which they were suspended, but still retained the lower joint, namely, that which passed through them; I then fixed them in their positive position, in such manner, that they could turn round on this joint, as a vertical axis. I also made an apparatus, by which I could turn them half way round, on these vertical axes, without opening the door of the room.

Having suffered the apparatus to remain in this manner for a day, I next morning observed the arm, and, having found it to be stationary, turned the weights half way round on their axes, but could not perceive any motion in the arm. Having suffered the weights to remain in this position for about an hour, I turned them back into their former position, but without its having any effect on the arm. This experiment was repeated on two other days, with the same result.

We may be sure, therefore, that the effect in question could not be produced by magnetism in the weights; for, if it was, turning them half round on their axes, would immediately have changed their magnetic attraction into repulsion, and have produced a motion in the arm.

As a further proof of this, I took off the leaden weights, and in their room placed two 10-inch magnets;■ the apparatus for turning them round being left as it was, and the magnets being placed horizontal, and pointing to the balls, and with their north poles turned to the north; but I could not find that any alteration was produced in the place of the arm, by turning them half round; which not only confirms the deduction drawn from the former experiment, but also seems to show, that in the experiments with the iron rods, the effect produced could not be owing to magnetism.

The sequence of tests described below mark Cavendish a most careful and thorough experimenter.

The next thing which suggested itself to me was that possibly the effect might be owing to a difference of temperature between the weights and the case; for it is evident, that if the weights were much warmer than the case, they would warm that side which was next to them, and produce a current of air, which would make the balls approach nearer to the weights. Though I thought it not likely that there should be sufficient difference, between the heat of the weights and case, to have any sensible effect, and though it seemed improbable that, in all foregoing experiments, the weights should happen to be warmer than the case, I resolved to examine into it, and for this purpose removed the apparatus used in the last experiments, and supported the weights by the copper rods, as before; and, having placed them in the midway position, I put a lamp under each, and placed a thermometer with its ball close to the outside of the case, near that part which one of the weights approached to in its positive position, and in such manner that I could distinguish the divisions by the telescope. Having done this, I shut the door, and some time after moved the weights to the positive position. At first, the arm was drawn aside only in its usual manner; but, in half an hour, the effect was so much increased that the arm was drawn 14 divisions aside, instead of about three, as it would otherwise have been, and the thermometer was raised near $1\frac{1}{2}°$; namely, from $61°$ to $62\frac{1}{2}°$.■ On opening the door, the weights were found to be no more heated, than just to prevent their feeling cool to my fingers.

As the effect of a difference of temperature appeared to be so great, I bored a small hole in one of the weights, about three quarters of an inch deep, and inserted the ball of a small thermometer, and then covered up the opening with cement. Another small thermometer was placed with its ball close to the case, and as near to that part to which the weight was approached as could be done with safety; the thermometers being so placed, that when the weights were in the negative position, both could be seen through one of the telescopes, by means of light reflected from a concave mirror.

In these three experiments, the effect of the weight appeared to increase from two to five tenths of a division, on standing an hour; and the thermometers showed, that the weights were three or five tenths of a degree warmer than the air close to the case. In the two last experiments, I put a lamp into the room, overnight, in hopes of making the air warmer than the weights, but without effect, as the heat of the weights exceeded that of the air more in these two experiments than in the former.

On the evening of October 17, the weights being placed in the midway position, lamps were put under them, in order to warm them; the door was then shut, and the lamps suffered to burn out. The next morning it was found, on moving the weights to the negative position, that they were $7\frac{1}{2}°$ warmer than the air near the case. After they had continued an hour in that position, they were found to have cooled $1\frac{1}{2}°$, so

EXPERIMENT VI. SEPT. 6. Weights in midway position

Extreme points	Divisions	Time		Point of rest	Thermometer in air	in weight
		h	′			
	18.9	9	43	—	55.5	
	18.85	10	3	18.85		
Weights moved to negative position						
13.1	—	10	12	—	55.5	55.8
18.4	—		18	15.82		
13.4	—		25			
missed						
13.6	—		39	—	55.5	55.8
17.6	—		46	15.65		
13.8	—		53	15.65		
17.4	—	11	0	15.65		
14.0	—		7	15.65		
17.2	—		14	—	55.5	
Weights moved to positive position						
25.8	—		23			
17.5	—		30	21.55		
25.4	—		37	21.6		
18.1	—		44	21.65		
25.0	—		51			
missed						
24.7	—	0	5			
19.	—		12	21.77		
24.4	—		19			

Motion of arm on moving weights from midway to − = 3.03
− to + = 5.9

as to be only 6° warmer than the air. They were then moved to the positive position; and in both positions the arm was drawn aside about four divisions more, after the weights had remained an hour in that position, than it was at first.

May 22, 1798. The experiment was repeated in the same manner, except that the lamps were made so as to burn only a short time, and only two hours were suffered to elapse before the weights were moved. The weights were now found to be scarcely 2° warmer than the case; and the arm was drawn aside about two divisions more, after the weights had remained an hour in the position they were moved to, than it was at first.

On May 23, the experiment was tried in the same manner, except that the weights were cooled by laying ice on them; the ice being confined in its place by tin plates, which, on moving the weights, fell to the ground, so as not to be in the way. On moving the weights to the negative position, they were found to be about 8° colder than the air, and their effect on the arm seemed now to diminish on standing, instead of increasing, as it did before; as the arm was drawn aside about 2½ divisions less, at the end of an hour after the motion of the weights, than it was at first.

It seems sufficiently proved, therefore, that the effect in question is produced, as above explained, by the difference of temperature between the weights and case; for, in the sixth, eighth, and ninth experiments, in which the weights were not much warmer than the case, their effect increased but little on standing; whereas, it increased much, when they were much warmer than the case, and decreased much, when they were much cooler.

Omitted are the results of several more measurements, and a discussion of the method employed to calculate the density of the earth, together with the corrections applied.

It must be observed, that in this apparatus, the box in which the balls play is pretty deep, and the balls hang near the bottom of it, which makes the effect of the current of air more sensible than it would otherwise be, and is a defect which I intend to rectify in some future experiments.

■ ....

This table summarizes all the results. The second and third columns give the motions of the weight and arm respectively. The fourth gives the motion of the arm corrected for the change in attraction caused by its motion (Cavendish concluded that this was the only significant correction). The next two columns give the period, as observed and corrected, again for the same reason that the motion of the arm required correction. The last column contains the values of density computed from these data.

CONCLUSION. The following table contains the result of the experiments.

Exper.	Mot. weight	Mot. arm	Do. corr.	Time vib.		Do. corr.	Dens.
				$'$	$''$		
1	m to +	14.32	13.42			—	5.5
	+ to m	14.1	13.17	14	55	—	5.61
2	m to +	15.87	14.69	—		—	4.88
	+ to m	15.45	14.14	14	42	—	5.07
3	+ to m	15.22	13.56	14	39	—	5.26
	m to +	14.5	13.28	14	54	—	5.55
	m to +	3.1	2.95			6.54	5.36
4	+ to −	6.18	—	7	1	—	5.29
	− to +	5.92	—	7	3	—	5.58
5	+ to −	5.9	—	7	5	—	5.65
	− to +	5.98	—	7	5	—	5.57
6	m to −	3.03	2.9	—		—	5.53
	− to +	5.9	5.71			—	5.62
7	m to −	3.15	3.03	7.4		6.57	5.29
	− to +	6.1	5.9	by mean			5.44
8	m to −	3.13	3.00	—		—	5.34
	− to +	5.72	5.54			—	5.79
9	+ to −	6.32	—	6	58	—	5.1
10	+ to −	6.15	—	6	59	—	5.27
11	+ to −	6.07	—	7	1	—	5.39
12	− to +	6.09	—	7	3	—	5.42
13	− to +	6.12	—	7	6	—	5.47
	+ to −	5.97	—	7	7	—	5.63
14	− to +	6.27	—	7	6	—	5.34
	+ to −	6.13	—	7	6	—	5.46
15	− to +	6.34	—	7	7	—	5.3
16	− to +	6.1	—	7	16	—	5.75
17	− to +	5.78	—	7	2	—	5.68
	+ to −	5.64	—	7	3	—	5.85

From this table it appears, that though the experiments agree pretty well together, yet the difference between them, both in the quantity of motion of the arm and in the time of vibration, is greater than can proceed merely from the error of observation. As to the difference in the motion of the arm, it may very well be accounted for, from the current of air produced by the difference of temperature; but, whether this can account for the difference in the time of vibration, is doubtful. If the current of air was regular, and of the same swiftness in all parts of the vibration of the ball, I think it could not; but, as there will most likely be much irregularity in the current, it may very likely be sufficient to account for the difference.

By a mean of the experiments made with the wire first used, the density of the earth comes out 5.48 times greater than that of water;[■] and by a mean of those made with the latter wire, it comes out the same; and the extreme difference of the results of the 23 observations made with this wire, is only .75; so that the extreme results do not differ from the mean by more than .38, or $\frac{1}{14}$ of the whole, and therefore the density should seem to be determined hereby, to great exactness. It, indeed, may be objected, that as the result appears to be influenced by the current of air, or some other cause, the laws of which we are not well acquainted with, this cause may perhaps act always, or commonly, in the same direction, and thereby make a considerable error in the result. But yet, as the experiments were tried in various weathers, and with considerable variety in the difference of temperature of the weights and air, and with the arm resting at different distances from the sides of the case, it seems very unlikely that this cause should act so uniformly in the same way, as to make the error of the mean result nearly equal to the difference between this and the extreme; and, therefore, it seems very unlikely that the density of the earth should differ from 5.48 by so much as $\frac{1}{14}$ of the whole.

Another objection, perhaps, may be made to these experiments, namely, that it is uncertain whether, in these small distances, the force of gravity follows exactly the same law as in greater distances. There is no reason, however, to think that any irregularity of this kind takes place, until the bodies come within the action of what is called the attraction of cohesion, and which seems to extend only to very minute distances. With a view to see whether the result could be affected by this attraction, I made the ninth, tenth, eleventh, and fifteenth experiments, in which the balls were made to rest as close to the sides of the case as they could; but there is no difference to be depended on, between the results under

Compare this result with that found by C. V. Boys (*Phil. Trans.*, 1895) using a similar technique, 5.5270. The agreement is within 1 percent.

By measuring the deflection of a plumb bob suspended near the hill. But to determine the density of the earth from this observation, one needs to know the mass of the mountain accurately.

An appendix, containing a calculation of the effect of the mahogany case on the balls (found to be negligible), has been omitted.

that circumstance, and when the balls are placed in any other part of the case.

According to the experiments made by Dr. Maskelyne, on the attraction of the hill Schehallien,■ the density of the earth is 4½ times that of water; which differs rather more from the preceding determination than I should have expected. But I forbear entering into any consideration of which determination is most to be depended on, till I have examined more carefully how much the preceding determination is affected by irregularities whose quantity I cannot measure.

■...

SUPPLEMENTARY READING

Dampier, W. C., *A History of Science* (New York: Cambridge University, 1946).

Lenard, P., *Great Men of Science* (New York: British Book Centre, Macmillan, 1934), pp. 145ff.

Magie, W. F., *A Source Book in Physics* (New York: McGraw-Hill, 1935), pp. 105ff.

Wolf, A., *A History of Science, Technology, and Philosophy in the 18th Century,* 2d ed. (London: Allen and Unwin, 1952), pp. 112f. and 242ff.

7

Thomas Young

1773 - 1829

The Interference of Light

AMONG the problems that concerned scientists of the seventeenth and eighteenth centuries, the nature of light must be counted among the most intriguing. Even in modern times it remains one of nature's greatest puzzles. The wave-particle controversy of the last few centuries has now been supplanted by a wave-particle dualism; but this apparent reconciliation of two opposing views has, in many ways, added to the complexity of the problem. While we are now able to account for most optical phenomena, it is not because we have a finished theory as much as the fact that we better understand the limitations of our current views. The question of the nature of light is as significant today as it was in Newton's time, 250 years ago.

Most physicists during the eighteenth century held some form of *corpuscular* or particle theory of light. Newton saw certain difficulties with a wave theory, particularly in the fact that diffraction was not as readily observed for light as it was for other wave phenomena, such as sound or water waves. Although he did not reject completely the idea of periodicity in connection with the propagation of light, he advanced a particle theory which, because of his great prestige, was widely accepted. It was essentially this view that prevailed through the eighteenth century. About the same time that Newton proposed his corpuscular theory, several contemporaries, notably Robert Hooke (1635–1703) and Christian Huygens (1629–1695), advocated wave theories, Hooke to account for the colors of thin films, Huygens to account for the finite velocity of light, recently determined by Olaus Roemer (1644–1710). Somewhat earlier, Francesco Grimaldi (1618–1663), on the strength of experiments which appeared to show the diffraction of light by small apertures, suggested that light had a wavelike character. The evidence was entirely qualitative, however, and offered no substantial proof of the incorrectness of Newton's view. As a result, the latter was not finally forced aside until a century later, when the weight of the evidence appeared to favor the wave hypothesis. The first experiment to cast serious doubt on the corpuscular view was a demonstration by Young, in 1803, of the interference of light. Young was a most remarkable individual: a practicing physician, he was at the same time a competent physicist and a noted Egyptologist, neither of which fields appears to have suffered much for his attention to the others.

Thomas Young was born into "comfortable circumstances" at Milverton, England, on June 13, 1773, toward the end of the period known as the *Intellectual Revolution*. He matured into the *Age of Romanticism* among such contemporaries as the poets Wordsworth and Shelley, composers Beethoven and Schubert, philosophers Hegel and Schopenhauer, and his own colleagues Fresnel, Avogadro, Oersted, and Faraday. Unlike most famous scientists, whose early lives were uneventful, Young was a precocious youngster who could read fluently at the age of two and who employed this extraordinary ability to read the classics placed before him by an admiring grandfather. At six he commenced Latin; he was tutored privately at first and later attended private schools. By the time he was sixteen he had acquired a thorough knowledge of Latin and Greek, and was acquainted with some eight other languages, both classical and modern. So proficient was he in classical studies that at the age of eighteen he was recognized by many in London as an accomplished scholar.

In 1792, then nineteen years old, Young decided to make his career in medicine. The following year he read a paper before the Royal Society in which he correctly attributed the accommodation of the eye to its muscular structure. This led to his election, one year later, to membership in the Society. After completing his medical studies at Edinburgh and Göttingen, he returned to London to practice, continuing, at the same time, his scholarly interests at Emmanuel College, Cambridge. The death of an uncle left him with independent means and the freedom to pursue his main desires. Some investigations on sound and light, which he conducted in 1798, apparently formed the starting point for his theory of interference several years later. His contributions to science and literature were so numerous that he took to making some anonymously in order to avoid the charge that he was neglecting his professional duties.

In 1801 Young was appointed professor of natural philosophy at the Royal Institution, which provided him the opportunity of presenting lectures to popular audiences. Apparently his lectures were not well suited to this kind of audience, being designed more for the specialist than the layman. He was appointed foreign secretary to the Royal Society in 1802, a post which he held to the end of his life. Feeling that his duties at the Royal Institution were interfering with his medical career he resigned his professorship in 1803. That same year he received the M.B. degree from Cambridge, and five years later, the degree of M.D.

It was during this period that Young found time to conduct his experimental investigations on light. He published his *Experiments on Sound and Light* in 1800 *(Phil. Trans.)*, and presented a detailed account of his theory of interference in his Bakerian Lecture, *On the Theory of Light and Colors (Phil. Trans.,* 1801).[1] In another Bakerian Lecture (*Phil. Trans.,* November 1803), entitled *Experiments and Calculations Relative to Physical Optics,* Young summarized his observations on interference and added several new phenomena. The importance of his work was not apparent to his contemporaries, with the result that the principle of interference remained more or less obscure for another fourteen years, when it was rediscovered by Fresnel. Young made other significant contributions to physical optics, particularly in the areas of double refraction and dispersion. His physical insight was not confined to optics, however. He was the first to assign the term *energy* to the product (mv^2) and to set the work expended (which he defined as the product of force and distance) proportional to the energy. He introduced absolute methods for the determination of elastic properties of materials (Young's modulus) and developed the most comprehensive theory of tides then available.

His contributions to archeology and philology were equally impressive, as were his researches in medicine. He could make himself at home in almost any scholarly activity and delight in the challenge it offered. Young retired from active practice in 1814 to devote full time to his scientific interests, continuing his productive work until his death in 1829. Sir Humphry Davy (1778–1829), his colleague at the Royal Institution, said of him: " . . . Had he limited himself to any one department of knowledge, he must have been the first in that department. But as a mathematician, a scholar, a hieroglyphist, he was eminent, and he knew so much that it was difficult to say what he did not know."[2]

The following extracts, on the interference of light, are taken from his Bakerian Lecture (read Nov. 24, 1803), *Philosophical Transactions of the Royal Society of London,* vol. 94 (1804), and from lecture 39 of his Course of Lectures. They may be found, as well, in *The Wave Theory of Light,* edited by H. Crew (New York: American Book, 1900).

---

[1] Young also published most of his results in his *Course of Lectures on Natural Philosophy and the Mechanical Arts,* 1807.

[2] *Life of Sir Humphry Davy,* by his brother, John Davy (London, 1839). Strangely enough, Young's mathematical writings are the most obscure of his works.

Young's Experiment

From the Bakerian Lecture (Nov. 24, 1803).

EXPERIMENTAL DEMONSTRATION OF THE GENERAL LAW OF THE INTERFERENCE OF LIGHT■

In making some experiments on the fringes of colors accompanying shadows, I have found so simple and so demonstrative a proof of the general law of the interference of two portions of light, which I have already endeavored to establish, that I think it right to lay before the Royal Society a short statement of the facts which appear to me so decisive. The proposition on which I mean to insist at present, is simply this, that fringes of colors are produced by the interference of two portions of light; and I think it will not be denied by the most prejudiced, that the assertion is proved by the experiments I am about to relate, which may be repeated with great ease, whenever the sun shines, and without any other apparatus than is at hand to every one.

Exper. 1. I made a small hole in a window shutter, and covered it with a piece of thick paper, which I perforated with a fine needle. For greater convenience of observation, I placed a small looking glass without■ the window shutter, in such a position as to reflect the sun's light, in a direction nearly horizontal, upon the opposite wall, and to cause the cone of diverging light to pass over a table, on which were several little screens of card paper. I brought into the sunbeam a slip of card, about one thirtieth of an inch in breadth, and observed its shadow, either on the wall, or on other cards held at different distances. Besides the fringes of colors on each side of the shadow, the shadow itself was divided by similar parallel fringes, of smaller dimensions, differing in number, according to the distance at which the shadow was observed, but leaving the middle of the shadow always white. Now these fringes were the joint effects of the portions of light passing on each side of the slip of cards, and inflected, or rather diffracted, into the shadow. For, a little screen being placed a few inches from the card, so as to receive either edge of the shadow on its margin, all the fringes which had before been observed in the shadow on the wall immediately disappeared, although the light inflected on the other side was allowed to retain its course, and although this light must have undergone any modification that the proximity of the other edge of the slip of card might have been capable of occasioning. When the interposed screen was more remote from the narrow card, it was necessary to plunge it more deeply into the shadow, in order to extinguish the parallel lines; for here the light, diffracted from the edge

Without: outside of

of the object, had entered further into the shadow, in its way towards the fringes. Nor was it for want of a sufficient intensity of light, that one of the two portions was incapable of producing the fringes alone; for, when they were both uninterrupted, the lines appeared, even if the intensity was reduced to one tenth or one twentieth.

Exper. 2. The crested fringes described by the ingenious and accurate Grimaldi,■ afford an elegant variation of the preceding experiment, and an interesting example of a calculation grounded on it. When a shadow is formed by an object which has a rectangular termination,■ besides the usual external fringes, there are two or three alternations of colors, beginning from the line which bisects the angle, disposed on each side of it, in curves, which are convex towards the bisecting line, and which converge in some degree towards it, as they become more remote from the angular point. These fringes are also the joint effect of the light which is inflected directly towards the shadow, from each of the two outlines of the object. For, if a screen be placed within a few inches of the object, so as to receive only one of the edges of the shadow, the whole of the fringes disappear. If, on the contrary, the rectangular point of the screen be opposed to the point of the shadow, so as barely to receive the angle of the shadow on its extremity, the fringes will remain undisturbed.

Physico-Mathesis de lumine, coloribus, et iride (Bologna, 1665).

That is, when the light is diffracted by a corner of the object.

COMPARISON OF MEASURES, DEDUCED FROM VARIOUS EXPERIMENTS

If we now proceed to examine the dimensions of the fringes, under different circumstances, we may calculate the differences of the lengths of the paths described by the portions of light, which have thus been proved to be concerned in producing those fringes; and we shall find, that where the lengths are equal, the light always remains white; but that, where either the brightest light, or the light of any given color, disappears and reappears, a first, a second, or a third time, the differences of the lengths of the paths of the two portions are in arithmetical progression, as nearly as we can expect experiments of this kind to agree with each other. I shall compare, in this point of view, the measures deduced from several experiments of Newton,■ and from some of my own.

In the eighth and ninth observations of the third book of Newton's *Optics,* some experiments are related, which, together with the third observation, will furnish us with the data necessary for the calculation. Two knives were placed, with their edges meeting at a very acute angle,

Keep in mind that Newton held a corpuscular view of light, as will be seen below.

in a beam of the sun's light, admitted through a small aperture; and the point of concourse∎ of the two first dark lines bordering the shadows of the respective knives, was observed at various distances. The results of six observations are expressed in the first three lines of the first table. On the supposition that the dark line is produced by the first interference of the light reflected from the edges of the knives, with the light passing in a straight line between them, we may assign, by calculating the difference of the two paths, the interval for the first disappearance of the brightest light, as it is expressed in the fourth line. The second table contains the results of a similar calculation, from Newton's observations on the shadow of a hair; and the third, from some experiments of my own, of the same nature: The second bright line being supposed to correspond to a double interval, the second dark line to a triple interval, and the succeeding lines to depend on a continuation of the progression. The unit of all the tables is an inch.

It appears, from five of the six observations of the first table, in which the distance of the shadow was varied from about 3 inches to 11 feet, and the breadth of the fringes was increased in the ratio of 7 to 1, that the difference of the routes constituting the interval of disappearance, varied but one eleventh at most; and that, in three out of the five, it agreed with the mean, either exactly, or within $\frac{1}{160}$ part. Hence we are warranted in inferring, that the interval appropriate to the extinction of the brightest light, is either accurately or very nearly constant.

TABLE I. *Obs.* 9. N.

Distance of the knives from the aperture∎............................ 101.

Distances of the paper from the knives.............. $1\frac{1}{2}$, $3\frac{1}{3}$, $8\frac{3}{8}$, 32, 96, 131.

Distances between the edges of the knives, opposite to the point of concourse.... .012, .020, .034, .057, .081, .087.

Interval of disappearance .0000122, .0000155, .0000182, .0000167, .0000166, .0000166.

TABLE II. *Obs.* 3. N.

Breadth of the hair..................................		$\frac{1}{280}$.
Distance of the hair from the aperture................		144.
Distances of the scale from the aperture..............	150,	252.
(Breadths of the shadow............................	$\frac{1}{64}$,	$\frac{1}{9}$.)
Breadth between the second pair of bright lines.........	$\frac{2}{47}$,	$\frac{4}{17}$.
Interval of disappearance, or half the difference of the paths	.0000151,	.0000173.
Breadth between the third pair of bright lines..........	$\frac{4}{73}$,	$\frac{3}{10}$.
Interval of disappearance, $\frac{1}{4}$ of the difference...........	.0000130,	.0000143.

TABLE III. *Exper.* 3.

Breadth of the object...	.434.
Distance of the object from the aperture.........................	125.
Distance of the wall from the aperture...........................	250.
Distance of the second pair of dark lines from each other.............	1.167.
Interval of disappearance, $\frac{1}{8}$ of the difference......................	.0000149.

Exper. 4.

Breadth of the wire......................	.083.
Distance of the wire from the aperture.......	32.
Distance of the wall from the aperture.......	250.
(Breadth of the shadow, by three measurements	.815, .826, or .827; mean, .823.)
Distance of the first pair of dark lines........	1.165, 1.170, or 1.160; mean, 1.165.
Interval of disappearance.................	.0000194.
Distance of the second pair of dark lines.....	1.402, 1.395, or 1.400; mean, 1.399.
Interval of disappearance.................	.0000137.
Distance of the third pair of dark lines.......	1.594, 1.580, or 1.585; mean, 1.586.
Interval of disappearance.................	.0000128.

But it may be inferred, from a comparison of all the other observations, that when the obliquity of the reflection is very great, some circumstance takes place, which causes the interval thus calculated to be somewhat greater: thus, in the eleventh line of the third table, it comes out one sixth greater than the mean of the five already mentioned. On the other hand, the mean of two of Newton's experiments and one of mine, is a result about one fourth less than the former. With respect to the nature of this circumstance, I cannot at present form a decided opinion; but I conjecture that it is a deviation of some of the light concerned, from the rectilinear direction assigned to it, arising either from its natural diffraction, by which the magnitude of the shadow is also enlarged, or from some other unknown cause.■ If we imagined the shadow of the wire, and the fringes nearest it, to be so contracted that the motion of the light bounding the shadow might be rectilinear, we should thus make a sufficient compensation for this deviation; but it is difficult to point out what precise track of the light would cause it to require this correction.

Possibly this was a geometrical factor not fully realized by Young.

The means of the three experiments which appear to have been least affected by this unknown deviation, gives .0000127 for the interval appropriate to the disappearance of the brightest light; and it may be inferred, that if they had been wholly exempted from its effects, the measure would have been somewhat smaller. Now the analogous interval, deduced from the experiments of Newton on thin plates, is .0000112, which is about one eighth less than the former result; and this appears to be a coincidence fully sufficient to authorize us to attribute these two classes of phenomena to the same cause. It is very easily shown, with respect to the colors of thin plates that each kind of light disappears and reappears, where the differences of the routes of two of its portions are in arithmetical progression; and we have seen, that the same law may be in general inferred from the phenomena of diffracted light, even independently of the analogy.

A discussion of the colors produced by interference, and an application to rainbows, are omitted.

■.......................................

ARGUMENTATIVE INFERENCE RESPECTING THE NATURE OF LIGHT

The experiment of Grimaldi, on the crested fringes within the shadow, together with several others of his observations, equally important, has been left unnoticed by Newton. Those who are attached to the Newtonian theory of light,■ or to the hypotheses of modern opticians, founded on views still less enlarged, would do well to endeavor to imagine anything like an explanation of these experiments, derived from their own doctrines; and, if they fail in the attempt, to refrain at least from idle declamation against a system which is founded on the accuracy of its application to all these facts, and to a thousand others of a similar nature.

From the experiments and calculations which have been premised, we may be allowed to infer, that homogeneous light, at certain equal distances in the direction of its motion, is possessed of opposite qualities, capable of neutralizing or destroying each other, and of extinguishing the light, where they happen to be united; that these qualities succeed each other alternately in successive concentric superficies,■ at distances which are constant for the same light, passing through the same medium. From the agreement of the measures, and from the similarity of the phenomena, we may conclude, that these intervals are the same as are concerned in the production of the colors of thin plates; but these are shown, by the experiments of Newton, to be the smaller, the denser the medium;■ and, since it may be presumed that their number must necessarily remain unaltered in a given quantity of light, it follows of course, that light moves more slowly in a denser, than in a rarer medium: and this being granted, it must be allowed, that refraction is not the effect of an attractive force directed to a denser medium.■ The advocates for the projectile hypothesis of light, must consider which link in this chain of reasoning they may judge to be the most feeble; for, hitherto, I have advanced in this paper no general hypothesis whatever. But, since we know that sound diverges in concentric superficies, and that musical sounds consist of opposite qualities,■ capable of neutralizing each other, and succeeding at certain equal intervals, which are different according to the difference of the note, we are fully authorized to conclude, that there must be some strong resemblance between the nature of sound and that of light.

I have not, in the course of these investigations, found any reason to suppose the presence of such an inflecting medium in the neighborhood of dense substances as I was formerly inclined to attribute to them; and, upon considering the phenomena of the aberration of the stars,■ I am disposed to believe, that luminiferous ether pervades the substance of

The Newtonian hypothesis was widely held at the time of Young's investigations.

Superficies: surfaces

That is, the wavelength decreases in denser media.

The means of accounting for refraction on the basis of Newton's corpuscular theory.

The *opposite qualities* here are the alternate compressions and rarefactions.

The apparent angular displacement of the stars caused by the motion of the earth, not by any diffraction effect.

all material bodies with little or no resistance, as freely perhaps as the wind passes through a grove of trees.■

The observations on the effects of diffraction and interference, may perhaps sometimes be applied to a practical purpose, in making us cautious in our conclusions respecting the appearances of minute bodies viewed in a microscope. The shadow of a fiber, however opaque, placed in a pencil of light admitted through a small aperture, is always somewhat less dark in the middle of its breadth than in the parts on each side. A similar effect may also take place, in some degree, with respect to the image on the retina, and impress the sense with an idea of a transparency which has no real existence: and, if a small portion of light be really transmitted through the substance, this may again be destroyed by its interference with the diffracted light,■ and produce an appearance of partial opacity, instead of uniform semitransparency. Thus, a central dark spot, and a light spot surrounded by a darker circle, may respectively be produced in the images of a semitransparent and an opaque corpuscle; and impress us with an idea of a complication of structure which does not exist. In order to detect the fallacy, we may make two or three fibers cross each other, and view a number of globules contiguous to each other; or we may obtain a still more effectual remedy by changing the magnifying power; and then, if the appearance remains constant in kind and in degree, we may be assured that it truly represents the nature of the substance to be examined. It is natural to inquire whether or not the figures of the globules of blood, delineated by Mr. Hewson in the *Phil. Trans.,* Vol LXIII, for 1773, might not in some measure have been influenced by a deception of this kind: but, as far as I have hitherto been able to examine the globules, with a lens of one fiftieth of an inch focus, I have found them nearly such as Mr. Hewson has described them.

■...................................

■The nature of light is a subject of no material importance to the concerns of life or to the practice of the arts, but it is in many other respects extremely interesting, especially as it tends to assist our views both of the nature of our sensations, and of the constitution of the universe at large. The examination of the production of colors, in a variety of circumstances, is intimately connected with the theory of their essential properties, and their causes; and we shall find that many of these phenomena will afford us considerable assistance in forming our opinion respecting the nature and origin of light in general.

■It is allowed on all sides, that light either consists in the emission of very minute particles from luminous substances, which are actually

The ether was the *medium* believed necessary for the propagation of light, assuming it to be a wave phenomenon. Following the work of Einstein the ether was later discarded.

Because of a difference in path.

Omitted are some remarks on the colors of bodies and on ultraviolet light (the dark rays of Ritter).

This extract is taken from Lecture 39 in Young's *Course of Lectures.*

The wave-particle controversy.

Newton's view, which contained some elements of a wave picture, is all the more interesting in the light of current optical theory, with its emphasis upon the apparent dual nature of light.

R. G. Boscovich, an eighteenth-century Jesuit who advocated the corpuscular theory.

Omitted is a detailed analysis of the two opposing views, as they apply in various cases.

The light must be *coherent*. Probably Young learned this experimentally.

Young's *double-slit experiment*.

projected, and continue to move, with the velocity commonly attributed to light, or in the excitation of an undulatory motion, analogous to that which constitutes sound, in a highly light and elastic medium pervading the universe; but the judgments of philosophers of all ages have been much divided with respect to the preference of one or the other of these opinions. There are also some circumstances which induce those, who entertain the first hypothesis, either to believe, with Newton,■ that the emanation of the particles of light is always attended by the undulations of an ethereal medium, accompanying it in its passage, or to suppose, with Boscovich,■ that the minute particles of light themselves receive, at the time of their emission, certain rotatory and vibratory motions, which they retain as long as their projectile motion continues. These additional suppositions, however necessary they may have been thought for explaining some particular phenomena, have never been very generally understood or admitted, although no attempt has been made to accommodate the theory in any other manner to those phenomena.

■..

Supposing the light of any given color to consist of undulations, of a given breadth, or of a given frequency, it follows that these undulations must be liable to those effects which we have already examined in the case of the waves of water, and the pulses of sound. It has been shown that two equal series of waves, proceeding from centers near each other, may be seen to destroy each other's effects at certain points, and at other points to redouble them; and the beating of two sounds has been explained from a similar interference. We are now to apply the same principles to the alternate union and extinction of colors (Fig. 1.).

In order that the effects of two portions of light may be thus combined, it is necessary that they be derived from the same origin, and that they arrive at the same point by different paths, in directions not much deviating from each other.■ This deviation may be produced in one or both of the portions by diffraction, by reflection, by refraction, or by any of these effects combined; but the simplest case appears to be, when a beam of homogeneous light falls on a screen in which there are two very small holes or slits, which may be considered as centers of divergence, from whence the light is diffracted in every direction.■ In this case, when the two newly formed beams are received on a surface placed so as to intercept them, their light is divided by dark stripes into portions nearly equal, but becoming wider as the surface is more remote from the apertures, so as to subtend very nearly equal angles from the apertures at all distances, and wider also in the same proportion as the apertures are closer to each other. The middle of the two portions is always light,

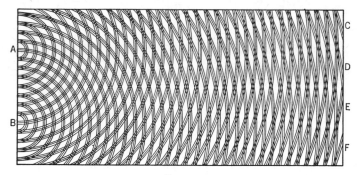

FIG. 1

and the bright stripes on each side are at such distances, that the light coming to them from one of the apertures, must have passed through a longer space than that which comes from the other, by an interval which is equal to the breadth of one, two, three, or more of the supposed undulations,■ while the intervening dark spaces correspond to a difference of half a supposed undulation, of one and a half, of two and a half, or more.■

That is, by an integral number of wavelengths.

From a comparison of various experiments, it appears that the breadth of the undulations constituting the extreme red light must be supposed to be, in air, about one 36 thousandth of an inch, and those of the extreme violet about one 60 thousandth; the mean of the whole spectrum, with respect to the intensity of light, being about one 45 thousandth.■ From these dimensions it follows, calculating upon the known velocity of light, that almost 500 millions of millions of the slowest of such undulations must enter the eye in a single second. The combination of two portions of white or mixed light, when viewed at a great distance exhibits a few white and black stripes, corresponding to this interval; although, upon closer inspection, the distinct effects of an infinite number of stripes of different breadths appear to be compounded together, so as to produce a beautiful diversity of tints, passing by degrees into each other. The central whiteness is first changed to a yellowish, and then to a tawny color, succeeded by crimson, and by violet and blue, which together appear, when seen at a distance, as a dark stripe; after this a green light appears, and the dark space beyond it has a crimson hue; the subsequent lights are all more or less green, the dark spaces purple and reddish;■ and the red light appears so far to predominate in all these effects, that the red or purple stripes occupy nearly the same place in the mixed fringes as if their light were received separately.

The comparison of the results of this theory with experiments fully establishes their general coincidence; it indicates, however, a slight cor-

An odd number of half wavelengths.

The actual range of visible light exceeds these limits only by small amounts. Young's observations were remarkably accurate.

The colors are formed as the different wavelengths satisfy the criterion for interference.

rection in some of the measures, on account of some unknown cause, perhaps connected with the intimate nature of diffraction, which uniformly occasions the portions of light, proceeding in a direction very nearly rectilinear, to be divided into stripes or fringes a little wider than the external stripes, formed by the light which is more bent.∎

When the parallel slits are enlarged, and leave only the intervening substance to cast its shadow, the divergence from its opposite margins still continues to produce the same fringes as before, but they are not easily visible, except within the extent of its shadow, being overpowered in other parts by a stronger light; but if the light thus diffracted be allowed to fall on the eye, either within the shadow, or in its neighborhood, the stripes will still appear; and in this manner the colors of small fibers are probably formed. Hence if a collection of equal fibers, for example a lock of wool, be held before the eye when we look at a luminous object, the series of stripes belonging to each fiber combine their effects, in such a manner, as to be converted into circular fringes or coronae. This is probably the origin of the colored circles or coronae sometimes seen round the sun and moon, two or three of them appearing together, nearly at equal distances from each other and from the luminary, the internal ones being, however, like the stripes, a little dilated. It is only necessary that the air should be loaded with globules of moisture,∎ nearly of equal size among themselves, not much exceeding one two thousandth of an inch in diameter, in order than a series of such coronae, at the distance of two or three degrees from each other, may be exhibited.

If, on the other hand, we remove the portion of the screen which separates the parallel slits from each other, their external margins will still continue to exhibit the effects of diffracted light in the shadow on each side; and the experiment will assume the form of those which were made by Newton on the light passing between the edges of two knives, brought very nearly into contact; although some of these experiments appear to show the influence of a portion of light reflected by a remoter part of the polished edge of the knives which indeed must unavoidably constitute a part of the light concerned in the appearance of fringes, wherever their whole breadth exceeds that of the aperture, or of the shadow of the fiber.

The edges of two knives, placed very near each other, may represent the opposite margins of a minute furrow, cut in the surface of a polished substance of any kind, which, when viewed with different degrees of obliquity, present a series of colors nearly resembling those which are exhibited within the shadows of the knives: in this case, however, the

See the explanation given by Fresnel (p. 118).

Or minute ice crystals.

paths of the two portions of light before their incidence are also to be considered, and the whole difference of these paths will be found to determine the appearance of color in the usual manner; thus when the surface is so situated, that the image of the luminous point would be seen in it by regular reflection, the difference will vanish, and the light will remain perfectly white, but in other cases various colors will appear, according to the degree of obliquity. These colors may easily be seen, in an irregular form, by looking at any metal, coarsely polished, in the sun-shine; but they become more distinct and conspicuous, when a number of fine lines of equal strength are drawn parallel to each other, so as to conspire in their effects.■

It sometimes happens that an object, of which a shadow is formed in a beam of light, admitted through a small aperture, is not terminated by parallel sides; thus the two portions of light which are diffracted from two sides of an object, at right angles with each other, frequently form a short series of curved fringes within the shadow, situated on each side of the diagonal, which were first observed by Grimaldi, and which are completely explicable from the general principle, of the interference of the two portions encroaching perpendicularly on the shadow.

But the most obvious of all the appearances of this kind is that of the fringes, which are usually seen beyond the termination of any shadow, formed in a beam of light, admitted through a small aperture: in white light three of these fringes are usually visible, and sometimes four;■ but in light of one color only, their number is greater; and they are always much narrower as they are remoter from the shadow. Their origin is easily deduced from the interference of the direct light with a portion of light reflected from the margin of the object which produces them, the obliquity of its incidence causing a reflection so copious as to exhibit a visible effect, however narrow that margin may be; the fringes are, however, rendered more obvious as the quantity of this reflected light is greater. Upon this theory it follows that the distance of the first dark fringe from the shadow should be half as great as that of the fourth, the difference of the lengths of the different paths of the light being as the squares of those distances; and the experiment precisely confirms this calculation, with the same slight correction only as is required in all other cases; the distances of the first fringes being always a little increased.■ It may also be observed, that the extent of the shadow itself is always augmented, and nearly in an equal degree with that of the fringes: the reason of this circumstance appears to be the gradual loss of light at the edges of every separate beam, which is so strongly analogous to the phenomena visible in waves of water. The same cause may also perhaps

The principle of the diffraction grating, which combines the effects from many slits.

Their sharpness suffers because of the overlapping of adjacent colors.

The spacing between fringes should be constant, provided that the screen is placed far from the slits compared with the distances of the fringes measured from the shadow. Otherwise, a geometric correction is required.

have some effect in producing the general modification or correction of the place of the first fringes, although it appears to be scarcely sufficient for explaining the whole of it.

A still more common and convenient method, of exhibiting the effects of the mutual interference of light, is afforded us by the colors of the thin plates of transparent substances. The lights are here derived from the successive partial reflections produced by the upper and under surface of the plate, or when the plate is viewed by transmitted light, from the direct beam which is simply refracted, and that portion of it which is twice reflected within the plate. The appearance in the latter case is much less striking than in the former, because the light thus affected is only a small portion of the whole beam, with which it is mixed; while in the former the two reflected portions are nearly of equal intensity, and may be separated from all other light tending to overpower them. In both cases, when the plate is gradually reduced in thickness to an extremely thin edge, the order of colors may be precisely the same as in the stripes and coronae already described; their distance only varying when the surfaces of the plate, instead of being plane, are concave, as it frequently happens in such experiments. The scale of an oxide, which is often formed by the effect of heat on the surface of a metal, in particular of iron, affords us an example of such a series formed in reflected light: this scale is at first inconceivably thin, and destroys none of the light reflected, it soon, however, begins to be of a dull yellow, which changes to red, and then to crimson and blue, after which the effect is destroyed by the opacity which the oxide acquires. Usually, however, the series of colors produced in reflected light follows an order somewhat different: the scale of oxide is denser than the air, and the iron below than the oxide, but where the mediums above and below the plate are either both rarer or both denser than itself, the different natures of the reflections at its different surfaces appear to produce a modification in the state of the undulations, and the infinitely thin edge of the plate becomes black instead of white, one of the portions of light at once destroying the other, instead of cooperating with it.■ Thus when a film of soapy water is stretched over a wine glass, and placed in a vertical position, its upper edge becomes extremely thin, and appears nearly black, while the parts below are divided by horizontal lines into a series of colored bands; and when two glasses, one of which is slightly convex, are pressed together with some force, the plate of air between them exhibits the appearance of colored rings,■ beginning from a black spot at the center, and becoming narrower and narrower, as the curved figure of the glass causes the thickness of the plate of air to increase more and more rapidly. The

Because of the reversal of phase upon reflection from a medium of greater index of refraction.

Newton's rings.

black is succeeded by a violet, so faint as to be scarcely perceptible; next to this is an orange yellow, and then crimson and blue. When water, or any other fluid, is substituted for the air between the glasses, the rings appear where the thickness is as much less than that of the plate of air, as the refractive density of the fluid is greater; a circumstance which necessarily follows from the proportion of the velocities with which light must, upon the Huygenian hypothesis, be supposed to move in different mediums. It is also a consequence equally necessary in this theory, and equally inconsistent with all others, that when the direction of the light is oblique, the effect of a thicker plate must be the same as that of a thinner plate, when the light falls perpendicularly upon it; the difference of the paths described by the different portions of light precisely corresponding with the observed phenomena.

■...

It is presumed, that the accuracy, with which the general law of the interference of light has been shown to be applicable to so great a variety of facts, in circumstances the most dissimilar, will be allowed to establish its validity in the most satisfactory manner. The full confirmation or decided rejection of the theory, by which this law was first suggested, can be expected from time and experience alone; if it be confuted, our prospects will again be confined within their ancient limits, but if it be fully established, we may expect an ample extension of our views of the operations of nature, by means of our acquaintance with a medium, so powerful and so universal, as that to which the propagation of light must be attributed.

Omitted are brief discussions of the colors of small bodies and of rainbows.

SUPPLEMENTARY READING

Crew, H. (ed.), *The Wave Theory of Light* (New York: American Book, 1900).

Dampier, W. C., *A History of Science* (New York: Cambridge University, 1946), pp. 219ff.

Lenard, P., *Great Men of Science* (New York: British Book Centre, 1934), pp. 196ff.

Mach, E., *The Principles of Physical Optics,* trans. by J. S. Anderson and A. F. A. Young (New York: Dover, 1953).

Magie, W. F., *A Source Book in Physics* (New York: McGraw-Hill, 1935), pp. 308ff.

Newton, I., *Opticks* (New York: Dover, 1952).

Wood, A., and Oldham, F., *Thomas Young* (New York: Cambridge University, 1954).

8

Augustin Fresnel
1788 - 1827

The Diffraction of Light

THOMAS YOUNG'S demonstration of the interference of light made little impression upon his colleagues in 1803. It remained for Fresnel, a decade later, to present such compelling evidence for the wave theory as to command its acceptance, even by the staunchest supporters of Newtonian theory. The dual significance of his achievement must not be overlooked. He was not the first to advocate a wave theory of light; others, notably Huygens and Young, preceded him. But Fresnel approached the problem so skillfully, and set forth such a clear and convincing solution, that he is generally regarded as the founder of the wave theory of light. At the same time he caused the first major break with the purely mechanical doctrines of Newtonian physics. It was understandably difficult at that time (nor do many find it much easier today) to conceive of light being propagated by some *nonmaterial* means. The corpuscular view found ready support not only because of the great prestige lent to it by Newton, but also because it fitted so well the philosophical outlook of the time. For the same reason, once the wave theory found acceptance, the *ether* was invented in an effort to retain some semblance of a mechanical view. It can be seen, then, how impressive Fresnel's theory must have been, if it could invite wide support in the face of such mental reservations. Not that Fresnel's explanation put an end to the matter; there remained certain gaps in his theory which were filled in later by George Stokes (1819–1903), Clerk Maxwell (1831–1879), and H. A. Lorentz (1858–1923), among others. Nevertheless, he accomplished the major part of it—and, what is more, set others to thinking along correct lines.

Augustin Jean Fresnel was born at Broglie, in the Normandy region of France, on May 10, 1788, the year before the start of the French Revolution.

His father was an architect; his influence may have decided young Fresnel on an engineering career. He was educated at Caen, making little impression until, at the age of sixteen, he entered the École Polytechnique in Paris. There he made rapid progress, particularly in mathematics, and attracted the attention of the famous mathematician Adrien Marie Legendre (1752–1833). Two years later he was admitted to the École des Ponts et Chausées, from which he graduated as an engineer in 1809. He became a government civil engineer shortly afterward, a post that, except for one brief period, remained his chief occupation for the rest of his life. After serving as a highway engineer in several parts of France he found himself assigned to the Drôme department, on the main highway between Spain and Italy. He was there when Napoleon returned from Elba in the spring of 1815 and the Hundred Days began. Having declared his allegiance to the Bourbons, Fresnel sought to join the troops that were to oppose Napoleon's entry into Paris. As a result he lost his appointment and spent the period of enforced leisure on his optical studies.

It appears that he began his researches about 1814 when he wrote a paper (unpublished) on the aberration of light. He turned to the problem of diffraction shortly afterward, working at first in total ignorance of Young's discovery. With the aid of some simple equipment he performed his first measurements on the fringes observed in the shadows cast by opaque objects, reporting his results to the Paris Academy of Sciences in the latter part of 1815 and early in 1816. He was faced with initial opposition no less than that which greeted Young's earlier experiments; such distinguished scientists as Laplace, Biot, and Poisson strongly criticized his views.* Poisson objected to Fresnel's theory on the ground that, if correct, the shadow of a round object should have a bright center, and this, of course, was not true! Fresnel agreed that this would be implied by his theory, whereupon he proceeded to demonstrate the effect in his laboratory. Fresnel had not yet reached a clear understanding of diffraction. He was familiar with Huygen's work, and by then had learned of Young's experiments, but he had yet to grasp the idea that the parts of a wave front giving rise to interference were limited areas[1] rather than points. Thus he accounted for the fringes, as did Young, by the interference of direct rays from the source with those *inflected* by the edge of the obstacle. By the middle of 1816, however, Fresnel saw clearly the explanation of diffraction on the basis of the wave hypothesis.

Following Napoleon's second abdication in 1815 Fresnel was restored to

---

* Pierre Laplace (1749–1827), known particularly for his work on the molecular forces in liquids and his great mathematical skill; Jean Biot (1774–1862), an active investigator in optics and electricity; Siméon Poisson (1781–1840), a mathematician noted for his application of mathematics to physics, especially in mechanics and heat.

[1] The so-called *Fresnel zones.*

his government post, and the next year, through the influence of his friend and colleague François Arago (1786–1853), he obtained an appointment in Paris, where he remained until his death eleven years later. While there he served for two years on the Oureq Canal and spent the remainder with the Lighthouse Commission. He also secured, toward the end of his life, the position of examiner in the École Polytechnique. Throughout this period he continued his investigations of optical phenomena. In 1818, in response to a competition proposed by the French Academy on the subject of diffraction, Fresnel submitted a paper that was "crowned" with the first prize (known as the *Crown Memoir*). The judges, including his earlier opponents Laplace, Biot, and Poisson, were unanimous in their decision. He was elected to membership in the Académie des Sciences in 1823, the same year that the first Fresnel Lenses were installed in a French lighthouse. In 1827, several days before his untimely death at the age of 39, he was awarded the Rumford Medal of the Royal Society for his discoveries in physical optics. These included his well-known studies of double refraction, and, with Arago, experiments on the interference of polarized light that showed it to be a transverse wave phenomenon.

His one consuming interest in life was experimental optics. As he wrote to Thomas Young, " . . . All the compliments that I have received from Arago, Laplace, and Biot never gave me as much pleasure as the discovery of a theoretic truth, or the confirmation of a calculation by experiment."

The following extract is taken from his prize memoir, *Mémoire Courronné* published in the *Mémoires de L'Academie Royale des Sciences de L'Institut de France,* vol. 5, 1826. The translation appears in *The Wave Theory of Light,* edited by H. Crew (New York: American Book Co., 1900).

Fresnel's Experiment

A general discussion comparing the emission-theory (corpuscular theory) with the wave theory is omitted.

Young's double-slit experiment.

■...

Grimaldi was the first to observe the effect which rays of light produce upon one another. Recently the distinguished Dr. Thomas Young has shown by a simple and ingenious experiment that the interior fringes are produced by the meeting of rays inflected at each side of the opaque body.■ This he proved by using a screen to intercept one of the two pencils of light; and in this way he was able to make the interior fringes completely vanish, whatever might be the form, mass, or nature of the screen, and whether he intercepted the luminous pencil before or after its passage into the shadow.

Brighter and sharper fringes may be produced by cutting two parallel slits close together in a piece of cardboard or a sheet of metal, and placing the screen thus prepared in front of the luminous point. We may then observe, by use of a magnifying glass between the opaque body and the eye, that the shadow is filled with a large number of very sharp-colored fringes so long as the light shines through both openings at the same time, but these disappear whenever the light is cut off from one of the slits.

If we allow two pencils of light, each coming from the same source and regularly reflected by two metallic mirrors, to meet under a very small angle, we obtain similar fringes, the colors of which are even purer and more brilliant than before.■ To obtain these bands, it is necessary to be very careful that in the region where the two mirrors come into contact, or at least throughout a portion of their line of contact, the surface of the one is not shifted sensibly past that of the other. This is necessary in order that the difference of path traversed by two reflected rays meeting in the area common to the two luminous* fields may be very small. I may remark in passing that the theory of interference alone will explain this experiment, and that the experiment calls for manipulation so delicate and effort so continued that it is almost impossible that one should strike upon it by accident.

Fresnel's double-mirror experiment. The two mirrors were placed adjacent to one another but inclined at a very small angle.

If we raise one of the mirrors or intercept the light which it reflects either before or after reflection, the fringes disappear as in the preceding case. This furnishes still further evidence that the fringes are produced, not by the action of the edges of the mirrors, but by the meeting of two pencils of light. For these fringes are always at right angles to the line which joins the two images of the luminous point, whatever be its inclination with respect to these edges, at least throughout the extent of the area which is common to the two regularly reflected pencils.†

---

* In the case of white light, or even in light as homogeneous as possible, the number of fringes which one can see is always limited, because even when the light has reached a degree of simplicity as great as possible without too far diminishing its intensity, it is still composed of rays which are heterogeneous; and since the bright and dark bands thus produced do not all have the same size, they encroach the one upon the other in proportion as their order increases, and finally they completely destroy each other; and this is why one does not see any fringes when the difference of paths becomes slightly sensible.■

† When the fringes extend outside, all their exterior portions resulting from the meeting of rays regularly reflected by one of the mirrors and rays inflected near the edge of the other should have different directions. If one observes this phenomenon carefully, he will see that the form and position of the fringes are in each case in accord with the theory of interference.

Sensible is used throughout in the sense of appreciable.

Since the fringes which one sees in the interior of the shadow of a very narrow body and those which one obtains by the use of two mirrors result evidently from the mutual influence of rays of light, analogy would indicate that the same thing ought to be true for the exterior fringes■ of the shadows of bodies illuminated by a point source. The first explanation which occurs to one is that these fringes are produced by the interference of direct rays with those which are reflected at the edge of the opaque body, while the interior fringes result from the combined action of rays inflected into the shadow from the two sides of the opaque body, these inflected rays having their origin either at the surface or at points indefinitely near it. This appears to be the opinion of Mr. Young, and it was at first my own opinion;■ but a closer examination of the phenomena convinced me of its falsity.

■..

In the first section of this memoir I have shown that the corpuscular theory, and even the principle of interference when applied only to direct rays and to rays *reflected or inflected at the very edge of the opaque screen,* is incompetent to explain the phenomena of diffraction. I now propose to show that we may find a satisfactory explanation and a general theory in terms of waves, without recourse to any auxiliary hypothesis, by basing everything upon the principle of Huygens and upon that of interference, both of which are inferences from the fundamental hypothesis.

Assuming that light consists in vibrations of the ether similar to sound waves,■ we can easily account for the inflection of rays of light at sensible distances from the diffracting body. For when any small portion of an elastic fluid undergoes condensation, for instance, it tends to expand in all directions;■ and if throughout the entire wave the particles are displaced only along the normal, the result would be that all points of the wave lying upon the same spherical surface would simultaneously suffer the same condensation or expansion, thus leaving the transverse pressures in equilibrium; but when a portion of the wave front is intercepted or retarded in its path by interposing an opaque or transparent screen, it is easily seen that this transverse equilibrium is destroyed and that various points of the wave may now send out rays along new directions.

To follow by analytical mechanics all the various changes which a wave front undergoes from the instant at which a part of it is intercepted by a screen would be an exceedingly difficult task, and we do not propose to derive the laws of diffraction in this manner, nor do we propose to inquire what happens in the immediate neighborhood of the

---

The exterior fringes are those formed immediately outside the shadow.

This was Fresnel's view until the middle of 1816.

A section showing the use of Young's theory is omitted.

However, Fresnel and Arago later showed light to be a transverse vibration.

Meaning that in such a fluid any compression or expansion is transmitted through the medium.

opaque body, where the laws are doubtless very complicated and where the form of the edge of the screen must have a perceptible effect upon the position and the intensity of the fringes. We propose rather to compute the relative intensities at different points of the wave front only after it has gone a large number of wavelengths beyond the screen. Thus the positions at which we study the waves are always to be regarded as separated from the screen by a distance which is very considerable compared with the length of a light wave.∎

This is almost always the case in practice.

We shall not take up the problem of vibrations in an elastic fluid from the point of view which the mathematicians have ordinarily employed—that is, considering only a single disturbance. Single vibrations are never met with in nature. Disturbances occur in groups, as is seen in the pendulum and in sounding bodies.∎ We shall assume that vibrations of luminous particles occur in the same manner—that is, one after another and series after series. This hypothesis follows not only from analogy, but as an inference from the nature of the forces which hold the particles of a body in equilibrium. To understand how a single luminous particle may perform a large series of oscillations all of which are nearly equal, we have only to imagine that its density is much greater than that of the fluid in which it vibrates∎—and, indeed, this is only what has already been inferred from the uniformity of the motions of the planets through this same fluid which fills planetary space. It is not improbable also that the optic nerve yields the sensation of sight only after having received a considerable number of successive stimuli.

Fourier's method of harmonic analysis had been recently developed to represent mathematically such periodic phenomena.

That is, the damping effect of the fluid is small. The attempt to assign physical properties to the ether is evident here.

However extended one may consider systems of wave fronts to be, it is clear that they have limits, and that in considering interference we cannot predicate of their extreme portions that which is true for the region in which they are superposed. Thus, for instance, two systems of equal wavelength and of equal intensity, differing in path by half a wave, interfere destructively only at those points in the ether where they meet, and the two extreme half wavelengths escape interference.

Nevertheless, we shall assume that the various systems of waves undergo the same change throughout their entire extent, the error introduced by this assumption being inappreciable; or, what amounts to the same thing, we shall assume in our discussion of interference that these series of lightwaves represent general vibrations of the ether, and are undefined as to their limits.

Such an assumption simplified the mathematical treatment of the theory, which is omitted here. Fresnel showed how to calculate the effect of superposing two or more wave trains.

∎···

Having determined the resultant of any number of trains of lightwaves, I shall now show how by the aid of these interference formulae and by the principle of Huygens alone it is possible to explain, and even

Huygen's
principle.

A straightforward
mechanical
concept.

The phase.

As in the case of
sound waves, but
see the footnote.

to compute, all the phenomena of diffraction. This principle, which I consider as a rigorous deduction from the basal hypothesis, may be expressed thus: ▪*The vibrations at each point in the wave front may be considered as the sum of the elementary motions which at any one instant are sent to that point from all parts of this same wave in any one of its previous* positions, each of these parts acting independently the one of the other.* It follows from the principle of the superposition of small motions that the vibrations produced at any point in an elastic fluid by several disturbances are equal to the resultant of all the disturbances reaching this point at the same instant from different centers of vibration,▪ whatever be their number, their respective positions, their nature, or the epoch▪ of the different disturbances. This general principle must apply to all particular cases. I shall suppose that all of these disturbances, infinite in number, are of the same kind, that they take place simultaneously, that they are contiguous and occur in the single plane or on a single spherical surface. I shall make still another hypothesis with reference to the nature of these disturbances, *viz.*, I shall suppose that the velocities impressed upon the particles are all directed in the same sense,▪ perpendicular to the surface of the sphere,† and, besides, that they are proportional to the compression, and in such a way that the particles have no retrograde motion. I have thus reconstructed a primary wave out of partial disturbances. We may, therefore, say that the vibrations at each point in the wave front can be looked upon as the resultant of all the secondary displacements which reach it at the same instant from all parts of this same wave in some previous position, each of these parts acting independently one of the other.

If the intensity of the primary wave is uniform, it follows from theoretical as well as from all other considerations that this uniformity will be maintained throughout its path, provided only that no part of the wave is intercepted or retarded with respect to its neighboring parts,

---

Single
disturbances, such
as shock waves or
electrical pulses,
do occur in nature
and generally are
treated as the
superposition of
many continuous
waves of the
proper
frequencies and
amplitudes.

* I am here discussing only an infinite train of waves, or the most general vibration of a fluid. It is only in this sense that one can speak of two light waves annulling one another when they are half a wavelength apart. The formulae of interference just given do not apply to the case of a single wave, not to mention the fact that such waves do not occur in nature.▪

† It is possible for light waves to occur in which the direction of the absolute velocity impressed upon the particles is not perpendicular to the wave surface. In studying the laws of interference of polarized light, I have become convinced since the writing of this memoir that light vibrations are at right angles to the rays or parallel to the wave surface. The arguments and computations contained in this memoir harmonize quite as well with this new hypothesis as with the preceding, because they are quite independent of the actual direction of the vibrations and presuppose only that the direction of these vibrations is the same for all rays belonging to any system of waves producing fringes.

because the resultant of the secondary displacements mentioned above will be the same at every point. But if a portion of the wave be stopped by the interposition of an opaque body, then the intensity of each point varies with its distance from the edge of the shadow, and these variations will be especially marked near the edge of the geometrical shadow.

Let C be the luminous point, AG the screen,■ AME a wave which has just reached A and is partly intercepted by the opaque body. Imagine it to be divided into an infinite number of small arcs—Am' $m'm$, mM, Mn, nn', $n'n''$, etc. In order to determine the intensity at any point P in any of the later positions of the wave BPD, it is necessary to find the resultant of all the secondary waves which each of these portions of the primitive wave would send to the point P, provided they were acting independently one of the other.

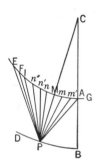

Since the impulse communicated to every part of the primitive wave was directed along the normal, the motion which each tends to impress upon the ether ought to be more intense in this direction than in any other; and the rays which would emanate from it, if acting alone, would be less and less intense as they deviated more and more from this direction.

The investigation of the law according to which their intensity varies about each center of disturbance is doubtless a very difficult matter; but, fortunately, we have no need of knowing it, for it is easily seen that the effects produced by these rays are mutually destructive when their directions are sensibly inclined towards the normal. Consequently, the rays which produce any appreciable effect upon the quantity of light received at any point P may be regarded as of equal intensity.*

Let us now consider the rays EP, FP, and IP, which are sensibly inclined and which meet at P, a point whose distance from the wave EA

---

* When the center of disturbance has been compressed, the force of expansion tends to thrust the particles in all directions; and if they have no backward motion,■ the reason is simply that their initial velocities forward destroy those which expansion tends to impress upon them towards the rear; but it does not follow that the disturbance can be transmitted only along the direction of the initial velocities, for the force of expansion in a perpendicular direction, for instance, combines with a primitive impulse without having its effect diminished. It is clear that the intensity of the wave thus produced must vary greatly at different points of its circumference, not only on account of the initial impulse, but also because the compressions do not obey the same law around the center of disturbance; but the variations of intensity in the resultant wave must follow the law of continuity, and may, therefore, be considered as vanishing throughout a small angle, especially along the normal to the primitive wave. For the initial velocities of the particles in any direction whatever are proportional to the cosine of the angle which this direction makes with that of the normal, so that these components vary much less rapidly than the angle so long as the angle is small.■

AG is an opaque obstacle in the path of the wave. See the diagram below.

The backward wave is destroyed by mutual interference.

The cosine of **very** small angles is nearly one, and is slowly varying.■

I shall suppose to include a large number of wavelengths. Take the two arcs *EF* and *FI* of such a length that the difference *EP* − *FP* and *FP* − *IP* shall be equal to a half wavelength. Since these rays are quite oblique, and since a half wavelength is very small compared with their length, these two arcs will be very nearly equal, and the rays which they send to the point *P* will be practically parallel; and since corresponding rays on the two arcs differ by half a wavelength, the two are mutually destructive.

We may then suppose that all the rays which various parts of the primary wave *AE* send to the point *P* are of equal intensity, since the only rays for which this assumption is not accurate produce no sensible effect upon the quantity of light which it receives.■ In the same manner, for the sake of simplifying the calculation of the resultant of all the elementary waves, we may consider their vibrations as taking place in the same direction, since the angles which these rays make with each other are very small; so that the problem reduces itself to the one which we have already solved—namely, *to find the resultant of any number of parallel trains of lightwaves of the same length, the intensities and relative positions being given.* The intensities are here proportional to the lengths of the illuminating arcs, and the relative positions of the wave trains are given by the differences of path traversed.

Properly speaking, we have considered up to this point only the section of the wave made by a plane perpendicular to the edge of the screen projected at *A*. We shall now consider it in its entirety, and shall think of it as divided by equidistant meridians perpendicular to the plane of the figure into infinitely thin spindles.■ We shall then be able to employ the same process of reasoning which we have just used for a section of the wave, and thus show that the rays which are quite oblique are mutually destructive.

In the case we are now considering these spindles are indefinitely extended in a direction parallel to the edge of the screen, for the wave is intercepted only on one side. Accordingly the intensity of the resultant of all the vibrations which they send to the point *P* would be the same for each of them; for, owing to the extremely small difference of path, the rays which emanate from these spindles must be considered as of equal intensity,■ at least throughout that region of the primitive wave which produces a sensible effect upon the light sent to *P*. Further, it is evident that each elementary resultant will differ in phase by the same quantity with respect to the ray coming from that point of the spindle nearest *P,* that is to say, from the point at which the spindle cuts the plane of the figure. The intervals between these elementary resultants

That is, only those rays from nearly adjacent regions of the wave front need be considered as far as interference is concerned.

Going to the three-dimensional case, Fresnel assumed for simplicity a cylindrical wave front perpendicular to the plane of the figure.

For the same reason given previously; the rays originate from nearly adjacent regions of the wave front.

will then be equal to the difference of path traversed by the rays, *AP,* *m'P, mP,* etc., all lying in the plane of the figure; and their intensities will be proportional to the arcs *Am', m'm, mM,* etc. In order now to obtain the intensity of the total resultant, we have to make the same calculation which we have already made, considering only the section of the wave by a plane perpendicular to the edge of the screen.*

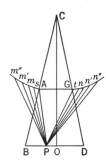

Before deriving the analytical expression for this resultant I propose to draw from the principle of Huygens some of the inferences which follow from simple geometrical considerations.

Let *AG* represent an opaque body sufficiently narrow for one to distinguish fringes in its shadow at the distance *AB.* Let *C* be the luminous point and *BD* be either the focal plane of the magnifying glass with which one observes these fringes or a white card upon which the fringes are projected.

Let us now imagine the original wave divided into small arcs—*Am,* *mm', m'm'',* etc., *Gn, nn', n'n'', n''n''',* etc.—in such a way that the rays drawn from the point *P* in the shadow to two consecutive points of division will differ by half a wavelength. All of the secondary waves sent to the point *P* by the elements of each of these arcs will completely interfere with those which emanate from the corresponding parts of the two arcs immediately adjoining it; so that, if all these arcs were equal, the rays which they would send to the point *P* would be mutually destructive, with the exception of the extreme arc *mA.* Half of the intensity of this arc would be left, for half the light sent by the arc *mm'* (with which *mA* is in complete discordance) would be destroyed by half of the preceding arc *m''m'.* As soon as the rays meeting at *P* are considerably inclined with respect to the normal,■ these arcs are practically equal. The resultant wave, therefore, corresponds in phase almost exactly to the middle of *mA,* the only arc which produces any sensible effect. It is thus seen that it differs in phase by one quarter of a wavelength from the element at the edge *A* of the opaque screen. Since the same thing takes place in the other part of the incident wave *Gn,* the interference

That is, the norma to *BD* at *P.*

---

* So long as the edge of the screen is rectilinear we can determine the position of the dark and bright bands and their relative intensities by considering only the section of the wave made by a plane which is perpendicular to the edge of the screen. But when the edge of the screen is curved or composed of straight edges inclined at an angle, it is then necessary to integrate■ along two directions at right angles to each other, or to integrate around the point under consideration. In some particular cases this latter method is simpler, as, for instance, when we have to calculate the intensity of the light in the center of the shadow produced by a screen or in the projection of a circular aperture.

Sum up the effects.

between these two vibrations occurring at the point P is determined by the difference of length between the two rays sP and tP, which take their rise at the middle of the arcs Am and Gn, or, what amounts to the same thing, by the difference between the two rays AP and GP coming from the very edge of the opaque body. It thus happens that when the interior fringes under consideration are rather distant from the edges of the geometrical shadow, we are able to apply practically without error the formula based upon the hypothesis that the inflected waves have their origin at the very edges of the opaque body; but in proportion as the point P approaches B the arc Am becomes greater in comparison with the arc mm', the arc mm' with respect to the arc $m'm''$, etc.; and likewise in the arc mA the elements in the immediate vicinity of the point A become sensibly greater than the elements which are situated near the point m, and which correspond to equal differences of path. It happens, therefore that the *effective** ray, sP, will not be the mean between the outside rays, mP and AP, but will more nearly approach the length of the latter. On the other side of the opaque body we have slightly different circumstances. The difference between the ray GP and the effective ray tP approximates more and more nearly a quarter of a wavelength as the point P moves farther and farther away from D, so that the difference of path traversed varies more rapidly between the effective rays sP and tP than between the rays AP and GP; consequently, the fringes in the neighborhood of the point B ought to be a little farther from the center of the shadow than would be indicated by the formula based upon the first hypothesis. ■

Compare this with Young's results, for which he was unable to account.

Having considered the case of fringes produced by a narrow body, I pass to the consideration of those which are caused by a small aperture.

Let AG be the aperture through which the light passes. I shall at first suppose that it is sufficiently narrow for the dark bands of the first order to fall inside the geometrical shadow of the screen, and at the same time to be fairly distant from the edges B and D. Let P be the darkest point in one of these two bands; it is then easily seen that this must correspond to a difference of one whole wavelength between the two extreme rays AP and GP. Let us now imagine another ray, PI, drawn in such a way that its length shall be a mean between the other two. Then, on

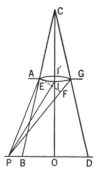

___

* I have given this name to the distance of the resultant wave from the original wave because the positions of the dark and bright bands are the same as they would be if these effective rays alone produced them.

account of its marked inclination to the arc *AIG,* the point *I* will fall almost exactly in the middle. We now have the arc divided into two parts, whose corresponding elements are almost exactly equal, and send to the point *P* vibrations in exactly opposite phases, so that these must annul each other.

By the same reasoning it is easily seen that the darkest points in the other dark bands also correspond to differences of an even number of half wavelengths between the two rays which come from the edges of the aperture; and, in like manner, the brightest points of the bright bands correspond to differences of an uneven number of half wavelengths— that is to say, their positions are exactly reversed as compared with those which are deduced from the interference of the limiting rays on the hypothesis that these alone are concerned in the production of fringes. This is true with the exception of the point at the middle, which, on either hypothesis, must be bright. The inferences deduced from the theory that the fringes result from the superposition of all of the disturbances from all parts of the arc *AG* are verified by experiments, which at the same time disprove the theory which looks upon these bands as produced only by rays inflected and reflected at the edges of the diaphragm. These are precisely the phenomena which first led me to recognize the insufficiency of this hypothesis, and suggested the fundamental principle of the theory which I have just explained—namely, the principle of Huygens combined with the principle of interference.

In the case which we have just considered, where, by virtue of a very small aperture, the dark bands of the first order fall at some distance from the edges of the geometrical shadow, it follows from theory, as well as from experiment, that the distance comprised between the darkest points is almost exactly double that of the other intervals between the middle points of two consecutive dark bands, and this is all the more nearly true in proportion as the aperture becomes smaller or more distant from the luminous point and from the focus of the magnifying glass with which one observes the fringes; for, by sufficiently increasing these distances one may produce the same effects with an aperture of any size whatever.

But when these distances are not very great, and when their aperture is too large for the rays producing the fringes to be very much inclined to the wavefront, *AG,* it follows that corresponding elements of the arcs into which we have supposed a wave to be divided can no longer be considered as each equal to the other, for they are sensibly larger on the side next to the band under consideration. Under these conditions we can

Omitted are a number of specific applications worked out by Fresnel in great detail.

rigorously deduce the positions of maximum and minimum intensity only by computing the resultant of all the small secondary waves which are sent out by the incident wave.

■ ....

SUPPLEMENTARY READING

Crew, H. (ed.), *The Wave Theory of Light* (New York: American Book, 1900), pp. 79ff.

———, "Portraits of Famous Physicists," *Scripta Mathematica*, Pictorial Mathematics, Portfolio No. 4, 1942.

Dampier, W. C., *A History of Science* (New York: Cambridge University, 1946).

Lenard, P., *Great Men of Science* (New York: British Book Centre, 1934), pp. 204ff.

Magie, W. F., *A Source Book in Physics* (New York: McGraw-Hill, 1935), pp. 318ff.

9

Hans Christian Oersted

1777 · 1851

Electromagnetism

BY THE END of the eighteenth century a considerable body of empirical knowledge concerning electric and magnetic phenomena had accumulated. Coulomb had discovered the inverse-square law of force; the electrical experiments of Benjamin Franklin (1706–1790) were widely known; Luigi Galvani (1737–1798) had found, but explained incorrectly as *animal electricity,* that dissimilar metals produce electrical effects upon contact,[1] and Alessandro Volta (1745–1827), at the turn of the century, correctly accounted for Galvani's results and invented the *voltaic cell* or *pile.* The essential link, then lacking, was the relationship between electricity and magnetism, putting aside the common problem of accounting for action at a distance. This highly significant connection, the basis for all electromagnetic phenomena, was discovered by the Danish physicist, Hans Christian Oersted, in 1820.

The son of an apothecary, Oersted was born in the town of Rudköbing, Denmark, on August 11, 1777. France was on the verge of revolution, and in America the War of Independence was already testing man's right to freedom. The Industrial Revolution had begun a decade or two earlier and *practical* inventions poured forth in a steady stream. It was a time of change that heralded new ways of life, new social and political ideas, a new philosophy, new concepts of art and literature; in short, it was the forerunner of the nineteenth-century Age of Romanticism, characterized chiefly by conservatism and idealism, by an apparent reaction to the eighteenth-century reliance upon science and reason. What influence the new age may have had on the development of physics cannot readily be determined. It must have touched, in some ways, the lives of those who made physics their

---

[1] Contact emf's.

careers. Fortunately, however, it apparently had little effect on the methods by which they sought answers to their questions about the physical universe.

Following an indifferent early education, Oersted was apprenticed, at the age of thirteen, to his father. It was here that he developed an interest in natural science and decided to continue in that walk of life. He was admitted to the University of Copenhagen in 1794, passed the philosophical examination the next year and the pharmaceutical examination in 1797. Two years later he received the Ph.D. degree for a thesis on metaphysics. Oersted was at that time regarded as a philosopher rather than a natural scientist, but shortly afterward he turned to experimental research in physics.

His first published paper[2] was on the identity of electrical and chemical forces, which formed the foundation of the electrochemical system developed more fully afterward by Jöns Jakob Berzelius (1779–1848). Oersted worked on a number of varied problems during the first two decades of the nineteenth century, including the compressibility of water and the use of electric currents to explode mines. But, while it appears that the connection between electricity and magnetism was a problem that had occupied him for some time, it was not until 1820 that he found the answer. He was at that time a professor of physics at the University of Copenhagen, an appointment which he had held since 1806, and his earliest observations were made in connection with his teaching duties there. It is not clear from his writings whether Oersted actually made the discovery during one of his lectures or whether he first demonstrated it on such an occasion. Assuming that he did observe the effect first during a lecture, Lenard[3] points out that the fact he had a voltaic pile and compass needle on the table indicates he was looking for such an effect, that the discovery was not pure accident. Oersted announced his discovery of the magnetic field accompanying an electric current on July 21, 1820, in a pamphlet[4] distributed privately to scientists and to scientific societies.

His stature as a physicist rose immediately. He was widely acclaimed and honored by membership in a number of scientific academies, including the Royal Society and l'Institut de France. In subsequent travels he met many of his contemporaries, including Davy and Faraday. The latter, no doubt, owed much of the inspiration for his induction experiments to these meetings with Oersted. As soon as André Marie Ampère (1775–1836) learned of Oersted's discovery, he began the investigations that led, within

---

[2] *Recherches sur l'identité des forces électriques et chimiques.*

[3] P. Lenard, *Great Men of Science* (New York: British Book Centre, 1934), p. 214.

[4] *Experimenta circa effectum conflictus electrici in acum magneticam.*

a short time, to his own important discovery of the force between two currents and the magnitude of the magnetic field due to a current.

Oersted founded the Society for Advancement of Natural Science, and established, in 1829, the Polytechnic Institution of Copenhagen, which he directed until his death on March 9, 1851, the year of the Great Exhibition. The purpose of the Society was to make science more readily available to the general public, an undertaking which Oersted considered extremely important and to which he devoted much of his time.

The following extract is a translation of the original pamphlet made by the Rev. J. E. Kempe. It appeared in the *Journal of the Society of Telegraph Engineers,* vol. V (1876), and included a brief memoir on Oersted by C. L. Madsen.

Oersted's Experiment

The first experiments on the subject which I undertake to illustrate were set on foot in the classes for electricity, galvanism, and magnetism, which were held by me in the winter just past.■ By these experiments it seemed to be shown that the magnetic needle was moved from its position by the help of a galvanic apparatus,■ and that, when the galvanic circuit was closed, but not when open, as certain very celebrated physicists in vain attempted several years ago. As, however, these experiments were conducted with somewhat defective apparatus, and, on that account, the phenomena which were produced did not seem clear enough for the importance of the subject, I got my friend Esmarch, the king's minister of justice, to join me, that the experiments might be repeated and extended with the great galvanic apparatus which we fitted up together. A distinguished man, Wleugel, knight of the Danish Order, and president of our Pilot Board, was also present at our experiments as a partner and a witness. Besides these there were as witnesses at these experiments that most excellent man, decorated by the king with the highest of honors, Hauch, whose acquaintance with natural science has long been celebrated; that most acute man Reinhardt, professor of natural history; Jacobsen, professor of medicine, a man of the utmost sagacity in conducting experiments; and the most experienced chemist, Zeise, doctor of philosophy. I have indeed somewhat frequently carried out by myself experiments relating to the matter proposed, but the phenomena which it thus befell me to disclose I repeated in the presence of these most learned men.

At the University of Copenhagen during the winter of 1819-1820.

A battery, as described below in more detail.

In reviewing my experiments I will pass over everything which, though they contributed to the discovery of the reason of the thing, cannot any further illustrate it. Those things, therefore, which clearly demonstrate the reason of the thing, let us take for granted.

The galvanic apparatus which we made use of consists of 20 rectangular copper receptacles, the length and height of which are alike 4 inches, the breadth, however, scarcely exceeding $2\frac{1}{2}$ inches.∎ Every receptacle is furnished with two copper plates, so inclined that they can carry a copper bar which supports a zinc plate in the water of the next receptacle. The water of the receptacles contains $\frac{1}{60}$ of its weight of sulphuric acid and likewise $\frac{1}{60}$ of its weight of nitric acid. The part of each plate which is immersed in the solution is square, the side being about 10 inches long. Even smaller apparatus may be used, provided they are able to make a metallic wire red hot.

Apparently the 20 Cu-Zn cells were placed in series, which would provide an emf of some 15 to 20 volts.

Let the opposite poles of the galvanic apparatus be joined by a metallic wire, which, for brevity, we will call hereafter the joining conductor or else the joining wire. To the effect, however, which takes place in this conductor and surrounding space, we will give the name of electric conflict.

Let the rectilinear part of this wire be placed in a horizontal position over the magnetic needle duly suspended, and parallel to it. If necessary, the joining wire can be so bent that the suitable part of it may obtain the position necessary for the experiment. These things being thus arranged, the magnetic needle will be moved, and indeed, under that part of the joining wire which receives electricity most immediately from the negative end of the galvanic apparatus, will decline towards the west.∎

Thus, if the current flows from south to north, the north pole of the needle will be deflected towards the west. Or, if the current flows from north to south, the south pole of the needle will be deflected in that direction.

If the distance of the joining wire from the magnetic needle does not exceed $\frac{3}{4}$ of an inch, the declination of the needle makes an angle of about 45°. If the distance is increased the angles decrease as the distances increase. The declination, however, varies according to the efficiency of the apparatus.∎

This is largely qualitative, inasmuch as the effect would depend upon the current in the wire, which probably is what Oersted means by the *efficiency*.

The joining wire can change its place either eastward or westward,∎ provided it keeps a position parallel to the needle, without any other change of effect than as respect magnitude; and thus the effect can by no means be attributed to attraction, for the same pole of the magnetic needle which approaches the joining wire while it is placed at the east side of it ought to recede from the same when it occupies a position at the west side of it if these declinations depended upon attractions or repulsions.

That is, the wire need not be placed directly above the magnetic needle.

The joining conductor may consist of several metallic wires or bands connected together. The kind of metal does not alter the effects, except, perhaps, as regards quantity.■ We have employed with equal success wires of platinum, gold, silver, copper, iron, bands of lead and tin, a mass of mercury. A conductor is not wholly without effect when water interrupts, unless the interruption embraces a space of several inches in length.■

The effects of the joining wire on the magnetic needle pass through glass, metal, wood, water, resin, earthenware, stones; for if a plate of glass, metal, or wood be interposed, they are by no means destroyed, nor do they disappear if plates of glass, metal,■ and wood be simultaneously interposed; indeed, they seem to be scarcely lessened. The result is the same if there is interposed a disc of amber, a plate of porphyry,■ an earthenware vessel, even if filled with water. Our experiments have also shown that the effects already mentioned are not changed if the magnetic needle is shut up in a copper box filled with water. It is unnecessary to state that the passing of the effects through all these materials in electricity and galvanism has never before been observed.■ The effects, therefore, which take place in electric conflict are as different as possible from the effects of one electric force on another.

If the joining wire is placed in a horizontal plane under the magnetic needle, all the effects are the same as in the plane over the needle, only in an inverse direction,■ for the pole of the magnetic needle under which is that part of the joining wire which receives electricity most immediately from the negative end of the galvanic apparatus will decline towards the east.

That these things may be more easily remembered let us use this formula: the pole *over* which negative electricity enters is turned towards the west, that *under* which it enters toward the east.■

If the joining wire is so turned in a horizontal plane as to form with the magnetic meridian a gradually increasing angle, the declination of the magnetic needle is increased if the motion of the wire tends towards the place of the disturbed needle, but is lessened if the wire goes away from this place.

The joining wire placed in the horizontal plane in which the magnetic needle moves balanced by means of a counterpoise,■ and parallel to the needle, disturbs the same neither eastward nor westward but only makes it quiver in the plane of inclination, so that the pole near which the negative electric force enters the wire is depressed when it is situated at the west side and elevated when at the east.

If the joining wire is placed perpendicular to the plane of the magnetic meridian, either above or below the needle, the latter remains at rest, unless the wire is very near to the pole, for then the pole is elevated

Since the resistance, and hence the current, depends upon the conductor.

Water with some mineral content is a fair conductor.

One would expect that an iron plate, however, would shield the magnetic field.

Porphyry: a form of igneous rock.

Except that the magnetic effect of the earth was known to pass through these materials. The magnetic field above the wire is opposite to that below. By *negative electricity* is meant electron flow, which is opposite to the conventional current. Check these against the modern right-hand rule: thumb pointing in direction of conventional current, fingers curling in direction of magnetic field. Apparently the magnetic needle was counterbalanced so as to annul the effect of inclination, or *dip*. Then, with the wire in the same horizontal plane the effect of its magnetic field is to rotate the needle in a *vertical* plane.

when the entrance is made from the western part of the wire and depressed when it is made from the eastern.■

When the joining wire is placed perpendicular to the pole of the magnetic needle,■ and the upper end of the wire receives electricity from the negative end of the galvanic apparatus, the pole is moved towards the east; but when the wire is placed opposite to a point situated between the pole and the middle of the needle it is driven towards the west.■ When the upper end of the wire receives electricity from the positive end reverse phenomena will occur.

If the joining wire is so bent that it is made parallel to itself at both parts of the bend, or forms two parallel legs, it repels or attracts the magnetic poles according to the different conditions of the case. Let the wire be placed opposite to either pole of the needle so that the plane of the parallel legs is perpendicular to the magnetic meridian, and let the eastern leg be joined with the negative end of the galvanic apparatus, the western with the positive, and when this is so arranged the nearest pole will be repelled either eastward or westward according to the position of the plane of the legs. When the eastern leg is joined with the positive end, and the western with the negative, the nearest pole is attracted. When the plane of the legs is placed perpendicular to a spot between the pole and the middle of the needle the same effects occur, only inverted.

A needle of copper, suspended like a magnetic needle, is not moved by the effect of a joining wire. Also needles of glass, or of so-called gumlac, subjected to the like experiments, remain at rest.■

From all this it may be allowable to adduce some considerations in explanation of these phenomena.

Electric conflict■ can only act upon magnetic particles of matter. All nonmagnetic bodies seem to be penetrable through electric conflict; but magnetic bodies, or rather their magnetic particles, seem to resist the passage of this conflict, whence it is that they can be moved by the impulse of contending forces.

That electric conflict is not inclosed in the conductor, but as we have already said is at the same time dispersed in the surrounding space, and that somewhat widely is clear enough from the observations already set forth.

In like manner it is allowable to gather from what has been observed that this conflict performs gyrations,■ for this seems to be a condition without which it is impossible that the same part of the joining wire, which, when placed beneath the magnetic pole, carries it eastward, drives it westward when placed above; for this is the nature of a gyration, that motions in opposite parts have an opposite direction. Moreover, motion

Actually, if the field, due to the current, is opposite to the earth's field, and greater, the needle should reverse.

That is, vertically in the magnetic meridian beyond either pole of the needle.

All of these effects are easily visualized with the aid of the *right-hand rule.*

All nonmagnetic materials.

It can be seen that the modern terminology for *electric conflict* would be *magnetic field.*

The *lines of force* associated with the magnetic field about a current form concentric circles about the conductor. Hence the impression of *gyrations* obtained by Oersted. The field does not have a spiral form!

by circuits combined with progressive motion, according to the length of the conductor, seems bound to form a cochlea or spiral line, which however, if I am not mistaken, contributes nothing to the explanation of phenomena hitherto observed.

All the effects on the northern pole, here set forth, are easily understood by stating that negatively electric force or matter runs through a spiral line bending to the right, and propels the northern pole, but does not act at all upon the southern.■ The effects on the southern pole are similarly explained if we attribute to force or matter positively electrified a contrary motion and the power of acting on the southern pole but not on the northern. The agreement of this law with nature will be better seen by the repetition of experiments than by a long explanation. To judge of the experiments, however, will be made much easier if the course of the electric force on the joining wire is indicated by marks, either painted or incised.

I will add this only to what has been said: that I have demonstrated in a book, published seven years ago, that heat and light are in electric conflict.■ From observations lately brought to bear we may now conclude that motion by gyrations also occurs in these effects; and I think that this does very much to illustrate the phenomena which they call the polarity of light.

> The same magnetic field acts upon both poles, but in opposite directions.

> Oersted correctly assumed a similarity between heat and light. However, the connection between these and the electro-magnetic field came later with the work of Faraday and Maxwell.

SUPPLEMENTARY READING

Dampier, W. C., *A History of Science* (New York: Cambridge University, 1946), pp. 217.

Lenard, P., *Great Men of Science* (New York: British Book Centre, 1934), pp. 213ff.

Magie, W. F., *A Source Book in Physics* (New York: McGraw-Hill, 1935), pp. 436ff.

Woodruff, L. L. (ed.), *The Development of the Sciences* (New Haven, Conn.: Yale University, 1923).

Michael Faraday
1791 - 1867

Electromagnetic Induction and Laws of Electrolysis

BY THE EARLY nineteenth century physics had advanced appreciably, particularly in the branch of mechanics, and one needed a strong background in mathematics to follow, professionally, this rapidly growing science. Yet the most prolific experimental physicist of that period, in fact, of any period in the history of physics, was a man whose formal schooling did not extend beyond the primary grades. He defied tradition in still another respect; by comparison with most of his predecessors, his greatest contributions were made relatively late in life. The two may bear some relation to each other, perhaps linked by a third characteristic; Faraday dedicated his adult life completely to the cause of science, allowing nothing to interfere with his singleness of purpose.

The son of a blacksmith who earned barely enough to support his family, Michael Faraday was born in Newington Butts, near London, on September 22, 1791. The Industrial Revolution was in full swing, as was another, more brutal transition, the French Revolution. It was largely because of Faraday's electrical discoveries that industry progressed so rapidly during the latter half of the nineteenth century. Faraday received what formal education he had between the ages of five and thirteen. By his own account this " . . . was of the most ordinary description, consisting of little more than the rudiments of reading, writing, and arithmetic at a common day-school." At thirteen he became errand boy for a bookbinder, and the following year was apprenticed to him for the usual seven-year period. He became deeply interested in books and took to educating himself by reading everything he could find, especially scientific books. These stimulated him so that he attended several lectures on chemistry delivered by Sir Humphry Davy (1778–1829). He was determined somehow to make science his career, and when his apprenticeship expired in 1812 he shortly applied to Davy for a position, presenting as evidence of his capabilities some notes he had taken of Davy's lectures. Davy, who was then president of the Royal Society, engaged Faraday as his assistant at a salary of twenty-five shillings per week plus living quarters at the top of the Royal Institution. Judging from all accounts, Faraday would have been happy to work without compensation, or with enough only to provide the barest essentials, so passionate was his desire to work in science.

His early work with Davy was largely in the field of chemistry; his first publication, in 1816, was an article describing an analysis he had made of a sample of caustic lime from Tuscany. During the next four years he published nearly forty articles and notes, climaxed by his discovery, in 1820, of two chlorides of carbon. He trained himself painstakingly, both in science and in the art of lecturing, becoming equally proficient in both. He began to lecture at the Royal Institution in 1827, a series that continued for more than thirty years, and marked him a brilliant popular lecturer. It has been suggested that Faraday's meager scientific background, particularly in mathematics, compelled him to seek simple explanations; hence his success in lecturing to the general public. This is perhaps an oversimplification. No doubt his lack of training had some influence on the manner in which he approached those problems on which he worked, but his *success*, either in research or on the lecture platform, could hardly be attributed to this deficiency.

Faraday became interested in electromagnetic phenomena following Oersted's discovery of the magnetic effects of a current in 1820 and Ampère's discovery, shortly afterward, of the action of currents upon one another. He performed experiments that led to the principle of the electric motor. These involved arranging magnets and current-carrying conductors so that either the conductors or the magnets were free to rotate continuously. Publication of his results involved Faraday in an unfortunate misunderstanding with Davy, who charged that Faraday had taken the idea from some work of William Wollaston (1766–1828) and Davy in the same area. Over Davy's opposition, Faraday was elected to fellowship in the Royal Society in 1824; yet the following year, on Davy's recommendation, he was appointed director of the laboratory at the Royal Institution. In his new position he was instrumental in reorganizing the Institution activities. He originated a series of evening meetings which became known as the Discourses, and which are still continued, and he started, in 1826–1827, the very popular

Christmas Courses of Lectures, which were designed for juvenile audiences and continue to attract them in large numbers.

Following his brief experience with electrical phenomena, Faraday returned to chemical investigations. He liquified chlorine, which resulted in another dispute with Davy, and discovered the compound now known as *benzene*. By then very well known, Faraday was offered a number of government positions. He became a lecturer at the Royal Military Academy, a member of the Scientific Advising Committee of the Admiralty, and scientific adviser to Trinity House. He returned to his electrical experiments from time to time in an effort to discover the induction of electricity, but failed to observe any effect until 1831, when he improved the coupling between his coils by means of an iron core. Faraday was forty years old when he discovered electromagnetic induction, yet he continued to make important contributions for many more years. Compared with many other notable contributors in physics, both before and after Faraday, this was a relatively advanced time of life for fundamental discovery.

Two years later he showed the identity of the different sorts of electricity: from voltaic cells, frictional effects, and electromagnetic induction. Faraday then turned to electrochemistry, where he extended the work of Davy, Berzelius, and others to formulate his well-known laws of electrolysis, together with a new terminology that is still in use. His fertile mind carried him from one problem to the next with remarkable agility. In 1834 he again investigated induction, including self-induction, which he discovered independently of Joseph Henry (1797–1878), and the induction of static electricity. He tried to account for the action at a distance of electromagnetic effects in terms of *lines of force,* fictional lines which proved useful in explaining these phenomena, and which, by emphasizing the role of the medium in such effects, led eventually to the concept of *field*.

Ill health kept him from active work for a number of years, but he resumed his researches in 1844 with a study on the liquefaction of gases. The following year found him investigating the effect of a magnetic field on light, during which he discovered the rotation of the plane of polarization of a beam of light in a magnetic field (Faraday effect) and diamagnetism. Year after year he poured out a steady stream of important discoveries. And the honors heaped upon him failed to slow his enormous pace. He put aside all activities that might interfere with his research. Twice he refused to accept the presidency of the Royal Society. He resigned his professorship in the Royal Institution in 1861, at the age of seventy, but continued his research for another year. His last investigation was a search for the splitting of a light beam in a magnetic field. He was unsuccessful in this, but the effect was later discovered by Pieter Zeeman (1865–1943). Faraday died in 1867, ending a career that would have taxed the energies of several ordinary men.

His electrical discoveries were reported to the Royal Society and later published as *Experimental Researches in Electricity*, both in book form and in the *Philosophical Transactions*. He was in the habit of numbering each paragraph consecutively throughout his work and of referring to earlier entries by paragraph number.

Of the extracts that follow, the first (through paragraph 120) was read before the Society on November 24, 1831, and appeared in *Philosophical Transactions* (1832), page 125. The second, on static electrical induction, was written in the form of a letter to R. Phillips, Esq., and appeared in *Philosophical Magazine*, 22 (1843), page 200. The last extract, beginning with paragraph 661, appeared in *Philosophical Transactions* (1834), page 77. All may be found, as well, in Faraday's *Experimental Researches in Electricity* (London: R. and J. E. Taylor, 1839–1855), volumes 1 and 2.

Faraday's Experiments

■1. The power which electricity of tension■ possesses of causing an opposite electrical state in its vicinity has been expressed by the general term Induction; which, as it has been received into scientific language, may also, with propriety, be used in the same general sense to express the power which electrical currents may possess of inducing any particular state upon matter in their immediate neighborhood, otherwise indifferent. It is with this meaning that I purpose using it in the present paper.

Philosophical Transactions (1832), page 125.

As used here, electricity of tension refers to static charges.

2. Certain effects of the induction of electrical currents have already been recognized and described: as those of magnetization; Ampère's experiments of bringing a copper disc near to a flat spiral; his repetition with electromagnets of Arago's extraordinary experiments, and perhaps a few others. Still it appeared unlikely that these could be all the effects induction by currents could produce; especially as, upon dispensing with iron,■ almost the whole of them disappear, whilst yet an infinity of bodies, exhibiting definite phenomena of induction with electricity of tension, still remain to be acted upon by the induction of electricity in motion.

The iron was needed to increase the magnetic coupling between circuits.

3. Further: Whether Ampère's beautiful theory■ were adopted, or any other, or whatever reservation were mentally made, still it appeared

Relating the magnetic field to the current causing it.

The magnetic lines of force are concentric circles about the current.

very extraordinary, that as every electric current was accompanied by a corresponding intensity of magnetic action at right angles to the current,■ good conductors of electricity, when placed within the sphere of this action, should not have any current induced through them, or some sensible effect produced equivalent in force to such a current.

4. These considerations, with their consequence, the hope of obtaining electricity from ordinary magnetism, have stimulated me at various times to investigate experimentally the inductive effect of electric currents. I lately arrived at positive results; and not only had my hopes fulfilled, but obtained a key which appeared to me to open out a full explanation of Arago's magnetic phenomena, and also to discover a new state, which may probably have great influence in some of the most important effects of electric currents.

5. These results I purpose describing, not as they were obtained, but in such a manner as to give the most concise view of the whole.

INDUCTION OF ELECTRIC CURRENTS

Adjacent turns.

6. About twenty-six feet of copper wire one twentieth of an inch in diameter were wound round a cylinder of wood as a helix, the different spires■ of which were prevented from touching by a thin interposed twine. This helix was covered with calico, and then a second wire applied in the same manner. In this way twelve helices were superposed, each containing an average length of wire of twenty-seven feet, and all in the same direction. The first, third, fifth, seventh, ninth, and eleventh of these helices were connected at their extremities end to end, so as to form one helix; the others were connected in a similar manner; and thus two principal helices were produced, closely interposed, having the same direction, not touching anywhere, and each containing one hundred and fifty-five feet in length of wire.■

This had the effect of two coils of wire inter-twined, but separated electrically.

7. One of these helices was connected with a galvanometer, the other with a voltaic battery of ten pairs of plates four inches square, with double coppers and well charged; yet not the slightest sensible deflection of the galvanometer needle could be observed.

Evidently Faraday thought the nature of the wire might influence the induction.

8. A similar compound helix, consisting of six lengths of copper and six of soft iron wire, was constructed.■ The resulting iron helix contained two hundred and fourteen feet of wire, the resulting copper helix two hundred and eight feet; but whether the current from the trough was passed through the copper or the iron helix, no effect upon the other could be perceived at the galvanometer.

9. In these and many other similar experiments no difference in action of any kind appeared between iron and other metals.

10. Two hundred and three feet of copper wire in one length were passed round a large block of wood; other two hundred and three feet of similar wire were interposed as a spiral between the turns of the first, and metallic contact everywhere prevented by twine. One of these helices was connected with a galvanometer, and the other with a battery of one hundred pairs of plates four inches square, with double coppers, and well charged.■ When the contact was made, there was a sudden and very slight effect at the galvanometer, and there was also a similar slight effect when the contact with the battery was broken. But whilst the voltaic current was continuing to pass through the one helix, no galvanometrical appearances of any effect like induction upon the other helix could be perceived, although the active power of the battery was proved to be great, by its heating the whole of its own helix, and by the brilliancy of the discharge when made through charcoal.

This would have provided a potential difference of the order of 100 volts, and hence a reasonably large current.

11. Repetition of the experiments with a battery of one hundred and twenty pairs of plates produced no other effects; but it was ascertained, both at this and the former time, that the slight deflection of the needle occurring at the moment of completing the connection, was always in one direction, and that the equally slight deflection produced when the contact was broken, was in the other direction; and also, that these effects occurred when the first helices were used (6,8).■

12. The results which I had by this time obtained with magnets led me to believe that the battery current through one wire, did, in reality, induce a similar current through the other wire, but that it continued for an instant only, and partook more of the nature of the electrical wave passed through from the shock of a common Leyden jar■ than of that from a voltaic battery, and therefore might magnetize a steel needle, although it scarcely affected the galvanometer.

With the larger battery. Numbers in parentheses refer to other paragraphs in Faraday's writings.

A Leyden jar, used for the storage of charge, could provide a very large potential difference.

13. This expectation was confirmed; for on substituting a small hollow helix, formed round a glass tube, for the galvanometer, introducing a steel needle, making contact as before between the battery and the inducing wire (7,10), and then removing the needle before the battery contact was broken, it was found magnetized.

14. When the battery contact was first made, then an unmagnetized needle introduced into the small indicating helix, and lastly the battery contact broken, the needle was found magnetized to an equal degree apparently to the first; but the poles were of the contrary kind.■

15. The same effects took place on using the large compound helices first described (6,8).

The magnetization was reversed because the induced current reversed direction when the contact was broken.

16. When the unmagnetized needle was put into the indicating helix, before contact of the inducing wire with the battery, and remained there until the contact was broken, it exhibited little or no magnetism; the first effect having been nearly neutralized by the second (13,14). The force of the induced current upon making contact was found always to exceed that of the induced current at breaking of contact; and if therefore the contact was made and broken many times in succession, whilst the needle remained in the indicating helix, it at last came out not unmagnetized, but a needle magnetized as if the induced current upon making contact had acted alone on it. This effect may be due to the accumulation (as it is called) at the poles of the unconnected pile, rendering the current upon first making contact more powerful than what it is afterwards, at the moment of breaking contact.■

Or it may have been due to the fact that it is difficult to break a current as abruptly as it can be made. A spark tends to prolong the current on breaking the contact; and the induced current is proportional to the rate of change of the current in the primary circuit.

Connections.

17. If the circuit between the helix or wire under induction and the galvanometer or indicating spiral was not rendered complete *before* the connection between the battery and the inducing wire was completed or broken, then no effects were perceived at the galvanometer. Thus, if the battery communications■ were first made, and then the wire under induction connected with the indicating helix, no magnetizing power was there exhibited. But still retaining the latter communications, when those with the battery were broken, a magnet was formed in the helix, but of the second kind, i.e., with poles indicating a current in the same direction to that belonging to the battery current, or to that always induced by that current in the first instance.

18. In the preceding experiments the wires were placed near to each other, and the contact of the inducing one with the battery made when the inductive effect was required; but as some particular action might be supposed to be exerted at the moments of making and breaking contact, the induction was produced in another way. Several feet of copper wire were stretched in wide zigzag forms, representing the letter W, on one surface of a broad board; a second wire was stretched in precisely similar forms on a second board, so that when brought near the first, the wires should everywhere touch, except that a sheet of thick paper was interposed. One of these wires was connected with the galvanometer, and the other with a voltaic battery. The first wire was then moved towards the second, and as it approached, the needle was deflected. Being then removed, the needle was deflected in the opposite direction. By first making the wires approach and then recede, simultaneously with the vibrations of the needle,■ the latter soon became very extensive; but when the wires ceased to move from or towards each other, the galvanometer needle soon came to its usual position.

That is, at the same frequency as the vibrations, so as to enhance the needle motion by a resonance effect.

19. As the wires approximated,■ the induced current was in the *contrary* direction to the inducing current. As the wires receded, the induced current was in the *same* direction as the inducing current. When the wires remained stationary, there was no induced current (54).

Approached each other.

20. When a small voltaic arrangement was introduced into the circuit between the galvanometer (10) and its helix or wire, so as to cause a permanent deflection of 30° or 40°, and then the battery of one hundred pairs of plates connected with the inducing wire, there was an instantaneous action as before (11); but the galvanometer needle immediately resumed and retained its place unaltered, notwithstanding the continued contact of the inducing wire with the trough: such was the case in whichever way the contacts were made (33).

21. Hence it would appear that collateral currents, either in the same or in opposite directions, exert no permanent inducing power on each other, affecting their quantity or tension.■

That is, a current already flowing in the secondary circuit had no effect upon the currents induced therein.

22. I could obtain no evidence by the tongue, by spark, or by heating fine wire or charcoal, of the electricity passing through the wire under induction;■ neither could I obtain any chemical effects, though the contacts with metallic and other solutions were made and broken alternately with those of the battery, so that the second effect of induction should not oppose or neutralize the first (13,16).

The induced currents evidently were quite feeble. A crude method of testing low voltages was to touch the two wires to the tongue.

23. This deficiency of effect is not because the induced current of electricity cannot pass fluids, but probably because of its brief duration and feeble intensity; for on introducing two large copper plates into the circuit on the induced side (20), the plates being immersed in brine, but prevented from touching each other by an interposed cloth, the effect at the indicating galvanometer, or helix, occurred as before. The induced electricity could also pass through the trough (20). When, however, the quantity of fluid was reduced to a drop, the galvanometer gave no indication.

24. Attempts to obtain similar effects to these by the use of wires conveying ordinary electricity■ were doubtful in the results. A compound helix similar to that already described (6) and containing eight elementary helices was used. Four of the helices had their similar ends bound together by wire, and the two general terminations thus produced connected with the small magnetizing helix contained an unmagnetized needle (13). The other four helices were similarly arranged, but their ends connected with a Leyden jar. On passing the discharge, the needle was found to be a magnet; but it appeared probable that a part of the electricity of the jar had passed off to the small helix, and so magnetized the needle. There was indeed no reason to expect that the electricity of

Static or *frictional* electricity.

High voltage.

a jar possessing as it does great tension,■ would not diffuse itself through all the metallic matter interposed between the coatings.

25. Still it does not follow that the discharge of ordinary electricity through a wire does not produce analogous phenomena to those arising from voltaic electricity; but as it appears impossible to separate the effects produced at the moment when the discharge begins to pass, from the equal and contrary effects produced when it ceases to pass (16), inasmuch as with ordinary electricity these periods are simultaneous, so there can be scarcely any hope that in this form of the experiment they can be perceived.■

With modern instruments this is easily observed.

26. Hence it is evident that currents of voltaic electricity present phenomena of induction somewhat analogous to those produced by electricity of tension, although, as will be seen hereafter, many differences exist between them. The result is the production of other currents (but which are only momentary) parallel, or tending to parallelism, with the inducing current. By reference to the poles of the needle formed in the indicating helix (13,14) and to the deflections of the galvanometer needle (11), it was found in all cases that the induced current, produced by the first action of the inducing current, was in the contrary direction to the latter,■ but that the current produced by the cessation of the inducing current was in the same direction. For the purpose of avoiding periphrasis,■ I propose to call this action of the current from the voltaic battery, *volta-electric induction*. The properties of the wire, after induction has developed the first current, and whilst the electricity from the battery continues to flow through its inducing neighbor (10,18) constitute a peculiar electric condition, the consideration of which will be resumed hereafter. All these results have been obtained with a voltaic apparatus consisting of a single pair of plates.

Compare with
Lenz's law,
Chapter 11,
page 162.

Wordiness.

EVOLUTION OF ELECTRICITY FROM MAGNETISM

27. A welded ring was made of soft round bar iron, the metal being seven eights of an inch in thickness, and the ring six inches in external diameter. Three helices were put round one part of this ring, each containing about twenty-four feet of copper wire one twentieth of an inch thick; they were insulated from the iron and each other, and superposed in the manner before described (6), occupying about nine inches in length upon the ring. They could be used separately or arranged together; the group may be distinguished by the mark *A* (Fig. 1). On

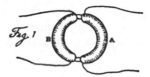

the other part of the ring about sixty feet of similar copper wire in two pieces were applied in the same manner, forming a helix *B*, which had the same common direction with the helices of *A*, but being separated from it at each extremity by about half an inch of the uncovered iron.

28. The helix *B* was connected by copper wires with a galvanometer three feet from the ring. The wires of *A* were connected end to end so as to form one long helix, the extremities of which were connected with a battery of ten pairs of plates four inches square. The galvanometer was immediately affected, and to a degree far beyond what has been described,■ when with a battery of tenfold power helices without iron were used (10); but though the contact was continued, the effect was not permanent, for the needle soon came to rest in its natural position, as if quite indifferent to the attached electromagnetic arrangement. Upon breaking the contact with the battery, the needle was again powerfully deflected, but in the contrary direction to that induced in the first instance.

The iron served to "couple" the two circuits more closely. That is, it increased the total flux threading the helix *B*.

29. Upon arranging the apparatus so that *B* should be out of use, the galvanometer be connected with one of the three wires of *A*, and the other two made into a helix through which the current from the trough (28) was passed; similar but rather more powerful effects were produced.

30. When the battery contact was made in one direction, the galvanometer needle was deflected on the one side; if made in the other direction, the deflection was on the other side. The deflection on breaking the battery contact was always the reverse of that produced by completing it. The deflection on making a battery contact always indicated an induced current in the opposite direction to that from the battery; but on breaking the contact the deflection indicated an induced current in the same direction as that of the battery. No making or breaking of the contact at *B* side, or in any part of the galvanometer circuit, produced any effect at the galvanometer. No continuance of the battery current caused any deflection of the galvanometer-needle. As the above results are common to all these experiments, and to similar ones with ordinary magnets to be hereafter detailed, they need not be again particularly described.

Omitted are several demonstrations employing the larger battery (100 plates). The effects were the same, except to a greater degree.

■.....................................

Both coils were wound upon a single cylinder of iron. However, the total flux is increased when the iron forms a closed path.

This helix was similar to that described in (6).

35. When the iron cylinder was replaced by an equal cylinder of copper, no effect beyond that of the helices alone was produced. The iron cylinder arrangement was not so powerful as the ring arrangement already described (27).■

36. Similar effects were then produced by *ordinary magnets:* thus the hollow helix■ just described (34) had all its elementary helices connected with the galvanometer by two copper wires, each five feet in length; the soft iron cylinder was introduced into its axis; a couple of bar magnets, each twenty-four inches long, were arranged with their opposite poles at one end in contact, so as to resemble a horseshoe magnet, and then contact made between the other poles and the ends of the iron cylinder, so as to convert it for the time into a magnet (Fig. 2): by breaking the magnetic contacts, or reversing them, the magnetism of the iron cylinder could be destroyed or reversed at pleasure.

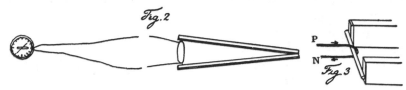

37. Upon making magnetic contact, the needle was deflected; continuing the contact, the needle became indifferent, and resumed its first position; on breaking the contact, it was again deflected, but in the opposite direction to the first effect, and then it again became indifferent. When the magnetic contacts were reversed, the deflections were reversed.

38. When the magnetic contact was made, the deflection was such as to indicate an induced current of electricity in the opposite direction to that fitted to form a magnet having the same polarity as that really produced by contact with the bar magnets. Thus when the marked and unmarked poles were placed as in Figure 3,■ the current in the helix was in the direction represented, *P* being supposed to be the end of the wire going to the positive pole of the battery, or that end towards which the zinc plates face, and *N* the negative wire. Such a current would have converted the cylinder into a magnet of the opposite kind to that formed by contact with the poles *A* and *B;* and such a current moves in the opposite direction to the currents which in M. Ampère's beautiful theory are considered as constituting a magnet in the position figured.*

Faraday marked his N-poles for identification. (Note upper pole in Fig. 3.) Thus, "marked pole" refers to a N-seeking pole.

---

* The relative position of an electric current and a magnet is by most persons found very difficult to remember, and three or four helps to the memory have been devised by M. Ampère and others. I venture to suggest the following as a very simple and effectual assistance in these and similar latitudes. Let the experimenter think he is looking down upon a dipping needle, or upon the pole of the

■...

42. These effects are simple consequences of the law hereafter to be described (114).

■...

Omitted are some further demonstrations of the transient nature of the effect.

114. The relation which holds between the magnetic pole, the moving wire or metal, and the direction of the current evolved, i.e., the law which governs the evolution of electricity by magneto-electric induction, is very simple, although rather difficult to express. If in Figure 4 *PN*

A number of additional demonstrations, plus observations on the *inertia* of induced currents, are omitted.

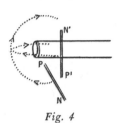

represent a horizontal wire passing by a marked magnetic pole, so that the direction of its motion shall coincide with the curved line proceeding from below upwards; or if its motion parallel to itself be in a line tangential to the curved line, but in the general direction of the arrows; or if it pass the pole in other directions, but so as to cut the magnetic curves† in the same general direction, or on the same side as they would

Fig. 4

be cut by the wire if moving along the dotted curved line;—then the current of electricity in the wire is from *P* to *N*. If it be carried in the reverse directions, the electric current will be from *N* to *P*. Or if the wire be in the vertical position, figured *P'N'*, and it be carried in similar directions, coinciding with the dotted horizontal curve so far, as to cut the magnetic curves on the same side with it, the current will be from *P'* to *N'*. If the wire be considered a tangent to the curved surface of the cylindrical magnet, and it be carried round that surface into any other position, or if the magnet itself be revolved on its axis, so as to bring any part opposite to the tangential wire,—still, if afterwards the wire be moved in the directions indicated, the current of electricity will be from *P* to *N;* or if it be moved in the opposite direction, from *N* to *P;* so that as regards the motions of the wire past the pole, they may be reduced to two, directly opposite to each other, one of which produces a current from *P* to *N*, and the other from *N* to *P*.■

115. The same holds true of the unmarked pole of the magnet, ex-

Note how much simpler this is by the use of Lenz's law (see p. 164). If the magnet revolves on its axis, another (rather complex) effect, *unipolar induction*, is brought into play.

The *right-hand rule* is readily applied to these examples. See Chapter 11.

---

earth, and then let him think upon the direction of the motion of the hands of a watch, or of a screw moving direct; currents in that direction round a needle would make it into such a magnet as the dipping needle, or would themselves constitute an electromagnet of similar qualities; or if brought near a magnet would tend to make it take that direction; or would themselves be moved into that position by a magnet so placed; or in M. Ampère's theory are considered as moving in that direction in the magnet.■ These two points of the position of the dipping needle and the motion of the watch hands being remembered, any other relation of the current and magnet can be at once deduced from it.

† By magnetic curves, I mean the lines of magnetic forces, however modified by the juxtaposition of poles, which would be depicted by iron filings; or those to which a very small magnetic needle would form a tangent.

cept that if it be substituted for the one in the figure, then, as the wires are moved in the direction of the arrows, the current of electricity would be from *N* to *P,* and as they move in the reverse direction, from *P* to *N.*

116. Hence the current of electricity which is excited in metal when moving in the neighborhood of a magnet, depends for its direction altogether upon the relation of the metal to the resultant of magnetic action, or to the magnetic curves, and may be expressed in a popular way thus; Let *AB* (Fig. 5) represent a cylinder magnet, *A* being the marked pole, and *B* the unmarked pole; let *PN* be a silver knife blade resting across the magnet with its edge upward, and with its marked or notched side

See the comment by Lenz on this experiment, Chapter 11, page 161.

towards the pole *A;*■ then in whatever direction or position this knife be moved edge foremost, either about the marked or the unmarked pole, the current of electricity produced will be from *P* to *N,* provided the in-

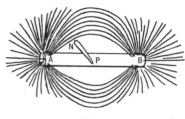

Fig. 5

tersected curves proceeding from *A* abut upon the notched surface of the knife, and those from *B* upon the unnotched side. Or if the knife be moved with its back foremost, the current will be from *N* to *P* in every possible position and direction, provided the intersected curves abut on the same surfaces as before. A little model is easily constructed, by using a cylinder of wood for a magnet, a flat piece for the blade, and a piece of thread connecting one end of the cylinder with the other, and passing through a hole in the blade, for the magnetic curves: this readily gives the result of any possible direction.

117. When the wire under induction is passing by an electromagnetic pole, as for instance one end of a copper helix traversed by the electric current (34), the direction of the current in the approaching wire is the same with that of the current in the parts or sides of the spirals nearest to it, and in the receding wire the reverse of that in the parts nearest to it.

118. All these results show that the power of inducing electric currents is circumferentially excited by a magnetic resultant or axis of power, just as circumferential magnetism is dependent upon and is exhibited by an electric current.■

Referring to the circular lines of force about a current.

119. The experiments described combine to prove that when a piece of metal (and the same may be true of all conducting matter) is passed either before a single pole, or between the opposite poles of a magnet,

or near electro-magnetic poles, whether ferruginous■ or not, electrical currents are produced across the metal transverse to the direction of motion; and which therefore, in Arago's experiments,■ will approximate towards the direction of radii. If a single wire be moved like the spoke of a wheel near a magnetic pole, a current of electricity is determined through it from one end towards the other. If the wheel be imagined, constructed of a great number of these radii, and this revolved near the pole, in the manner of the copper disc (85), each radius will have a current produced in it as it passes by the pole. If the radii be supposed to be in contact laterally, a copper disc results, in which the directions of the currents will be generally the same, being modified only by the coaction which can take place between the particles, now that they are in metallic contact.

120. Now that the existence of these currents is known, Arago's phenomena may be accounted for without considering them as due to the formation in the copper of a pole of the opposite kind to that approximated, surrounded by a diffuse polarity of the same kind (82) ;■ neither is it essential that the plate should acquire and lose its state in a finite time; nor on the other hand does it seem necessary that any repulsive force should be admitted as the cause of the rotation (82).

■...................................

Note.—In consequence of the long period which has intervened between the reading and printing of the foregoing paper, accounts of the experiments have been dispersed, and, through a letter of my own to M. Hachette, have reached France and Italy. The letter was translated (with some errors), and read to the Academy of Sciences at Paris, 26th December, 1831. A copy of it in *Le Temps* of the 28th December quickly reached Signor Nobili, who with Signor Antinori, immediately experimented upon the subject, and obtained many of the results mentioned in my letter; others they could not obtain or understand, because of the brevity of my account. These results by Signori Nobili and Antinori have been embodied in a paper dated 31st January 1832, and printed and published in the number of the *Antologia* dated November 1831, (according at least to the copy of the paper kindly sent me by Signor Nobili). It is evident the work could not have been then printed; and though Signor Nobili, in his paper, has inserted my letter as the text of his experiments, yet the circumstances of back date has caused many here, who have heard of Nobil's experiments by report only, to imagine his results were anterior to, instead of being dependent upon, mine.

I may be allowed under these circumstances to remark, that I experimented on this subject several years ago, and have published results.

Pertaining to iron. Electromagnetic poles would result from a helix, for example, even without an iron core.

Arago's experiments essentially showed the magnetic fields associated with the induced currents. If a copper plate were rotated near a magnet free to move, it was found that the magnet tended to follow the motion of the plate, or the plate followed the motion of the magnet, as the case might be.

The *eddy currents* produce a *drag* on the magnet. See Lenz, Chapter 11, page 162.

Omitted are descriptions of several efforts to devise electric generators.

(See *Quarterly Journal of Science* for July 1825, p. 338.) The following also is an extract from my notebook, dated November 28, 1825: "Experiments on induction by connecting wire of voltaic battery:—a battery of four troughs, ten pairs of plates, each arranged side by side—the poles connected by a wire about four feet long, parallel to which was another similar wire separated from it only by two thicknesses of paper, the ends of the latter were attached to a galvanometer:—exhibited no action, &c. &c. &c.—Could not in any way render any induction evident from the connecting wire." The cause of failure at that time is now evident (79). —M. F., April 1832.

Letter on static electric induction. *Phil. Mag.* 22 (1843), page 200.

Generally known as the "ice-pail" experiment.

■Perhaps you may think the following experiments worth notice; their value consists in their power to give a very precise and decided idea to the mind respecting certain principles of inductive electrical action, which I find are by many accepted with a degree of doubt or obscurity that takes away much of their importance: they are the expression and proof of certain parts of my view of induction.* Let A in the diagram represent an insulated pewter ice pail■ ten and a half inches high and seven inches diameter, connected by a wire with a delicate gold-leaf electrometer E, and let C be a round brass ball insulated by a dry thread of white silk, three or four feet in length, so as to remove the influence of the hand holding it from the ice pail below. Let A (Fig. 6) be perfectly discharged, then let C be charged at a distance by a machine or Leyden jar, and introduced into A as in the figure. If C be positive, E also will diverge positively; if C be taken away, E will collapse perfectly, the apparatus being in good order. As C enters the vessel A the divergence of E will increase until C is about three inches below the edge of the vessel, and will remain quite steady and unchanged for any lower distance. This shows that at that distance the inductive action of C is entirely exerted upon the interior of A, and not in any degree directly upon external objects. If C be made to touch the bottom of A, *all* its charge is communicated to A; there is no longer any inductive action between C and A, and C, upon being withdrawn and examined, is found perfectly discharged.

These are all well-known and recognized actions, but being a little varied, the following conclusions may be drawn from them. If C be merely suspended in A, it acts upon it by induction, evolving electricity of its own kind on the outside of A;■ but if C touch A its electricity is

And of opposite kind on the inside of A.

---

* See *Experimental Researches*, Par. 1295, &c., 1667, &c., and Answer to Dr. Hare, *Philosophical Magazine*, 1840, S. 3, vol. xvii, p. 56, viii.

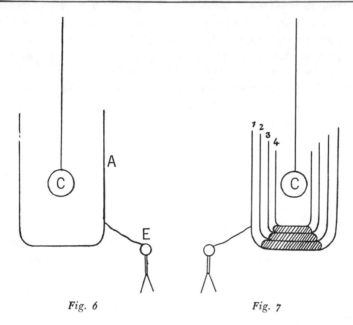

Fig. 6   Fig. 7

then communicated to it, and the electricity that is afterwards upon the outside of *A* may be considered as that which was originally upon the carrier *C*. As this change, however, produces no effect upon the leaves of the electrometer, it proves that the electricity *induced* by *C* and the electricity *in C* are accurately equal in amount and power.

Again, if *C* charged be held equidistant from the bottom and sides of *A* at one moment, and at another be held as close to the bottom as possible without discharging to *A*, still the divergence remains absolutely unchanged, showing that whether *C* acts at a considerable distance or at the very smallest distance, the amount of its force is the same. So also if it be held eccentric and near to the side of the ice pail in one place, so as to make the inductive action take place in lines expressing almost every degree of force in different directions, still the sum of their forces is the same constant quantity as that obtained before; for the leaves alter not. Nothing like expansion or coercion of the electric force appears under these varying circumstances.

As virtually all lines of force from C end on the inside of A, its position would have little effect.

I can now describe experiments with many concentric metallic vessels arranged as in the diagram (Fig. 7), where four ice pails are represented insulated from each other by plates of shellac on which they respectively stand. With this system the charged carrier *C* acts precisely as with the single vessel, so that the intervention of many conducting plates causes no difference in the amount of inductive effect. If *C* touch the inside of vessel 4, still the leaves are unchanged. If 4 be taken out by a silk thread,

The shellac served as the insulator.

the leaves perfectly collapse; if it be introduced again, they open out to the same degree as before. If 4 and 3 be connected by a wire let down between them by a silk thread, the leaves remain the same, and so they still remain if 3 and 2 be connected by a similar wire; yet all the electricity originally on the carrier and acting at a considerable distance, is now on the outside of 2, and acting through only a small nonconducting space. If at last it be communicated to the outside of 1, still the leaves remain unchanged.

Again, consider the charged carrier C in the center of the system, the divergence of the electrometer measures its inductive influence; this divergence remains the same whether 1 be there alone, or whether all four vessels be there; whether these vessels be separate as to insulation, or whether 2, 3, and 4 be connected so as to represent a very thick metallic vessel, or whether all four vessels be connected.

Again, if in place of the metallic vessels 2, 3, 4, a thick vessel of shellac or of sulphur be introduced, or if any other variation in the character of the substance within the vessel 1 be made, still not the slightest change is by that caused upon the divergence of the leaves.

If in place of one carrier many carriers in different positions are within the inner vessel, there is no interference of one with the other; they act with the same amount of force outwardly as if the electricity were spread uniformly over one carrier, however much the distribution on each carrier may be disturbed by its neighbors. If the charge of one carrier be by contact given to vessel 4 and distributed over it, still the others act through and across it with the same final amount of force; ■ and no state of charge given to any of the vessels 1, 2, 3, or 4, prevents a charged carrier introduced within 4 acting with precisely the same amount of force as if they were uncharged. If pieces of shellac, slung by white silk thread and excited, be introduced into the vessel, they act exactly as the metallic carriers, except that their charge cannot be communicated by contact to the metallic vessels. ■

The charges act independently of one another.

Thus a certain amount of electricity acting within the center of the vessel A exerts exactly the same power externally, whether it act by induction through the space between it and A, or whether it be transferred by conduction to A, so as absolutely to destroy the previous induction within. Also, as to the inductive action, whether the space between C and A be filled with air, or with shellac or sulphur, having above twice the specific inductive capacity ■ of air; or contain many concentric shells of conducting matter; or be nine tenths filled with conducting matter, or be metal on one side and shellac on the other; or whatever other means be taken to vary the forces, either by variation of distance or substance,

Charged insulators give up their charge on contact with conductors only at the very point of contact; hence, they can be discharged completely only by contacting the entire surface.

Dielectric constant.

or actual charge of the matter in this space, still the amount of action is precisely the same.

Hence if a body be charged, whether it be a particle or a mass, there is nothing about its action which can at all consist with the idea of exaltation or extinction; the amount of force is perfectly definite and unchangeable: or to those who in their minds represent the idea of the electric force by a fluid,■ there ought to be no notion of the compression or condensation of this fluid within itself, or of its coercibility, as some understand that phrase. The only mode of affecting this force is by connecting it with force of the same kind, either in the same or the contrary direction. If we oppose to it force of the contrary kind, we may *by discharge* neutralize the original force, or we may *without discharge* connect them by the simple laws and principles of static induction; but away from induction, which is *always of the same kind,* there is no other state of the power in a charged body; that is, there is no state of static electric force corresponding to the terms of *simulated* or *disguised* or *latent* electricity away from the ordinary principles of inductive action; nor is there any case where the electricity is *more latent* or *more disguised* than when it exists upon the charged conductor of an electrical machine and is ready to give a powerful spark to any body brought near it.■

The several *fluid theories* of electric charge.

A curious consideration arises from this perfection of inductive action. Suppose a thin uncharged metallic globe two or three feet in diameter, insulated in the middle of a chamber, and then suppose the space within this globe occupied by myriads of little vesicles or particles charged alike with electricity (or differently), but each insulated from its neighbor and the globe; their inductive power would be such that the outside of the globe would be charged with a force equal to the sum of *all* their forces, and any part of this globe (not charged of itself) would give as long and powerful a spark to a body brought near it as if the electricity of all the particles near and distant were on the surface of the globe itself. If we pass from this consideration to the case of a cloud, then, though we cannot altogether compare the external surface of the cloud to the metallic surface of the globe, yet the previous inductive effects upon the *earth* and its buildings are the same; and when a charged cloud is over the earth, although its electricity may be diffused over every one of its particles, and no important part of the *inductric* charge■ be accumulated upon its under surface yet the induction upon the earth will be as strong as if all that portion of force which is directed towards the earth *were* upon that surface; and the state of the earth and its tendency to discharge to the cloud will also be as strong in the former as in the latter case. As to whether lightning discharge begins first at the cloud or at the

Here Faraday sought to dispute those theories of electricity which had as their basis the notion of a *latent* charge. However, these experiments prove nothing about latent charge.

Inducing charge.

Faraday was correct in concluding that this is a complex phenomenon. Actually the discharge appears to originate in both places at the same time.

Philosophical Transactions (1834), page 77. This extract contains Faraday's most important electrochemical researches.

earth, that is a matter far more difficult to decide than is usually supposed;* theoretical notions would lead me to expect that in most cases, perhaps in all, it begins at the earth.■

PRELIMINARY■

661. The theory which I believe to be a true expression of the facts of electrochemical decomposition, and which I have therefore detailed in a former series of these researches, is so much at variance with those previously advanced, that I find the greatest difficulty in stating results, as I think correctly, whilst limited to the use of terms which are current with a certain accepted meaning. Of this kind is the term pole, with its prefixes of positive and negative, and the attached ideas of attraction and repulsion. The general phraseology is that the positive pole *attracts* oxygen, acids, &c., or more cautiously, that it *determines* their evolution upon the surface; and that the negative pole acts in an equal manner upon hydrogen, combustibles, metals, and bases. According to my view, the determining force is *not* at the poles, but *within* the decomposing

By *decomposing body* Faraday meant the electrolytic solution.

body;■ and the oxygen and acids are rendered at the *negative* extremity of the body, whilst hydrogen, metals, &c., are evolved at the *positive* extremity (518, 524).■

The use of "negative extremity" and "positive extremity" is a confusing terminology which Faraday wishes to abandon.

662. To avoid, therefore, confusion and circumlocution, and for the sake of greater precision of expression than I can otherwise obtain, I have deliberately considered the subject with two friends, and with their assistance and concurrence in framing them, I purpose henceforward using certain other terms, which I will now define. The poles, as they are usually called, are only the doors or ways by which the electric current passes into and out of the decomposing body (556); and they of course, when in contact with that body, are the limits of its extent in the direction of the current. The term has been generally applied to the metal surfaces in contact with the decomposing substance; but whether philosophers generally would also apply it to the surfaces of air (465, 471) and water (493), against which I have effected electrochemical decomposition, is subject to doubt. In place of the term pole, I propose using that of *electrode,* and I mean thereby that substance, or rather

Generally metal.

surface, whether of air, water, metal,■ or any other body, which bounds the extent of the decomposing matter in the direction of the electric current.

* *Experimental Researches,* Pars. 1370, 1410, 1484.

663. The surfaces at which, according to the common phraseology, the electric current enters and leaves a decomposing body, are most important places of action, and require to be distinguished apart from the poles, with which they are mostly, and the electrodes, with which they are always, in contact. Wishing for a natural standard of electric direction to which I might refer these, expressive of their difference and at the same time free from all theory, I have thought it might be found in the earth. If the magnetism of the earth be due to electric currents passing round it,■ the latter must be in a constant direction, which, according to present usage of speech, would be from east to west, or, which will strengthen this help to the memory, that in which the sun appears to move. If in any case of electro-decomposition we consider the decomposing body as placed so that the current passing through it shall be in the same direction, and parallel to that supposed to exist in the earth, then the surfaces at which the electricity is passing into and out of the substance would have an invariable reference, and exhibit constantly the same relations of powers.■ Upon this notion we purpose calling that towards the east the *anode,* and that towards the west the *cathode;* and whatever changes may take place in our views of the nature of electricity and electrical action, as they must affect the natural standard referred to in the same direction, and to an equal amount with any decomposing substances to which these terms may at any time be applied, there seems no reason to expect that they will lead to confusion, or tend in any way to support false views. The *anode* is therefore that surface at which the electric current, according to our present expression, enters; it is the negative extremity of the decomposing body; it is where oxygen, chlorine, acids, &c., are evolved;■ and is against or opposite the positive electrode. The *cathode* is that surface at which the current leaves the decomposing body, and is its positive extremity;■ the combustible bodies, metals, alkalies, and bases, are evolved there, and it is in contact with the negative electrode.

664. I shall have occasion in these researches, also, to class bodies together according to certain relations derived from their electrical actions (822); and wishing to express those relations without at the same time involving the expression of any hypothetical views, I intend using the following names and terms. Many bodies are decomposed directly by the electric current, their elements being set free, these I propose to call *electrolytes.*■ Water, therefore, is an electrolyte. The bodies which, like nitric or sulphuric acids, are decomposed in a secondary manner (752, 757),■ are not included under this term. Then for *electrochemically decomposed,* I shall often use the term *electrolyzed,* derived in the same

Imagine the earth to be circled with currents parallel to the magnetic equator. This is not the case, in practice.

This description may seem unnecessarily complex in view of the definitions of anode and cathode given below.

That is, where the negative ions are attracted.

Where positive ions are evolved.

When a small quantity of acid is added to make them conductors.

These molecules dissociate in water, forming positive and negative ions.

way, and implying that the body spoken of is separated into its components under the influence of electricity: it is analogous in its sense and sound to *analyze*, which is derived in a similar manner. The term *electrolytical* will be understood at once. Muriatic acid is electrolytical, boracic acid is not.■

665. Finally, I require a term to express those bodies which can pass to the *electrodes*, or, as they are usually called, the poles. Substances are frequently spoken of as being *electronegative*, or *electropositive*, according as they go under the supposed influence of a direct attraction to the positive or negative pole. But these terms are much too significant for the use to which I should have to put them; for though the meanings are perhaps right, they are only hypothetical, and may be wrong; and then, through a very imperceptible, but still very dangerous, because continual, influence, they do great injury to science, by contracting and limiting the habitual views of those engaged in pursuing it. I propose to distinguish these bodies by calling those *anions* which go to the *anode* of the decomposing body; and those passing to the *cathode, cations;* and when I have occasion to speak of these together, I shall call them *ions.* Thus, the chloride of lead is an *electrolyte*, and when *electrolyzed* evolves the two *ions*, chlorine and lead, the former being an anion, and the latter a cation.

666. These terms■ being once well defined, will, I hope, in their use enable me to avoid much periphrasis and ambiguity of expression. I do not mean to press them into service more frequently than will be required, for I am fully aware that names are one thing and science another.*

667. It will be well understood that I am giving no opinion respecting the nature of the electric current now, beyond what I have done on a former occasion (283,517) ;■ and that though I speak of the current as proceeding from the parts which are positive to those which are negative (663), it is merely in accordance with the conventional, though in some degree tacit, agreement entered into by scientific men, that they may have a constant, certain, and definite means of referring to the direction of the forces of that current.

■..

Muriatic acid is hydrochloric acid, HCl; Boric acid, H_3BO_3. The first forms ions in water; the second does not.

Most of these terms continue to be used at the present time.

Except for the fact that Franklin's *one-fluid theory* was in good agreement with experiment, there could be no more detailed description of electricity prior to the discovery of the electron.

Some observations on the electro-chemical properties of several compounds are omitted.

---

* Since this paper was read, I have changed some of the terms which were first proposed, that I might employ only such as were at the same time simple in their nature, clear in their reference, and free from hypothesis.

ON A NEW MEASURER OF VOLTA ELECTRICITY

704. I have already said, when engaged in reducing common and voltaic electricity to one standard of measurement (377),■ and again when introducing my theory of electrochemical decomposition (504, 505, 510), that the chemical decomposing action of a current *is constant for a constant quantity of electricity*, notwithstanding the greatest variations in its sources, in its intensity, in the size of the *electrodes* used, in the nature of the conductors (or nonconductors, 307) through which it is passed, or in other circumstances.■ The conclusive proofs of the truth of these statements shall be given almost immediately (783, etc.).

705. I endeavored upon this law to construct an instrument which should measure out the electricity passing through it, and which, being interposed in the course of the current used in any particular experiment, should serve at pleasure, either as a *comparative standard* of effect, or as a *positive measurer* of this subtile agent.■

706. There is no substance better fitted, under ordinary circumstances, to be the indicating body in such an instrument than water; for it is decomposed with facility when rendered a better conductor by the addition of acids or salts; its elements may in numerous cases be obtained and collected without any embarrassment from secondary action, and, being gaseous, they are in the best physical condition for separation and measurement. Water, therefore, acidulated by sulphuric acid, is the substance I shall generally refer to, although it may become expedient in peculiar cases or forms of experiment to use other bodies (843).

707. The first precaution needful in the construction of the instrument was to avoid the recombination of the evolved gases, an effect which the positive electrode has been found so capable of producing (571). For this purpose various forms of decomposing apparatus were used. The first consisted of straight tubes, each containing a plate and wire of platina■ soldered together by gold, and fixed hermetically in the glass at the closed extremity of the tube (Fig. 8). The tubes were about eight inches long, 0.7 of an inch in diameter, and graduated. The platina plates were about an inch long, as wide as the tubes would permit, and adjusted as near to the mouths of the tubes as was consistent with the safe collection of the gases evolved. In certain cases, where it was required to evolve the elements upon as small a surface as possible, the metallic extremity, instead of being a plate, consisted of the wire bent into the form of a ring (Fig. 9). When these tubes were used as measurers, they were filled with the dilute sulphuric acid, and inverted in a basin of the same liquid (Fig. 10), being placed in an inclined position,

Common electricity was the term applied to static electricity.

Actually, Faraday's first law of electrolysis.

A fairly sensitive, though inconvenient form of ammeter, or rather, coulometer, since it measures total charge.

Platinum.

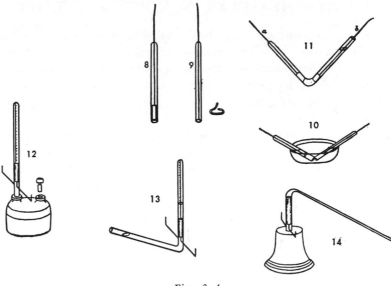

Figs. 8-14

with their mouths near to each other, that as little decomposing matter should intervene as possible; and also, in such a direction that the platina plates should be in vertical planes (720).

708. Another form of apparatus was that delineated (Fig. 11). The tube is bent in the middle; one end is closed; in that end is fixed a wire and plate, *a,* proceeding so far downwards, that, when in the position figured, it shall be as near to the angle as possible, consistently with the collection, at the closed extremity of the tube, of all the gas evolved against it. The plane of this plate is also perpendicular (720). The other metallic termination, *b,* is introduced at the time decomposition is to be effected, being brought as near the angle as possible, without causing any gas to pass from it towards the closed end of the instrument. The gas evolved against it is allowed to escape.■

The gas collected in the closed tube being a measure of the total charge.

709. The third form of apparatus contains both electrodes in the same tube; the transmission, therefore, of the electricity, and the consequent decomposition, is far more rapid than in the separate tubes. The resulting gas is the sum of the portions evolved at the two electrodes, and the instrument is better adapted than either of the former as a measurer of the quantity of voltaic electricity transmitted in ordinary cases. It consists of a straight tube (Fig. 12) closed at the upper extremity, and graduated, through the sides of which pass the platina wires (being fused into the glass), which are connected with two plates within. The tube is fitted by grinding into one mouth of a double-necked bottle. If the latter

be one-half or two-thirds full of the dilute sulphuric acid, it will, upon inclination of the whole, flow into the tube and fill it. When an electric current is passed through the instrument, the gases evolved against the plates collect in the upper portion of the tube, and are not subject to the recombining power of the platina.

710. Another form of the instrument is given at Figure 13.

711. A fifth form is delineated (Fig. 14). This I have found exceedingly useful in experiments continued in succession for days together, and where large quantities of indicating gas were to be collected. It is fixed on a weighted foot, and has the form of a small retort containing the two electrodes: the neck is narrow, and sufficiently long to deliver gas issuing from it into a jar placed in a small pneumatic trough. The electrode chamber, sealed hermetically at the part held in the stand, is five inches in length, and 0.6 of an inch in diameter; the neck about nine inches in length, and 0.4 of an inch in diameter internally. The figure will fully indicate the construction.

712. It can hardly be requisite to remark, that in the arrangement of any of these forms of apparatus, they, and the wires connecting them with the substance, which is collaterally subjected to the action of the same electric current, should be so far insulated as to ensure a certainty that all the electricity which passes through the one shall also be transmitted through the other.

■..

729. Although not necessary for the practical use of the instrument I am describing, yet as connected with the important point of constant electrochemical action upon water, I now investigated the effects produced by an electric current passing through aqueous solutions of acids, salts, and compounds, exceedingly different from each other in their nature, and found them to yield astonishingly uniform results. But many of them which are connected with a secondary action will be more usefully described hereafter (778).

Omitted are some details of the operation of these instruments.

730. When solutions of caustic potassa or soda, or sulphate of magnesia, or sulphate of soda, were acted upon by the electric current, just as much oxygen and hydrogen was evolved from them as from the diluted sulphuric acid, with which they were compared. When a solution of ammonia, rendered a better conductor by sulphate of ammonia (554), or a solution of subcarbonate of potassa was experimented with, the *hydrogen* evolved was in the same quantity as that set free from the diluted sulphuric acid with which they were compared. Hence *changes in the nature of the solution do not alter the constancy of electrolytic action upon water.*

731. I have already said, respecting large and small electrodes, that change of order caused no change in the general effect (715). The same was the case with different solutions, or with different intensities; and however the circumstances of an experiment might be varied, the results came forth exceedingly consistent, and proved that the electrochemical action was still the same.

732. I consider the foregoing investigation as sufficient to prove the very extraordinary and important principle with respect to water, *that when subjected to the influence of the electric current, a quantity of it is decomposed exactly proportionate to the quantity of electricity which has passed,* notwithstanding the thousand variations in the conditions and circumstances under which it may at the time be placed; and further, that when the interference of certain secondary effects (742, etc.),■ together with the solution or recombination of the gas and the evolution of air, are guarded against, *the products of the decomposition may be collected with such accuracy, as to afford a very excellent and valuable measurer of the electricity concerned in their evolution.*

For example, the liberated gas may react with the material of the electrode.

■···

821. All these facts combine into, I think, an irresistible mass of evidence, proving the truth of the important proposition which I at first laid down, namely, *that the chemical power of a current of electricity is in direct proportion to the absolute quantity of electricity which passes* (377, 783). They prove, too, that this is not merely true with one substance, as water, but generally with all electrolytic bodies; and, further, that the results obtained with any *one substance* do not merely agree amongst themselves, but also with those obtained from *other substances,* the whole combining together into *one series of definite electrochemical actions* (505). I do not mean to say that no exceptions will appear: perhaps some may arise, especially amongst substances existing only by weak affinity; but I do not expect that any will seriously disturb the result announced. If, in the well-considered, well-examined, and, I may surely say, well-ascertained doctrines of the definite nature of ordinary chemical affinity, such exceptions occur, as they do in abundance, yet, without being allowed to disturb our minds as to the general conclusion, they ought also to be allowed if they should present themselves at this, the opening of a new view of electrochemical action; not being held up as obstructions to those who may be engaged in rendering that view more and more perfect, but laid aside for a while, in hopes that their perfect and consistent explanation will finally appear.

A discussion of experiments performed with various substances, and the precautions to be observed, is omitted.

822. The doctrine of *definite electrochemical action* just laid down, and, I believe, established, leads to some new views of the relations and

classifications of bodies associated with or subject to this action. Some of these I shall proceed to consider.

823. In the first place, compound bodies may be separated into two great classes, namely, those which are decomposable by the electric current, and those which are not.∎ Of the latter, some are conductors, others nonconductors, of voltaic electricity.* The former do not depend for their decomposability upon the nature of their elements only; for, of the same two elements, bodies may be formed, of which one shall belong to one class and another to the other class; but probably on the proportions also (697). It is further remarkable, that with very few, if any, exceptions (414, 691), these decomposable bodies are exactly those governed by the remarkable law of conduction I have before described (394); for that law does not extend to the many compound fusible substances that are excluded from this class. I propose to call bodies of this, the decomposable class, *electrolytes* (664).

> Electrovalent and covalent, respectively.

824. Then, again, the substances into which these divide, under the influence of the electric current, form an exceedingly important general class. They are combining bodies; are directly associated with the fundamental parts of the doctrine of chemical affinity; and have each a definite proportion, in which they are always evolved during electrolytic action. I have proposed to call these bodies generally *ions,* or particularly *anions* and *cations,* according as they appear at the *anode* or *cathode* (665); and the numbers representing the proportions in which they are evolved electrochemical equivalents.∎ Thus hydrogen, oxygen, chlorine, iodine, lead, tin, are *ions;* the three former are anions, the two metals are cations, and 1, 8, 36, 125, 104, 58, are their *electrochemical equivalents* nearly.

> The term is used here in the sense of *chemical equivalent,* which is the ratio of the atomic weight of a substance to its valence. *Electrochemical equivalent* is now used for the mass deposited per-unit charge.

825. A summary of certain points already ascertained respecting *electrolytes, ions,* and *electrochemical* equivalents may be given in the following general form of propositions, without, I hope, including any serious error.

826. i. A single ion, i.e., one not in combination with another,∎ will have no tendency to pass to either of the electrodes, and will be perfectly indifferent to the passing current, unless it be itself a compound of more elementary ions, and so subject to actual decomposition. Upon this fact is founded much of the proof adduced in favor of the new theory of electrochemical decomposition, which I put forth in a former series of these Researches (518, etc.).

> Here Faraday apparently meant a neutral molecule which did not dissociate, or perhaps an atom.

---

* I mean here by voltaic electricity, merely electricity from a most abundant source, but having very small intensity.

827. ii. If one *ion* be combined in right proportions (697) with another strongly opposed to it in its ordinary chemical relations, i.e., if an anion be combined with a *cation,* then both will travel, the one to the *anode,* the other to the *cathode,* of the decomposing body (530, 542, 547).

828. iii. If, there, an *ion* pass towards one of the electrodes, another *ion* must also be passing simultaneously to the other electrode, although, from secondary action, it may not make its appearance (743). ■

829. iv. A body decomposable directly by the electric current, i.e., an *electrolyte,* must consist of two ions, and must also render them up during the act of decomposition.

830. v. There is but one electrolyte composed of the same two elementary *ions;* at least such appears to be the fact (697), dependent upon a law, that *only single electrochemical equivalents of elementary ions can go to the electrodes, and not multiples.*

831. vi. A body not decomposable when alone, as boracic acid, is not directly decomposable by the electric current when in combination (780). It may act as an *ion,* going wholly to the *anode* or *cathode,* but does not yield up its elements, except occasionally by a secondary action. Perhaps it is superfluous for me to point out that this proposition has *no relation* to such cases as that of water, which, by the presence of other bodies, is rendered a better conductor of electricity, and *therefore* is more freely decomposed.

832 vii. The nature of the substance of which the electrode is formed, provided it be a conductor, causes no difference in the electro-decomposition, either in kind or degree (807, 813); but it seriously influences, by secondary action (744), the state in which the *ions* finally appear. Advantage may be taken of this principle in combining and collecting such *ions* as,■ if evolved in their free state, would be unmanageable.*

833. viii. A substance which, being used as the electrode, can combine altogether with the *ion* evolved against it, is also, I believe, an *ion,* and combines, in such cases, in the quantity represented by its *electrochemical equivalent.* All the experiments I have made agree with this view; and it seems to me, at present, to result as a necessary consequence. Whether, in the secondary actions that take place, where the *ion* acts, not upon the matter of the electrode, but on that which is around it in

---

* It will often happen that the electrodes used may be of such a nature as, with the fluid in which they are immersed, to produce an electric current,■ either according with or opposing that of the voltaic arrangement used, and in this way, or by direct chemical action, may sadly disturb the results. Still, in the midst of all these confusing effects, the electric current, which actually passes in any direction through the decomposing body, will produce its own definite electrolytic action,

the liquid (744), the same consequence follows, will require more extended investigation to determine.

834. ix. Compound *ions* are not necessarily composed of electrochemical equivalents of simple *ions*. For instance, sulphuric acid, boracic acid, phosphoric acid, are *ions,* but not *electrolytes,* i.e., not composed of electrochemical equivalents of simple *ions*.

835. x. Electrochemical equivalents are always consistent; i.e., the same number which represents the equivalent of a substance *A* when it is separating from a substance *B,* will also represent *A* when separating from a third substance *C*. Thus, 8 is the electrochemical equivalent of oxygen, whether separating from hydrogen, or tin, or lead; and 103.5 is the electrochemical equivalent of lead, whether separating from oxygen, or chlorine, or iodine.

836. xi. Electrochemical equivalents coincide, and are the same, with ordinary chemical equivalents.

■...

851. A very valuable use of electrochemical equivalents will be to decide, in cases of doubt, what is the true chemical equivalent, or definite proportional, or atomic number of a body; for I have such conviction that the power which governs electro-decomposition and ordinary chemical attractions is the same; and such confidence in the overruling influence of those natural laws which render the former definite, as to feel no hesitation in believing that the latter must submit to them also. Such being the case, I can have no doubt that, assuming hydrogen as 1, and dismissing small fractions for the simplicity of expression, the equivalent number or atomic weight of oxygen is 8, of chlorine 36, of bromine 78.4, of lead 103.5, of tin 59, etc., notwithstanding that a very high authority doubles several of these numbers.■

An analysis of the chemical equivalents of various substances is omitted.

ON THE ABSOLUTE QUANTITY OF ELECTRICITY ASSOCIATED WITH THE PARTICLES OR ATOMS OF MATTER

The atomic weights of oxygen, lead and tin are double the values given. Evidently Faraday was misled by the different valences.

852. The theory of definite electrolytical or electrochemical action appears to me to touch immediately upon the *absolute quantity* of electricity or electric power belonging to different bodies. It is impossible, perhaps, to speak on this point without committing oneself beyond what present facts will sustain; and yet it is equally impossible, and perhaps would be impolitic, not to reason upon the subject. Although we know nothing of what an atom is, yet we cannot resist forming some idea of a small particle, which represents it to the mind; and though we are in

equal, if not greater, ignorance of electricity, so as to be unable to say whether it is a particular matter or matters, or mere motion of ordinary matter, or some third kind of power or agent, yet there is an immensity of facts which justify us in believing that the atoms of matter are in some way endowed or associated with electrical powers, to which they owe their most striking qualities, and amongst them their mutual chemical affinity.■ As soon as we perceive, through the teaching of Dalton, that chemical powers are, however varied the circumstances in which they are exerted, definite for each body, we learn to estimate the relative degree of force which resides in such bodies: and when upon that knowledge comes the fact, that the electricity, which we appear to be capable of loosening from its habitation for a while, and conveying from place to place, *whilst it retains its chemical force,* can be measured out, and, being so measured, is found to be *as definite in its action* as any of *those portions* which, remaining associated with the particles of matter, give them their *chemical relation;* we seem to have found the link which connects the proportion of that we have evolved to the proportion of that belonging to the particles in their natural state.

853. Now it is wonderful to observe how small a quantity of a compound body is decomposed by a certain portion of electricity. Let us, for instance, consider this and a few other points in relation to water. *One grain* of water acidulated to facilitate conduction, will require an electric current to be continued for three minutes and three quarters of time to effect its decomposition, which current must be powerful enough to retain a platina wire $\frac{1}{104}$ of an inch in thickness,* red hot, in the air during the whole time; and if interrupted anywhere by charcoal points, will produce a very brilliant and constant star of light.■ If attention be paid to the instantaneous discharge of electricity of tension, as illustrated in the beautiful experiments of ■Mr. Wheatstone,† and to what I have

Faraday showed remarkable perception in this conclusion.

The measure of current employed here was the usual semi-quantitative heating effect.

Sir Charles Wheatstone (1802-1875). Generally regarded as the *practical* founder of modern telegraphy. The *Wheatstone bridge,* used for the measurement of resistance, while not conceived by him, was developed by him.

Probably because of cooling effects at the ends.

---

* I have not stated the length of wire used, because I find by experiment, as would be expected in theory, that it is indifferent. The same quantity of electricity which, passed in a given time, can heat an inch of platina wire of a certain diameter red hot, can also heat a hundred, a thousand, or any length of the same wire to the same degree, provided the cooling circumstances are the same for every part in both cases. This I have proved by the volta-electrometer. I found that whether half an inch or eight inches were retained at one constant temperature of dull redness, equal quantities of water were decomposed in equal times in both cases. When the half-inch was used, only the center portion of wire was ignited.■ A fine wire may even be used as a rough but ready regulator of a voltaic current; for if it be made part of the circuit, and the larger wire communicating with it be shifted nearer to or further apart, so as to keep the portion of wire in the circuit sensibly at the same temperature, the current passing through it will be nearly uniform.

† *Literary Gazette,* 1833, March 1 and 8. *Philosophical Magazine,* 1833, p. 204. *L'Institute,* 1833, p. 261.

said elsewhere on the relation of common and voltaic electricity (371, 375), it will not be too much to say, that this necessary quantity of electricity is equal to a very powerful flash of lightning.■ Yet we have it under perfect command; can evolve, direct, and employ it at pleasure; and when it has performed its full work of electrolyzation, it has only separated the elements of a single grain of water.

This is rather unlikely, despite the fact that large quantities of charge could be stored in Leyden jars, etc.

854. On the other hand, the relation between the conduction of the electricity and the decomposition of the water is so close, that one cannot take place without the other. If the water is altered only in that small degree which consists in its having the solid instead of the fluid state, the conduction is stopped, and the decomposition is stopped with it. Whether the conduction be considered as depending upon the decomposition, or not (413, 703), still the relation of the two functions is equally intimate and inseparable.

855. Considering this close and twofold relation, namely, that without decomposition transmission of electricity does not occur; and, that for a given definite quantity of electricity passed, an equally definite and constant quantity of water or other matter is decomposed; considering also that the agent, which is electricity, is simply employed in overcoming electrical powers in the body subjected to its action; it seems a probable, and almost a natural consequence, that the quantity which passes is the *equivalent* of, and therefore equal to, that of the particles separated; i.e., that if the electrical power which holds the elements of a grain of water in combination, or which makes a grain of oxygen and hydrogen in the right proportions unite into water when they are made to combine, could be thrown into the condition of *a current,* it would exactly equal the current required for the separation of that grain of water into its elements again.

A further discussion of what Faraday considered the *large* quantity of charge required to electrolyze a grain of water is omitted.

■··

SUPPLEMENTARY READING

Crew, H., "Portraits of Famous Physicists," *Scripta Mathematica,* Pictorial Mathematics, Portfolio No. 4, 1942.

Crowther, J. G., *Men of Science* (New York: Norton, 1936).

Dampier, W. C., *A History of Science* (New York: Cambridge University, 1946), pp. 216ff.

Faraday, M., *Diary,* vols. 1–7 (London: G. Bell and Sons, 1932–1936).

Jones, H. B., *The Life and Letters of Faraday* (London: Longmans, Green, 1870–1876).

Knedler, J. W. (ed.), *Masterworks of Science* (New York: Doubleday, 1947), pp. 447ff.

Kondo, H., "Michael Faraday," in *Lives in Science* by the Editors of Scientific American (New York: Simon and Schuster, 1957), pp. 127ff.

Lenard, P., *Great Men of Science* (New York: British Book Centre, 1934), pp. 247ff; 339ff.

Magie, W. F., *A Source Book in Physics* (New York: McGraw-Hill, 1935), pp. 472ff.

Thompson, S. P., *Michael Faraday, His Life and Work* (London: Cassell, 1898).

Tyndall, J., *Faraday as a Discoverer* (New York: Appleton, 1873).

Woodruff L. L. (ed.), *The Development of the Sciences* (New Haven, Conn.: Yale University, 1923).

Heinrich Lenz
1804 - 1865

Lenz's Law

ALONG the frontiers of science it is not surprising that many investigators seek solutions to the same problems, and, in the days of poor communications, that they might arrive independently at similar results. Thus Faraday discovered the mutual induction of electricity in England in 1831. Shortly afterward, in the United States, Joseph Henry (1797–1878) observed the same phenomenon independently, and in Russia, with no knowledge of Henry's work but with some basic information about Faraday's, Heinrich Lenz extended both with his discovery of the principle of induction known by his name. Announced in 1834, Lenz's law was a deceptively simple statement relating to the direction taken by an induced current. Perhaps it is for this reason that Lenz has remained relatively obscure. He is little known beyond the physical law bearing his name and the available biographical material is extremely sketchy. By comparison, Henry is well known, yet Lenz made a highly significant contribution, for his observation was essentially a statement of energy conservation many years prior to its formal development. Not until James Joule (1818–1889) first showed the equivalence of work and heat about a decade later, and von Helmhotz (1821–1894) published his great paper on the conservation of energy in 1847, were these concepts clearly established.

While Lenz did not arrive at a general statement of energy conservation, no doubt his experimental observations had some influence on contemporary thought along these lines. From our present vantage point Lenz's law seems perfectly obvious. An induced current represents energy, which must be produced at the expense of the inducing agency, such as a moving magnet or a transient current in a neighboring circuit. Hence *the direction of the induced current must be such as to oppose, by its own magnetic field, the action tending to induce the current.* The justification for this state-ment, which is essentially Lenz's law, is our concept of energy conservation, for if the induced current were in the direction to aid the inducing agency, it is clear that an experiment could be devised which would violate the principle of conservation of energy. However apparent this may seem, it must be recalled that Lenz did not have the concept of energy conservation to guide him and that his conclusions were based solely upon experimental findings.

Heinrich Friedrich Emil Lenz was born at Dorpat, Russia (later Estonia), on February 12, 1804. Apparently very little is known of his early environ-ment or education, except that he may have studied theology for a while before turning to science. Upon completion of his academic course he re-ceived the degree of doctor of philosophy. His interests evidently were not confined to physics, for at the age of twenty he made a trip around the world as a naturalist, reporting his observations upon his return to the Imperial Academy of Sciences at St. Petersburg. In recognition of this he was elected an associate member of the Academy in 1828. Following his initial experiments on electromagnetism in 1834, he was elected to fellow-ship. At the same time he became professor of physics at the University of St. Petersburg, and sometime professor of physics at the Pedagogical Insti-tute. It would seem that he was highly regarded by his academic associates, for he served for a time as rector of the university.

In the period 1835–1838, Lenz devised an expression for the temperature dependence of electrical resistance of a conductor. He was a prolific writer, but except for a *Handbook of Physics* in 1864, he published little beyond reports and scientific papers. During a journey to Italy for his health in 1865 his condition grew worse and he died in Rome on Febru-ary 10.

The following extract is taken from his paper, "On the Determination of the Direction of Galvanic Currents Caused by Electrodynamic Induc-tion," *Annalen der Physik und Chemie,* vol. 31 (1834), page 483.

Lenz's Experiment

Faraday published many papers under the same general title. Probably the one referred to here appeared in the

In his *Experimental Researches in Electricity,* which contain the discovery of the so-called electrodynamic induction,■ Faraday determines the direction of the galvanic currents produced by it in the following manner: (1) a galvanic current will induce a current in the opposite direction in a wire which is approaching parallel to this galvanic current;

in a receding wire, however, the current will flow in the same direction, and (2) a magnet induces a current in a conductor moving in its vicinity whose sense depends upon the direction in which the conductor cuts the magnetic lines of force in its movement. Apart from the fact that two completely different rules are given for one and the same phenomenon (since the magnet, according to Ampère's beautiful theory,■ may be considered as a system of circular galvanic currents), the rule is not entirely adequate, as it does not include some cases at all; for example, the case in which a conductor is oriented perpendicular to a current and is moved along it; and finally, according to my conviction, the second statement does not have the desired simplicity, so that it cannot be applied easily to individual cases, and I believe that other readers of this otherwise remarkable dissertation will agree with me in this, if they recall paragraph 116,■ where Faraday seeks to illustrate the above rule by the movement of a knife blade toward a magnet; even Faraday himself mentions the difficulty in explaining clearly the direction of the current.

Nobili■ (in his memoir, *Poggendorff's Annalen* 1833, No. 3) proceeds from Faraday's first rule that when a conductor approaches parallel to a galvanic current, a current in the opposite sense is induced in it, and when it recedes the current is in the same direction, and tries by means of this alone to explain the appearance and directions of all currents produced by electrodynamic induction. While I find this work so valuable in other respects, it does not seem to me to offer in some points the degree of evidence which one may expect to find in a dissertation in physics, particularly in the explanation of those currents which are induced in a conductor placed perpendicular to a galvanic current and moved along it. Faraday is surely correct in objecting to the theory of the Italian physicist,■ that in the case of the rotation of a magnet about its own axis, with proper connection of the test wires, a galvanic current is produced, even though here there is no approach nor withdrawal of the conductors, since everything remains in the same relative positions.

On reading the dissertation of Faraday it appeared to me that all the experiments on electrodynamic induction could be reduced to the laws of electrodynamic movements, so that if these are considered known, the others may be determined thereby; and since this view has been confirmed by many experiments, I shall present it in the following pages and prove it partly by known observations and partly by experiments which I have especially devised for the purpose.

The law whereby the reduction of the magneto-electric effects to the electromagnetic effects is accomplished is the following:

■Whenever a metallic conductor is moved in the vicinity of a galvanic

Philosophical Transactions of 1832.

Ampère showed that the same magnetic field as obtained from a magnet could be derived from a system of currents flowing in small circular paths.

See Faraday's description of this effect. Page 140.

Leopoldo Nobili (1784-1835), professor of physics in Florence.

Nobili. See the note following paragraph 120 in Faraday's account. page 141.

The statement of Lenz's principle.

current or of a magnet, a galvanic current will be induced, which has a direction such that it would have caused the wire, if it were at rest, to move in a direction opposite to the one given, provided that the wire were movable only in the direction of the motion and in the opposite direction.

In order to explain the sense of the direction of the current produced by electrodynamic induction in the movable wire, we consider what the direction of the current must be according to electromagnetic laws, in order to cause this movement; the current will then be induced in the opposite direction. As an example, let us consider the well-known rotation experiment due to Faraday in which a movable conductor hanging vertically carries a galvanic current from top to bottom, and therefore would encircle the N pole of the magnet placed beneath it in the direction from north through east to south;▪ now if we were not to allow the current to pass through the movable conductor, but give it that motion through mechanical means, then according to our laws a current will be induced in the same, which, as opposed to the previous situation, will flow through the movable wire from the bottom to the top, as can be shown by connecting top and bottom end of the same to a galvanometer.

When we now reconsider the above law, we see that it follows from the same, that to each case of electromagnetic movement there corresponds a case of electrodynamic induction; one merely has to cause the movement produced electromagnetically by a different means, and thus induce a current in the movable conductor which is opposite to that of the electromagnetic experiment.▪ I shall show in the following several such corresponding phenomena, and shall follow each electromagnetic phenomenon with the corresponding magneto-electric and shall denote the first by a capital letter, and the latter by the corresponding lower-case letter. This will serve at the same time to show the essential correctness of our law. Furthermore the accompanying diagrams have been labelled with the same letters for greater clarification, to which I add the following: The arrows indicate the direction of the movement as well as that of the current, while I have distinguished between them by using the arrow o⟶ to indicate movement and the arrow ≫⟶ to mark the current; furthermore, the solid arrows will refer to the expected motion or direction of current, as opposed to the arrows of identical form drawn with broken lines, which indicate the experimental result. Keeping these in mind one should have no trouble in understanding the figures. I now turn to the experiments:

A. A straight conductor through which a galvanic current is flowing will attract a parallel movable wire if the current through the latter is

The magnet is situated vertically, with its N pole uppermost.

It will be noted that Lenz determined the direction of the induced current in each case by visualizing the inverse experiment in which the electromagnetic interaction causes motion, and then taking the current in the opposite direction. He gave no reasons for this rule, except that it agreed with experiment.

flowing in the same direction; it will repel it however, as soon as the direction of the current in the movable conductor is opposite to the current in the fixed conductor (Ampère).■

(a) If one of two straight, mutually parallel conductors is carrying a current, and if we were to move the other parallel to and towards the first, then during the motion there will be induced a current in the moving conductor which is opposite in direction to the current in the fixed conductor; if it is moving away, however, then the induced current is in the same direction as the inducing current (Faraday).

B. If one has two vertical circular conductors, of nearly equal diameter, which are perpendicular with respect to their planes and have a common vertical diameter as axis, about which both (or only one) can be turned, and if one allows a galvanic current to flow through both, then they will arrange themselves so that the direction of the currents is the same (Ampère).■

(b) If one of the two circular conductors as above carries a galvanic current, and we were to turn the other quickly from a perpendicular to a parallel position, a current will be induced in the second which is opposite to that in the first conductor (Lenz).

This last experiment I have carried out with two circular conductors each of which was made of 20 turns of copper wire; one of them was connected to a 2 square foot zinc-copper couple,■ the other to a sensitive Nobili galvanometer.

C. If, in the vicinity of a straight infinite conductor there is another straight conductor, perpendicular to the first, movable and limited in such manner as to lie completely on one side of the first conductor, and if both conductors carry galvanic currents, then the movable conductor will tend to move along the infinite conductor in the direction of the current in the same if its own current is flowing away from the infinite conductor, but against the current if its current flows towards the infinite conductor.

The term *infinite* and *finite* current must be understood to have the meanings usually given them in the texts on electromagnetism.■

(c) If a finite conductor, which is perpendicular to an infinite conductor carrying a galvanic current, moves along the same and in the direction of the current, a current will be induced in the finite conductor which flows toward the infinite conductor; if, however, the finite conductor moves opposite to the direction of the current in the infinite conductor, then the direction of the current caused by induction is away from the infinite current (Nobili, *Poggend. Annalen* 1833, No. 3, p. 407).

In the preceding the chief examples in which a galvanic current

---

Where a name follows a particular experiment it is that of the individual who first made the observation.

That is, they will move into the same plane.

A wet cell.

Here an *infinite* conductor is one sufficiently long so that its ends have little influence upon the phenomenon under study.

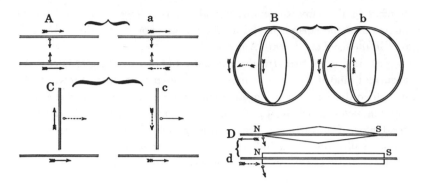

affects another have been considered; in the following we shall, in like manner, examine the cases where there is a mutual interaction between a galvanic current and a magnet. By an analysis first given by Ampère one can easily proceed as far as electromagnetic effects of this kind are concerned; it consists, as is well known, in assigning to the current a head and a foot, a right and left hand, or even better, imagine oneself to be the current, so that it enters at the feet and exits at the head, while the face is turned toward the N pole of the magnet; which will then be moved to the left because of the current, or the current (and with it the observer) will move to the right of the N pole.■

This is essentially the right-hand rule, where the thumb points in the direction of the current and the fingers then curve in the direction of the magnetic field surrounding the current.

Starting from our universal law of the relation of the magneto-electric to the electromagnetic effects, we shall be able rather easily to derive a similar rule as follows:

A galvanic current will be induced in a conductor moved in front of the N pole of a magnet because of the electrodynamic induction, which will flow through one from head to foot if one imagines himself to be the conductor and turns his face toward the N pole and moves with the conductor. This rule will be found to hold in all the following experiments.

D. Whenever a current passes over a freely suspended magnetic needle, which is oriented according to the earth's field, so that it moves over it from south to north, then the N pole of the magnetic needle will be turned to the west;■ if, however, the current flows from N to S the resulting deflection will be to the east. If the wire is placed underneath the needle, then in the first case there results a deflection to the east, in the second to the west (Oersted).

The N pole is urged in the direction of the magnetic field.

(d) If a conductor is situated above and parallel to a magnet, oriented according to the earth's field, and the magnet is suddenly turned about its center with the N pole toward the west, then a current from N to S will be caused to flow in the conductor; if the magnet be turned toward the east the current flow will be from S to N. If the conductor be placed underneath the magnet then the current flows in the first case from S to N, in the second from N to S (Lenz).

For this experiment I used as a conductor the foot-long side of a square which consisted of several turns of copper wire covered with silk; the side I moved toward the 5-inch long magnet was so close to it that the electrodynamic effect of the magnet on the three other sides could be considered negligible as compared to the effect on this side. To find the direction of the induced current according to the rule just given, one may think of the magnet as being at rest and the conductor in the first case turned to the east, in the second case to the west.■

■..

It is only the relative motion that is significant.

Several further examples along similar lines are omitted.

SUPPLEMENTARY READING

Holton, G., and Roller, D. H. D., *Foundations of Modern Physical Science* (Reading, Mass.: Addison-Wesley, 1958), pp. 529–530.

Magie, W. F., *A Source Book in Physics* (New York: McGraw-Hill, 1935), pp. 511ff.

Stine, W. M., *The Contributions of H. F. E. Lenz to Electromagnetism* (New York: Acorn, 1923).

12

James Joule
1818-1889

The Mechanical Equivalent of Heat

AMONG GALILEO'S many scientific interests was the quantitative study of thermal phenomena, for which he invented the first practical instrument, a sensitive but inaccurate form of gas thermometer. During the seventeenth and eighteenth centuries substantial advances were made in the field of thermometry, notably by Gabriel Fahrenheit (1686–1736), Ferchault de Reaumur (1683–1757), and Anders Celsius (1701–1744). These experimenters developed several improved types of liquid thermometers and established various *fixed points* (such as the temperature of melting ice) from which thermometric scales could be devised. The difference between *temperature* and *quantity of heat*, frequently confused with one another, was made clear in the latter half of the eighteenth century by Joseph Black (1728–1799). He introduced the science of *calorimetry*, by which he was led to the idea of *specific heat*,[1] and he discovered the concept of *latent heat:* in changes of state such as ice to water or water to steam, he found that large quantities of heat were absorbed without change in temperature.

Despite these advances, the nature of heat was not understood. It was generally regarded as an imponderable fluid, called *caloric,* made up of minute particles which repelled one another but were attracted by matter. The *caloric* was supposed to flow from hotter to colder bodies, and the development of heat by friction, for example, was explained as due to the fact that friction removed some of the *caloric* and thereby made the body appear warmer. The melting of ice was accounted for in terms of the combination of *caloric* with the ice to form water, somewhat in the nature of a chemical compound. Even Black adhered to the caloric theory, although a number

---

[1] Which he called "capacity for heat."

of earlier scientists, including Newton and Boyle, had inclined to the view that heat was related to the motions of the particles of bodies.

At the end of the eighteenth century, Benjamin Thompson, Count Rumford (1753–1814), an American who served as minister of war for the elector of Bavaria, conducted what appeared only later to have been convincing experiments regarding the nature of heat. As he pointed out,[2]

> Being engaged, lately, in superintending the boring of cannon, in the workshops of the military arsenal at Munich, I was struck with the very considerable degree of heat which a brass gun acquires, in a short time, in being bored; and with the still more intense heat (much greater than that of boiling water, as I found by experiment) of the metallic chips separated from it by the borer.
>
> The more I meditated on these phenomena, the more they appeared to me to be curious and interesting. A thorough investigation of them seemed even to bid fair to give a farther insight into the hidden nature of heat; and to enable us to form some reasonable conjectures respecting the existence, or nonexistence, of an *igneous fluid:* a subject on which the opinions of philosophers have, in all ages, been much divided.
>
> In order that the Society may have clear and distinct ideas of the speculations and reasonings to which these appearances gave rise in my mind, and also of the specific objects of philosophical investigation they suggested to me, I must beg leave to state them at some length, and in such manner as I shall think best suited to answer this purpose.
>
> From *whence comes* the heat actually produced in the mechanical operation above mentioned?
>
> Is it furnished by the metallic chips which are separated by the borer from the solid mass of metal?
>
> If this were the case, then, according to the modern doctrines of latent heat, and of caloric, the *capacity for heat* of the parts of the metal, so reduced to chips, ought not only to be changed, but the change undergone by them should be sufficiently great to account for all the heat produced.
>
> But no such change had taken place; for I found, upon taking equal quantities, by weight, of these chips, and of thin slips of the same block of metal separated by means of a fine saw, and putting them, at the same temperature (that of boiling water) into equal quantities of cold water (that is to say, at the temperature of $59\frac{1}{2}°F$) the portion of water into which the chips were put was not, to all appearance, heated either less or more than the other portion, in which the slips of metal were put.

---

[2] Rumford's *Collected Works,* vol. II, essay IX. Read before the Royal Society on January 25, 1798. Later published in *Philosophical Transactions,* vol. 88 (1798), page 80, and reprinted in *The Complete Works of Count Rumford* (American Academy of Arts and Sciences, 1870), vol. I, pages 471ff.

This experiment being repeated several times, the results were always so nearly the same, that I could not determine whether any, or what change, had been produced in the metal, *in regard to its capacity for heat*, by being reduced to chips by the borer.

From hence it is evident, that the heat produced could not possibly have been furnished at the expense of the latent heat of the metallic chips.

Rumford's experiments were disregarded, for the most part, until the middle of the nineteenth century, when two quite independent developments took place that led to general acceptance of the mechanical theory of heat. The first was a suggestion by Julius Mayer (1814–1878), in 1842, that heat and work were equivalent and could be converted one to the other,[3] and the second was the actual measurement by Joule, during the decade between 1840 and 1850, of the mechanical equivalent of heat (or more appropriately, the heat equivalent of work).

James Prescott Joule was born near Manchester, England, on December 24, 1818, the son of a prosperous brewery owner. It was still the Age of Romanticism and of the first Industrial Revolution, but new social and economic doctrines were in the wind. Among his contemporaries were the economic philosophers John Stuart Mill (1806–1873) and Karl Marx (1818–1883). The growing use of labor-saving devices had caused serious dislocations of the labor supply, and the factories of Manchester were particularly notorious for the apalling conditions accorded its workers. The cultural atmosphere was no less suffocating than the bad air in the factories and tenements that housed the working-class families. In this environment, hardly one designed to inspire cultural pursuits, but shielded from it by wealth and position, Joule spent the major part of his life.

He was educated at his home by resident tutors until the age of sixteen, when he was sent, together with his brother, to study under the famous chemist, John Dalton (1766–1844), who supported himself in part by such tutoring. This arrangement lasted but a short time, owing to the illness of the teacher, but probably it contributed immeasurably to Joule's interest in science. So ended his "formal" education, although he did receive, for a brief period in 1839, some private lessons in chemistry from one John Davies. Because of his financial independence he needed no further conventional training; his experimental researches were a form of entertainment for him. In 1838 Joule converted one of the rooms of his father's home into a laboratory and started his experimental investigations. In the same year he published his first short paper, but it was not until 1840 that

---

[3] Actually the principle of conservation of energy, later formalized by Helmholtz in 1847.

he presented an important paper[4] to the Royal Society. In this he showed that the rate at which heat is generated by an electric current in a conductor is proportional to the square of the current, the constant relating the two being the resistance of the conductor. For the next ten years Joule continued his thermal experiments, refining his measurements time and again, and reporting his results at frequent intervals to the Royal Society. In 1850 he published a memoir in the *Philosophical Transactions* which contained his most precise value of the mechanical equivalent of heat, including the famous *paddle-wheel experiment.* Following the publication of this paper he was elected to fellowship in the Royal Society, and his reputation as a scientist was firmly established.

Joule's researches after 1850, while numerous and significant, did not rank in importance with his measurements of the heat equivalent of work. During this period he performed the well-known porous-plug experiment with William Thomson (1824–1907)[5] to show the cooling effect of a gas due to the separation of its molecules upon expansion. Economic misfortune overtook him toward the end of his life; his investments had declined to the point where his income no longer permitted him to carry on research at his own expense. In 1878 he was granted a pension by the Government of £ 200 per annum, which was continued until his death in 1889. Of the many honors that came to Joule during his lifetime, probably none was more eloquent than the decision of the second International Congress to use his name for the practical unit of energy.

Of the extracts given below, the first was a letter to the editor entitled *On the Existence of an Equivalent Relation beween Heat and the Ordinary Forms of Mechanical Power,* published in the *Philosophical Magazine,* vol. 27, Series 3 (1845), page 205. The second is taken from the memoir of 1850, *Philosophical Transactions,* vol. 140, page 61.

---

[4] *Philosophical Magazine,* 1841, vol. 19, p. 260.
[5] Later Lord Kelvin.

Joule's Experiment

■The principal part of this letter was brought under the notice of the British Association at its last meeting at Cambridge. I have hitherto hesitated to give it further publication, not because I was in any degree doubtful of the conclusions at which I had arrived, but because I intended to make slight alteration in the apparatus calculated to give still

A letter to the editors of the *Philosophical Magazine,* vol. 27 (1845), page 205.

greater precision to the experiments. Being unable, however, just at present to spare the time necessary to fulfill this design, and being at the same time most anxious to convince the scientific world of the truth of the positions I have maintained, I hope you will do me the favor of publishing this letter in your excellent magazine.

The apparatus exhibited before the Association consisted of a brass paddle wheel working *horizontally* in a can of water.■ Motion could be communicated to this paddle by means of weights, pulleys, &c., exactly in the manner described in a previous paper.*

The paddle moved with great resistance in the can of water, so that the weights (each of four pounds) descended at the slow rate of about one foot per second. The height of the pulleys from the ground was twelve yards, and consequently, when the weights had descended through that distance, they had to be wound up again in order to renew the motion of the paddle. After this operation had been repeated sixteen times, the increase of the temperature of the water was ascertained by means of a very sensible■ and accurate thermometer.

A series of nine experiments was performed in the above manner, and nine experiments were made in order to eliminate the cooling or heating effects of the atmosphere. After reducing the result to the capacity for heat of a pound of water,■ it appeared that for each degree of heat evolved by the friction of water, a mechanical power equal to that which can raise a weight of 890 lbs to the height of one foot, had been expended.

The equivalents I have already obtained are,—first, 823 lbs, derived from magneto-electrical experiments;† second, 795 lbs, deduced from the cold produced by the rarefaction of air;‡ and third, 774 lbs from experiments (hitherto unpublished) on the motion of water through narrow tubes. This last class of experiments being similar to that with the paddle wheel, we may take the mean of 774 and 890, or 832 lbs, as the equivalent derived from the friction of water. In such delicate experiments, where one hardly ever collects more than half a degree of heat,■ greater accordance of the results with one another than that above exhibited could hardly have been expected. I may therefore conclude that the existence of an equivalent relation between heat and the ordinary forms

The famous *paddle-wheel* apparatus, described in detail in the following paper.

sensible: sensitive

That is, the specific heat, or amount of heat required to raise the temperature of one pound of substance by one degree Fahrenheit; this is readily determined experimentally. For water, the value is 1.0 Btu/lb - deg F.

The change in temperature during an experiment was very small.

So that work could be done on the paddle wheel. If there were no opposition to the rotation of the *floats* (paddles) work could not be performed.

---

* *Phil. Mag.*, vol. xxiii (1843), p. 436. The paddle wheel used by Rennie in his experiments on the friction of water (*Phil. Trans.*, 1831, plate xi, fig. 1) was somewhat similar to mine. I employed, however, a greater number of "floats," and also a corresponding number of stationary floats, in order to prevent the rotatory motion of the water in the can.■

† *Phil. Mag.*, vol. xxiii (1843), pp. 263, 347.

‡ *Phil Mag.*, vol. xxvi, series 3 (1845), p. 369.

of mechanical power is proved; and assume 817 lbs, the mean of the results of three distinct classes of experiments, as the equivalent, until still more accurate experiments shall have been made.■

Any of your readers who are so fortunate as to reside amid the romantic scenery of Wales or Scotland, could, I doubt not, confirm my experiments by trying the temperature of the water at the top and at the bottom of a cascade. If my views be correct, a fall of 817 feet will of course generate one degree of heat; and the temperature of the river Niagara■ will be raised about one fifth of a degree by its fall of 160 feet.

Admitting the correctness of the equivalent I have named, it is obvious that the *vis viva*■ of the particles of a pound of water at (say) 51° is equal to the *vis viva* possessed by a pound of water at 50° plus the *vis viva* which would be acquired by a weight of 817 lbs after falling through the perpendicular height of one foot.

Assuming that the expansion of elastic fluids on the removal of pressure is owing to the centrifugal force of revolving atmospheres of electricity, we can easily estimate the absolute quantity of heat in matter. For in an elastic fluid the pressure will be proportional to the square of the velocity of the revolving atmospheres; and the *vis viva* of the atmospheres will also be proportional to the square of their velocity; consequently the pressure will be proportional to the *vis viva*.■ Now the ratio of the pressures of elastic fluids at the temperatures 32° and 33° is 480:481, consequently the zero of temperature must be 480° below the freezing point of water.■

We see then what an enormous quantity of *vis viva* exists in matter. A single pound of water at 60° must possess 480° + 28° = 508° of heat, in other words, it must possess a *vis viva* equal to that acquired by a weight of 415036 lbs after falling through the perpendicular height of one foot. The velocity with which the atmospheres of electricity must revolve in order to present this enormous amount of *vis viva,* must of course be prodigious, and equal probably to the velocity of light in the planetary space, or to that of an electric discharge as determined by the experiments of Wheatstone.■

■In accordance with the pledge I gave the Royal Society some years ago, I have now the honor to present it with the results of the experiments I have made in order to determine the mechanical equivalent of heat with exactness. I will commence with a slight sketch of the progress of the mechanical doctrine, endeavoring to confine myself, for the sake of conciseness, to the notice of such researches as are immediately connected with the subject. I shall not therefore be able to review the valu-

The correct value is 778 foot-pounds. Joule's value differs from this by about 5 percent.

Niagara Falls.

The capacity for work, or energy.

This strange concept of the pressure of fluids preceded by several years Joule's more convincing explanation on the basis of kinetic theory.

Actually 492° F below the freezing point, or −460° F.

See page 156.

Following is the 1850 paper, in which Joule reviewed all his earlier work.

James D. Forbes (1809-1868), professor of physics at Edinburgh.

Rumford (page 167) showed that the heat evolved was related in some manner to the work done on the cannon, but did not obtain the actual relationship.

2^h30^m = 2 hours, 30 minutes

Later taken as the definition of the horsepower.

Note that even as late as 1850, Joule did not make use of scientific notation.

Rather, the work represented by. . . . The use of the term *force* was frequently ambiguous, sometimes signifying work, and other times having its conventional meaning.

able labors of Mr. Forbes■ and other illustrious men, whose researches on radiant heat and other subjects do not come exactly within the scope of the present memoir.

For a long time it had been a favorite hypothesis that heat consists of "a force or power belonging to bodies,"* but it was reserved for Count Rumford to make the first experiments decidedly in favor of that view. That justly celebrated natural philosopher demonstrated by his ingenious experiments that the very great quantity of heat excited by the boring of cannon could not be ascribed to a change taking place in the calorific capacity of the metal;■ and he therefore concluded that the motion of the borer was communicated to the particles of metal, thus producing the phenomena of heat:—"It appears to me," he remarks, "extremely difficult, if not quite impossible, to form any distinct idea of anything, capable of being excited and communicated, in the manner the heat was excited and communicated in these experiments, except it be motion."†

One of the most important parts of Count Rumford's paper, though one to which little attention has hitherto been paid, is that in which he makes an estimate of the quantity of mechanical force required to produce a certain amount of heat. Referring to his third experiment, he remarks that the "total quantity of ice-cold water which, with the heat actually generated by friction, and accumulated in 2^h 30^m,■ might have been heated 180°, or made to boil, = 26.58 lbs."‡ In the next page he states that "the machinery used in the experiment could easily be carried round by the force of one horse (though, to render the work lighter, two horses were actually employed in doing it)." Now the power of a horse is estimated by Watt at 33,000 foot-pounds per minute,■ and therefore if continued for two hours and a half will amount to 4,950,000 foot-pounds,■ which, according to Count Rumford's experiment, will be equivalent to 26.58 lbs of water raised 180°. Hence the heat required to raise a lb of water 1° will be equivalent to the force represented by 1034 foot-pounds.■ This result is not very widely different from that which I have deduced from my own experiments related in this paper, viz., 772 foot-pounds; and it must be observed that the excess of Count Rumford's equivalent is just such as might have been anticipated from the circumstance, which he himself mentions, that "no estimate was made of the heat accumulated in the wooden box, nor of that dispersed during the experiment."

---

* Crawford on *Animal Heat,* p. 15.
† "An Inquiry concerning the Source of the Heat which is excited by Friction," *Phil. Trans.,* abridged, vol. xviii, p. 286.
‡ "An Inquiry concerning the Source of the Heat which is excited by Friction," *Phil. Trans.,* abridged, vol. xviii, p. 283.

About the end of the last century Sir Humphry Davy communicated a paper to Dr. Beddoes' West Country Contributions, entitled, "Researches on Heat, Light, and Respiration," in which he gave ample confirmation to the views of Count Rumford. By rubbing two pieces of ice against one another in the vacuum of an air pump, part of them was melted, although the temperature of the receiver■ was kept below the freezing point. This experiment was the more decisively in favor of the doctrine of the immateriality of heat, inasmuch as the capacity of ice for heat is much less than that of water.■ It was therefore with good reason that Davy drew the inference that "the immediate cause of the phenomena of heat is motion, and the laws of its communication are precisely the same as the laws of the communication of motion."*

The researches of Dulong■ on the specific heat of elastic fluids were rewarded by the discovery of the remarkable fact that "equal volumes of all the elastic fluids,■ taken at the same temperature, and under the same pressure, being compressed or dilated suddenly to the same fraction of their volume, disengage or absorb the same *absolute quantity of heat*."† This law is of the utmost importance in the development of the theory of heat, inasmuch as it proves that the calorific effect is, under certain conditions, proportional to the force expended.

In 1834 Dr. Faraday demonstrated the "Identity of the Chemical and Electrical Forces."■ This law, along with others subsequently discovered by that great man, showing the relations which subsist between magnetism, electricity, and light, have enabled him to advance the idea that the so-called imponderable bodies are merely the exponents of different forms of force. Mr. Grove and M. Mayer have also given their powerful advocacy to similar views.■

My own experiments in reference to the subject were commenced in 1840, in which year I communicated to the Royal Society my discovery of the law of the heat evolved by voltaic electricity, a law from which the immediate deductions were drawn,—first, that the heat evolved by any voltaic pair is proportional, *caeteris paribus*,■ to its intensity or electromotive force;** and second, that the heat evolved by the combustion of a body is proportional to the intensity of its affinity for oxygen.†† I thus succeeded in establishing relations between heat and chemical affinity. In 1843 I showed that the heat evolved by magneto-electricity is proportional to the force absorbed;■ and that the force of the electro-

The vacuum chamber.

The specific heat of ice is about half that of water, 0.55 compared to 1.0 cal per gm per °C.

Pierre Dulong (1785-1838), professor of physics at the Polytechnic School in Paris.

The reference here is to gases, and states simply that the heat evolved or absorbed is proportional to the work done, since the latter is the same in each case.

By his experiments on electrolysis.

Sir William Grove (1811-1896), professor of experimental philosophy at the London Institution. Julius R. Mayer (1814-1878) developed a mechanical theory of heat, published one year before Joule's paper, for which he claimed priority over Joule.

All other things being equal.

Here the term force should be replaced by energy.

---

* *Elements of Chemical Philosophy*, p. 94.
† *Memoires de l'Académie des Sciences*, t. x., p. 188.
** *Phil. Mag.*, vol. xix, p. 275.
†† *Ibid.*, vol. xx, p. 111.

magnetic engine is derived from the force of chemical affinity in the battery, a force which otherwise would be evolved in the form of heat: from these facts I considered myself justified in announcing "that the quantity of heat capable of increasing the temperature of a lb of water by one degree of Fahrenheit's scale, is equal to, and may be converted into, a mechanical force capable of raising 838 lbs to the perpendicular height of one foot."‡■

The correct value is 778 foot-pounds.

In a subsequent paper, read before the Royal Society in 1844, I endeavored to show that the heat absorbed and evolved by the rarefaction and condensation of air is proportional to the force■ evolved and absorbed in those operations.§ The quantitative relation between force and heat deduced from these experiments, is almost identical with that derived from the electromagnetic experiments just referred to, and is confirmed by the experiments of M. Séguin■ on the dilatation of steam.¶

Read *force* as *energy.*

Marc Séguin (1786-1875).

From the explanation given by Count Rumford of the heat arising from the friction of solids, one might have anticipated, as a matter of course, that the evolution of heat would also be detected in the friction of liquid and gaseous bodies. Moreover there were many facts, such as, for instance, the warmth of the sea after a few days of stormy weather, which had long been commonly attributed to fluid friction. Nevertheless the scientific world, preoccupied with the hypothesis that heat is a substance,■ and following the deductions drawn by Pictet■ from experiments not sufficiently delicate, have almost unanimously denied the possibility of generating heat in that way. The first mention, so far as I am aware, of experiments in which the evolution of heat from fluid friction is asserted, was in 1842 by M. Mayer,‖ who states that he has raised the temperature of water from 12°C to 13°C, by agitating it, without however indicating the quantity of force■ employed, or the precautions taken to secure a correct result. In 1843 I announced the fact that "heat is evolved by the passage of water through narrow tubes,"* and that each degree of heat per lb of water required for its evolution in this way a mechanical force represented by 770 foot-pounds. Subsequently in 1845,† and 1847,** I employed a paddle wheel to produce the fluid friction, and obtained the equivalents 781ʹ5, 782ʹ1 and 787ʹ6, respectively, from the agitation of water, sperm oil, and mercury. Results so closely coinciding

An imponderable fluid.

Marc Pictet (1752-1825).

Here, read *force* as *work.*

---

‡ *Ibid.,* vol. xxiii, p. 441.
§ *Ibid.,* vol. xxvi, pp. 375, 379.
¶ *Comptes Rendus,* tome 25, p. 421.
‖ *Annalen of Waehler and Liebig,* May 1842.
* *Phil. Mag.,* vol. xxiii, p. 442.
† *Ibid.,* vol. xxvii, p. 205.
** *Ibid.,* vol. xxxi, p. 173, and *Comptes Rendus,* tome xxv, p. 309.

with one another, and with those previously derived from experiments with elastic fluids and the electromagnetic machine, left no doubt on my mind as to the existence of an equivalent relation between force and heat; but still it appeared of the highest importance to obtain that relation with still greater accuracy. This I have attempted in the present paper.

Description of Apparatus

The thermometers employed had their tubes calibrated and graduated according to the method first indicated by M. Regnault.■ Two of them, which I shall designate by *A* and *B*, were constructed by Mr. Dancer of Manchester; the third, designated by *C*, was made by M. Fastre of Paris. The graduation of these instruments was so correct, that when compared together their indications coincided to about $\frac{1}{100}$ of a degree Fahr.■ I also possessed another exact instrument made by Mr. Dancer, the scale of which embraced both the freezing and boiling points. The latter point in this standard thermometer was obtained, in the usual manner, by immersing the bulb and stem in the steam arising from a considerable quantity of pure water in rapid ebullition.■ During the trial the barometer stood at 29.94 inches, and the temperature of the air was 50°; so that the observed point required very little correction to reduce it to 0.760 metre and 0°C, the pressure used in France, and I believe the Continent generally, for determining the boiling point, and which has been employed by me on account of the number of accurate thermometrical researches which have been constructed on that basis.* The values of the scales of thermometers *A* and *B* were ascertained by plunging them along with the standard in large volumes of water kept constantly at various temperatures. The value of the scale of thermometer *C* was determined by comparison with *A*. It was thus found that the number of divisions corresponding to 1° Fahr in the thermometers *A, B,* and *C* were 12.951, 9.829 and 11.647, respectively. And since constant practice had enabled me to read off with the naked eye to $\frac{1}{20}$th of a division, it followed that $\frac{1}{200}$th of a degree Fahr was an appreciable temperature.

Henri Victor Regnault (1810-1878) employed a constant-volume gas thermometer as the standard.

Fahrenheit.

Boiling water.

---

* A barometrical pressure of 30 inches of mercury at 60° is very generally employed in this country, and fortunately agrees almost exactly with the Continental standard. In the "Report of the Committee appointed by the Royal Society to consider the best method of adjusting the Fixed Points of Thermometers," *Phil. Trans.*, abridged, vol. xiv, p. 258, the barometrical pressure 29.8 is recommended, but the temperature is not named—a remarkable omission in a work so exact in other respects.

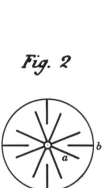

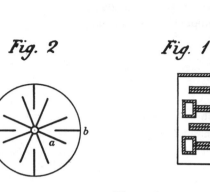

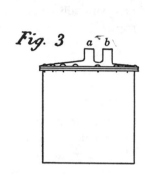

Fig. 2

Fig. 1

Fig. 3

The figures are drawn to scale: 1 inch = 1 foot.

Figure 1 represents a vertical, and Figure 2 a horizontal plan of the apparatus■ employed for producing the friction of water, consisting of a brass paddle wheel furnished with eight sets of revolving arms, *a, a,* & *c,* working between four sets of stationary vanes, *b, b,* & *c,* affixed to a framework also in sheet brass. The brass axis of the paddle wheel worked freely, but without shaking, on its bearings at *c, c,* and at *d* was divided into two parts by a piece of boxwood intervening, so as to prevent the conduction of heat in that direction.

Figure 3 represents the copper vessel into which the revolving apparatus was firmly fitted: it had a copper lid, the flange of which, furnished with a very thin washer of leather saturated with white lead, could be screwed perfectly watertight to the flange of the copper vessel. In the lid there were two necks, *a, b,* the former for the axis to revolve in without touching, the latter for the insertion of the thermometer.

■ ....

Omitted are several diagrams (and descriptions) pertaining to apparatus designed for similar experiments on mercury and on solid materials.

Figure 4 is a perspective view of the machinery employed to set the frictional apparatus just described in motion; *a a* are wooden pulleys, 1 foot in diameter and 2 inches thick, having wooden rollers, *bb, bb,* 2 inches in diameter, and steel axles, *cc, cc,* one quarter of an inch in diameter. The pulleys were turned perfectly true and equal to one another. Their axles were supported by brass friction wheels *dddd, dddd,* the steel axles of which worked in holes drilled into brass plates attached to a very strong wooden framework firmly fixed into the walls of the apartment.*

* This was a spacious cellar, which had the advantage of possessing an uniformity of temperature far superior to that of any other laboratory I could have used.

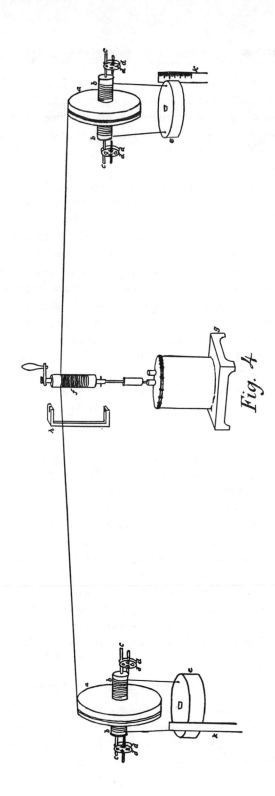

Fig. 4.

The leaden weights *e, e,* which in some of the ensuing experiments weighed about 29 lbs and in others about 10 lbs a piece, were suspended by string from the rollers *bb, bb;* and fine twine attached to the pulleys *aa* connected them with the central roller *f,* which, by means of a pin, could with facility be attached to, or removed from, the axis of the frictional apparatus.

The wooden stool *g,* upon which the frictional apparatus stood, was perforated by a number of transverse slits, so cut out that only a very few points of wood came in contact with the metal, whilst the air had free access to almost every part of it.■ In this way the conduction of heat to the substance of the stool was avoided.

A large wooden screen (not represented in the figure) completely obviated the effects of radiant heat from the person of the experimenter.

The method of experimenting was simply as follows: The temperature of the frictional apparatus having been ascertained and the weights wound up with the assistance of the stand *h,* the roller was refixed to the axis. The precise height of the weights above the ground having then been determined by means of the graduated slips of wood, *k, k,* the roller was set at liberty and allowed to revolve until the weights reached the flagged floor of the laboratory, after accomplishing a fall of about 63 inches. The roller was then removed to the stand, the weights wound up again, and the friction renewed. After this had been repeated twenty times, the experiment was concluded with another observation of the temperature of the apparatus. The mean temperature of the laboratory was determined by observations made at the commencement, middle and termination of each experiment.

Previously to, or immediately after each of the experiments, I made trial of the effect of radiation and conduction of heat to or from the atmosphere, in depressing or raising the temperature of the frictional apparatus. In these trials, the position of the apparatus, the quantity of water contained by it, the time occupied, the method of observing the thermometers, the position of the experimenter, in short everything, with the exception of the apparatus being at rest, was the same as in the experiments in which the effect of friction was observed.

First Series of Experiments

Friction of Water. Weight of the leaden weights along with as much of the string in connexion with them as served to increase the pressure, 203066 grs.■ and 203086 grs. Velocity of the weights in descending, 2.42 inches per second. Time occupied by each experiment, 35 minutes.

Probably the conduction of heat to the wooden stool would have been substantially less than that carried off by convection in the air. The heat loss was small primarily because the temperature difference was small.

grs. = grains

Thermometer employed for ascertaining the temperature of the water, A.
Thermometer for registering the temperature of the air, B.

No. of experiment and cause of change of temperature.	Total fall of weights in inches.	Mean temperature of air.	Difference between mean of Columns 5 and 6 and Column 3.	Temperature of apparatus.		Gain or loss of heat during experiment.
				Commencement of experiment.	Termination of experiment.	
1 Friction....	1256·96	57·698°	2·252° −	55·118°	55·774°	0·656° gain
1 Radiation...	0	57·868	2·040 −	55·774	55·882	0·108 gain
2 Friction....	1255·16	58·085	1·875 −	55·882	56·539	0·657 gain
2 Radiation...	0	58·370	1·789 −	56·539	56·624	0·085 gain
3 Friction.......	1253·66	60·788	1·596 −	58·870	59·515	0·645 gain
3 Radiation...	0	60·926	1·373 −	59·715	59·592	0·077 gain
4 Friction....	1252·74	61·001	1·110 −	59·592	60·191	0·599 gain
4 Radiation...	0	60·890	0·684 −	60·191	60·222	0·031 gain
5 Friction.......	1251·81	60·940°	0·431° −	60·222°	60·797°	0·575° gain
5 Radiation...	0	61·035	0·237 −	60·797	60·799	0·002 gain
6 Radiation...	0	59·675	0·125 +	59·805	59·795	0·010 loss
6 Friction....	1254·71	59·919	0·157 +	59·795	60·357	0·562 gain
7 Radiation...	0	59·888	0·209 −	59·677	59·681	0·004 gain
7 Friction....	1254·02	60·076	0·111 −	59·681	60·249	0·568 gain
8 Radiation...	0	59·240	0·609 +	58·871	58·828	0·043 loss
8 Friction....	1251·22	58·237	0·842 +	58·828	59·330	0·502 gain
9 Friction....	1253·92	55·328	0·070 +	55·118	55·678	0·560 gain
9 Radiation...	0	55·528	0·148 +	55·678	55·674	0·004 loss
10 Radiation...	0	54·941	0·324 −	54·614	54·620	0·006 gain
10 Friction....	1257·96	54·985	0·085 −	54·620	55·180	0·560 gain
11 Radiation...	0	55·111	0·069 +	55·180	55·180	0·000
11 Friction....	1258·59	55·229	0·227 +	55·180	55·733	0·553 gain
12 Friction....	1258·71	55·433	0·238 +	55·388	55·954	0·566 gain
12 Radiation...	0	55·687	0·265 +	55·954	55·950	0·004 loss
13 Friction....	1257·91	55·677	0·542 +	55·950	56·488	0·538 gain
13 Radiation...	0	55·674	0·800 +	56·488	56·461	0·027 loss
14 Radiation...	0	55·579	0·583 −	54·987	55·006	0·019 gain
14 Friction....	1259·69	55·864	0·568 −	55·006	55·587	0·581 gain
15 Radiation...	0	56·047	0·448 −	55·587	55·612	0·025 gain
15 Friction....	1259·89	56·182	0·279 −	55·612	56·195	0·583 gain
1	2	3	4	5	6	7

■..................................

A part of the
table is omitted.

Mean Friction	1260.248	....	0.305075° −	....	....	0.575250° gain
Mean Radiation	0	....	0.322950 −	....	....	0.012975 gain
1	2	3	4	5	6	7

From the various experiments in the above table in which the effect
of radiation was observed, it may be readily gathered that the effect of
the temperature of the surrounding air upon the apparatus was, for each

degree of difference between the mean temperature of the air and that of the apparatus, 0.04654°. Therefore, since the excess of the temperature of the atmosphere over that of the apparatus was 0.32295° in the mean of the radiation experiments, but only 0.305075° in the mean of the friction experiments, it follows that 0.000832° must be added to the difference between 0.57525° and 0.012975°, and the result, 0.563107°, will be the proximate heating effect of the friction.■ But to this quantity a small correction must be applied on account of the mean of the temperatures of the apparatus at the commencement and termination of each friction experiment having been taken for the true mean temperature, which was not strictly the case, owing to the somewhat less rapid increase of temperature towards the termination of the experiment when the water had become warmer. The mean temperature of the apparatus in the friction experiments ought therefore to be estimated 0.002184° higher, which will diminish the heating effect of the atmosphere by 0.000102°. This, added to 0.563107°, gives 0.563209° as the true mean increase of temperature due to the friction of water.*

In order to ascertain the absolute quantity of heat evolved, it was necessary to find the capacity for heat of the copper vessel and brass paddle wheel. That of the former was easily deduced from the specific heat of copper according to M. Regnault. Thus, capacity of 25541 grs.† of copper × 0.09515 = capacity of 2430.2 grs. of water.■ A series of seven very careful experiments with the brass paddle wheel gave me 1783 grs. of water as its capacity, after making all the requisite corrections for the heat occasioned by the contact of the water with the surface of the metal, &c. But on account of the magnitude of these corrections, amounting to one thirtieth of the whole capacity, I prefer to avail myself to M. Regnault's law, viz., *that the capacity in metallic alloys is equal to the sum of the capacities of their constituent metals.*** Analysis

----

* This increase of temperature was, it is necessary to observe, a mixed quantity, depending partly upon the friction of the water, and partly upon the friction of the vertical axis of the apparatus upon its pivot and bearing, cc, Figure 1. The latter source of heat was however only equal to about 1/80th of the former. Similarly also, in the experiments on the friction of solids hereafter detailed, the cast-iron discs revolving in mercury rendered it impossible to avoid a very small degree of friction among the particles of that fluid. But since it was found that the quantity of heat evolved was the same, for the same quantity of force expended, in both cases, i.e., whether a minute quantity of heat arising from friction of solids was mixed with the heat arising from the friction of a fluid, or whether, on the other hand, a minute quantity of heat arising from the friction of a fluid was mingled with the heat developed by the friction of solids, I thought there could be no inpropriety in considering the heat as if developed from a simple source—in the one case entirely from the friction of a fluid, and in the other entirely from the friction of a solid body.■

† The washer, weighing only 38 grs., was reckoned as copper in this estimate.

** *Ann. de Ch.*, 1841, t.i.

Margin notes:

It seems very unlikely that the extreme precision indicated by these temperature differences were warranted by the conditions of the experiment.

The specific heat of copper is 0.093 cal/gm—°C. Regnault was in error here (and in the case of zinc, below) by some 2 or 3 percent. However, because of the large mass of water compared to the metal, the effect of this error on the final result would have been negligible.

Joule was, of course, correct in this, provided only that the heat evolved went into raising the temperature of the water or other fluid employed.

of a part of the wheel proved it to consist of a very pure brass containing 3933 grs. of zinc to 14968 grs. of copper. Hence

Cap. 14968 grs. copper × 0.09515 = cap. 1424.2 grs. water.
Cap. 3933 grs. zinc × 0.09555■ = cap. 375.8 grs. water.
    Total cap brass wheel = cap. 1800 grs. water.

The specific heat of zinc is 0.092 cal/gm—°C.

The capacity of a brass stopper which was placed in the neck *b* (Fig. 3) for the purpose of preventing the contact of air with the water as much as possible, was equal to that of 10.3 grs. of water: the capacity of the thermometer had not to be estimated, because it was always brought to the expected temperature before immersion.■ The entire capacity of the apparatus was therefore as follows:

The thermal capacity of the thermometer probably was small enough to be neglected by this technique, provided the expected temperature was estimated fairly closely.

Water 93229.7
Copper as water................................ 2430.2
Brass as water.................................... 1810.3
    Total 97470.2

So that the total quantity of heat evolved was 0.563209° in 97470.2 grs. of water, or, in other words, 1° Fahr. in 7.842299 lbs of water.

The estimate of the force applied in generating this heat may be made as follows:

The weights amounted to 406152 grs., from which must be subtracted the friction arising from the pulleys and the rigidity of the string; which was found by connecting the two pulleys with twine passing round a roller of equal diameter to that employed in the experiments. Under these circumstances, the weight required to be added to one of the leaden weights in order to maintain them in equable■ motion was found to be 2955 grs. The same result, in the opposite direction, was obtained by adding 3055 grs. to the other leaden weight. Deducting 168 grs., the friction of the roller on its pivots, from 3005, the mean of the above numbers, we have 2837 grs. as the amount of friction in the experiments, which, subtracted from the leaden weights, leaves 403315 grs. as the actual pressure applied.

equable: uniform

The velocity with which the leaden weights came to the ground, viz., 2.42 inches per second, is equivalent to an altitude of 0.0076 inch. This, multiplied by 20, the number of times the weights were wound up in each experiment, produces 0.152 inch, which, subtracted from 1260.248, leaves 1260.096 as the corrected mean height from which the weights fell.■

Here again, it should be evident that Joule tended to assign greater precision to his measurements than were warranted by the experiment.

This fall, accompanied by the above-mentioned pressure, represents a force equivalent to 6050.186 lbs through one foot; and 0.8464 × 20 = 16.928 lbs added to it, for the force developed by the elasticity of the string after the weights had touched the ground, gives 6067.144 foot-pounds as the mean corrected force.▪

Hence $\frac{6067.114}{7.842299} = 773.64$ foot-pounds, will be the force which, according to the above experiments on the friction of water, is equivalent to 1° Fahr. in a lb of water.

▪..

The following table contains a summary of the equivalents derived from the experiments above detailed. In its fourth column I have supplied the results with the correction necessary to reduce them to a vacuum.

No. of series	Material employed	Equivalent in air	Equivalent in vacuo	Mean
1	Water.................	773.640	772.692	772.692
2	Mercury...............	773.762	772.814 ⎫	
3	Mercury...............	776.303	775.352 ⎬	774.083
4	Cast iron..............	776.997	776.045 ⎫	
5	Cast iron..............	774.880	773.930 ⎭	774.987

It is highly probable that the equivalent from cast iron was somewhat increased by the abrasion of particles of the metal during friction, which could not occur without the absorption of a certain quantity of force in overcoming the attraction of cohesion. But since the quantity abraded was not considerable enough to be weighed after the experiments were completed, the error from this source cannot be of much moment. I consider that 772.692, the equivalent derived from the friction of water, is the most correct, both on account of the number of experiments tried, and the great capacity of the apparatus for heat. And since, even in the friction of fluids, it was impossible entirely to avoid vibration and the production of a slight sound, it is probable that the above number is slightly in excess.▪ I will therefore conclude by considering it as demonstrated by the experiments contained in this paper,

1st. *That the quantity of heat produced by the friction of bodies, whether solid or liquid, is always proportional to the quantity of force expended.* And,

2d. *That the quantity of heat capable of increasing the temperature of a pound of water (weighed in vacuo, and taken at between 55° and 60°) by 1° Fahr., requires for its evolution the expenditure of a mechanical force represented by the fall of 772 lbs through the space of one foot.*

SUPPLEMENTARY READING

Conant, J. B. (ed.), *The Early Development of the Concepts of Temperature and Heat,* Harvard Case Histories in Experimental Science (Cambridge, Mass.: Harvard University, 1950), pp. 119ff.

Crew, H., "Portraits of Famous Physicists," *Scripta Mathematica,* Pictorial Mathematics, Portfolio No. 4, 1942.

Crowther, J. G., *Men of Science* (New York: Norton, 1936).

Dampier, W. C., *A History of Science* (New York: Cambridge University, 1946), pp. 226ff.

Lenard, P., *Great Men of Science* (New York: British Book Centre, 1934), pp. 286ff.

Magie, W. F., *A Source Book in Physics* (New York: McGraw-Hill, 1935), pp. 203ff.

Wolf, A., *A History of Science, Technology, and Philosophy in the 18th Century,* 2d ed. (London: Allen and Unwin, 1952).

13

Heinrich Hertz
1857 - 1894

Electromagnetic Waves

THE BRANCH of physics known as electromagnetism has a most remarkable history. It was born and brought to full maturity in the relatively short span of seven decades. There were no doubt some who witnessed its entire evolution. It began with Oersted's discovery in 1820 of the connection between electricity and magnetism; Faraday, Lenz, and Ampère contributed greatly to its growth during the 1830's; Clerk Maxwell (1831–1879), a Scot, put these ideas into mathematical form with his theory of electromagnetic waves in 1870, and seventeen years later Hertz found the waves predicted by Maxwell. All else, wireless telegraphy, radio, television, in a sense were anticlimactic—the *practical* results of basic discovery.

Maxwell's contributions to this field stagger the imagination.[1] Just as Newton's laws form the starting point of every argument in dynamics so do Maxwell's equations serve this function in electromagnetic theory. They summarize the fundamental relations between electricity and magnetism—and electromagnetic radiation, doing so in the most elegant mathematical fashion. In a few equations Maxwell said everything: Coulomb's laws of electric and magnetic force, Oersted's discovery, Ampère's rule, Faraday's law of induction, even Ohm's law[2]; nothing was omitted. They contained as well Maxwell's suggestion of the existence of *electric waves* propagated through dielectric media, based largely upon Faraday's *lines of force*. It was this suggestion that Hertz put to experimental test, with the result that he discovered electromagnetic radiation and thereby confirmed Maxwell's theory.

---

[1] A chapter on Maxwell will be found in the Appendix, page 283.
[2] George Simon Ohm (1789–1854) discovered the relationship between the current flowing in a conductor and the potential difference across it, now known by his name.

184

Maxwell had predicted that radiation should proceed from electric oscillations and be propagated through free space with the speed of light. Hertz started with Maxwell's ideas. He was a most ingenious experimenter who was forced to devise entirely new techniques for the detection of Maxwell's electric waves. From his success grew the communications industry, and perhaps more important, a clearer picture of the wave nature of light.

Heinrich Rudolf Hertz was born at Hamburg, Germany, on February 22, 1857, the son of a lawyer and senator of the city. Europe was then passing through a period of intense nationalism. Louis Napoleon Bonaparte had so distorted the office of president of France as to make himself virtual dictator, assuming the title of Napoleon III, Emperor of France. Austria and Hungary, following unsuccessful attempts at liberal government, were heading toward the dual monarchy *(Ausgleich)* that was to survive only until the first World War. And in Prussia Bismarck had started on his master plan to form a united nation of German states. When Hertz was still a young boy the ambitions of Bismarck and Napoleon clashed, the Franco-Prussian War erupted, and the German empire was established.

In this atmosphere Hertz matured. He inherited from his father a love for the humanities and from his father he also learned many languages. A grandfather, who studied natural science as a hobby, gave Hertz, when he was a boy, some apparatus which evidently stimulated his interest in science. He attended private school until he was fifteen, when he left to continue his studies with a tutor because the school emphasized practical subjects rather than the classics. During this period he conducted simple experiments in chemistry and physics in a small laboratory which he had fitted out in his home. At seventeen he entered the *Gelehrtenschule* of the Johanneums in Hamburg where he completed his classical studies and received his diploma in 1875. Thinking that he was better suited for the study of engineering than pure science, Hertz studied for a "practical year" with a firm of engineers before entering the Technical High School of Dresden. He remained at Dresden for six months, after which he was called into military service for a period of one year. By this time, he was convinced that his interests were in the field of pure science and he went to Munich to continue his studies in mathematics and mechanics. In 1878 Hertz, then twenty-one, transferred to the University of Berlin to study under von Helmholtz (1821–1894) and Gustav Kirchhoff (1824–1887). There he won a prize for his first independent research in physics, a paper on the inertia of electricity based upon a problem set by the physics faculty. He received his doctorate *magna cum laude* in 1880, for a dissertation on *Induction in Rotating Spheres,* and that same year was appointed a demonstrator in physics.

In 1883 Hertz left Berlin to become a lecturer *(privat-docent)* in theoretical physics at the University of Kiel, where he devoted himself to the study of electromagnetism, including Maxwell's recent work. Two years later he became professor of experimental physics at the Technische Hochschule at Karlsruhe, where he conducted his famous experiments on *electric waves.* The Prussian Academy of Sciences had offered a prize for the experimental confirmation of the relation between electromagnetic actions and the polarization of a dielectric. Hertz did not enter the competition, since he knew of no way to accomplish the assignment, but it was this experiment that provided the stimulus for his later accomplishments. He discovered the propagation of electromagnetic effects through space. He measured the speed of propagation and the wavelength of the waves, and showed their transverse nature through experiments on reflection, refraction, and polarization. In short, he completely established the electromagnetic character of light in accordance with Maxwell's theory.

He completed his work in 1889, and in that same year was appointed professor of physics at the University of Bonn. There, with his colleague Philipp Lenard (1862–1947), he studied gas discharges. Finding that cathode rays could pass through thin sheets of metal he concluded, in agreement with the German school of thought, that the rays had a wave character. Several years later J. J. Thomson was to demolish this view with his identification of the electron (see page 216).

Hertz worked in several branches of physics but in none did he distinguish himself to such a great degree as in his research on *electric waves.* Among his many honors was the coveted Rumford Medal of the Royal Society; were it not for his untimely death in 1894, at the age of 37, he most surely would have won a Nobel prize. (The first Nobel prize was awarded in 1901.)

The extract that follows is taken from his paper *On Electric Radiation.* It was published originally in *Sitzungsber. d. Berlin Akad. d. Wiss.,* December 13, 1888, and in the *Annalen der Physik,* vol. 36 (1889), page 769. In 1893 Hertz collected his papers in the form of a book which was translated by Jones[3] in 1900. The Jones translation is given here.

---

[3] D. E. Jones (trans.), *Electric Waves by Heinrich Hertz* (London: Macmillan, 1900).

Hertz's Experiment

■As soon as I had succeeded in proving that the action of an electric oscillation spreads out as a wave into space, I planned experiments with the object of concentrating this action and making it perceptible at greater distances by putting the primary conductor in the focal line of a large concave parabolic mirror.■ These experiments did not lead to the desired result, and I felt certain that the want of success was a necessary consequence of the disproportion between the length (4–5 meters) of the waves used and the dimensions which I was able, under the most favorable circumstances, to give to the mirror.■ Recently I have observed that the experiments which I have described can be carried out quite well with oscillations of more than ten times the frequency, and with waves less than one tenth the length of those which were first discovered. I have, therefore, returned to the use of concave mirrors, and have obtained better results than I had ventured to hope for. I have succeeded in producing distinct rays of electric force, and in carrying out with them the elementary experiments which are commonly performed with light and radiant heat. The following is an account of these experiments.

> In earlier papers Hertz had discussed various aspects of oscillating circuits and the application of Maxwell's theory to the production of *electric waves.*
>
> The *primary* conductor was the radiating antenna.
>
> For appreciable reflection, the diameter of the mirror should be at least several wavelengths.

THE APPARATUS

The short waves were excited by the same method which we used for producing the longer waves. The primary conductor used may be most simply described as follows: Imagine a cylindrical brass body,* 3 cm in diameter and 26 cm long, interrupted midway along its length by a spark gap whose poles on either side are formed by spheres of 2 cm radius. The length of the conductor is approximately equal to the half wavelength of the corresponding oscillation in straight wires; from this we are at once able to estimate approximately the period of oscillation. It is essential that the pole surfaces of the spark gap should be frequently repolished, and also that during the experiments they should be carefully protected from illumination by simultaneous side discharges; otherwise the oscillations are not excited.■ Whether the spark gap is in a satisfactory state can always be recognized by the appearance and sound of the sparks. The discharge is led to the two halves of the conductor

> Because of photoelectric effects.

---

* See Figures 1 and 2 and the description of them at the end of this paper.

> Figures 1 and 2, page 195.

Heinrich Ruhmkorff (1803-1877), a manufacturer of physical apparatus.

The sparks were produced in the high-voltage secondary circuit of the induction coil. The accumulators, or batteries, provided the current for the primary circuit of the coil.

Thus the *detector* used by Hertz was simply a small (resonant) circuit in which the induced currents were made evident by the minute sparks produced.

by means of two gutta-percha-covered wires which are connected near the spark gap on either side. I no longer made use of the large Ruhmkorff,■ but found it better to use a small induction-coil by Keiser and Schmidt; the longest sparks, between points, given by this were 4.5 cm long.■ It was supplied with current from three accumulators, and gave sparks 1–2 cm long between the spherical knobs of the primary conductor. For the purpose of the experiments the spark gap was reduced to 3 mm.

Here, again, the small sparks induced in a secondary conductor were the means used for detecting the electric forces in space. As before, I used partly a circle which could be rotated within itself and which had about the same period of oscillation as the primary conductor. It was made of copper wire 1 mm thick, and had in the present instance a diameter of only 7.5 cm. One end of the wire carried a polished brass sphere a few millimeters in diameter; the other end was pointed and could be brought up, by means of a fine screw insulated from the wire, to within an exceedingly short distance from the brass sphere.■ As will be readily understood, we have here to deal only with minute sparks of a few hundredths of a millimeter in length; and after a little practice one judges more according to the brilliancy than the length of the sparks.

The circular conductor gives only a differential effect, and is not adapted for use in the focal line of a concave mirror. Most of the work was therefore done with another conductor arranged as follows: Two straight pieces of wire, each 50 cm long and 5 mm in diameter, were adjusted in a straight line so that their near ends were 5 cm apart. From these ends two wires, 15 cm long and 1 mm in diameter, were carried parallel to one another and perpendicular to the wires first mentioned to a spark gap arranged just as in the circular conductor. In this conductor the resonance action was given up, and indeed it only comes slightly into play in this case. It would have been simpler to put the spark gap directly in the middle of the straight wire; but the observer could not then have handled and observed the spark gap in the focus of the mirror without obstructing the aperture. For this reason the arrangement above described was chosen in preference to the other which would in itself have been more advantageous.

THE PRODUCTION OF THE RAY

If the primary oscillator is now set up in a fairly large free space, one can, with the aid of the circular conductor, detect in its neighborhood on a smaller scale all those phenomena which I have already observed and described as occurring in the neighborhood of a larger oscillation. The greatest distance at which sparks could be perceived in the secondary conductor was 1.5 meters, or, when the primary spark gap was in very good order, as much as 2 meters. When a plane reflecting plate is set up at a suitable distance on one side of the primary oscillator, and parallel to it, the action on the opposite side is strengthened.■ To be more precise: If the distance chosen is either very small, or somewhat greater than 30 cm, the plate weakens the effect; it strengthens the effect greatly at distances of 8–15 cm, slightly at a distance of 45 cm, and exerts no influence at greater distances. We have drawn attention to this phenomenon in an earlier paper, and we conclude from it that the wave in air corresponding to the primary oscillation has a half wavelength of about 30 cm. We may expect to find a still further reinforcement if we replace the plane surface by a concave mirror having the form of a parabolic cylinder, in the focal line of which the axis of the primary oscillation lies. The focal length of the mirror should be chosen as small as possible, if it is properly to concentrate the action. But if the direct wave is not to annul immediately the action of the reflected wave,■ the focal length must not be much smaller than a quarter wavelength. I therefore fixed on 12½ cm as the focal length, and constructed the mirror by bending a zinc sheet 2 meters long, 2 meters broad, and ½ mm thick into the desired shape over a wooden frame of the exact curvature. The height of the mirror was thus 2 meters, the breadth of its aperture 1.2 meters, and its depth 0.7 meter. The primary oscillator was fixed in the middle of the focal line. The wires which conducted the discharge were led through the mirror; the induction coil and the cells were accordingly placed behind the mirror so as to be out of the way. If we now investigate the neighborhood of the oscillator with our conductors, we find that there is no action behind the mirror or at either side of it; but in the direction of the optical axis of the mirror the sparks can be perceived up to a distance of 5–6 meters.■ When a plane conducting surface was set up so as to oppose the advancing waves at right angles, the sparks could be detected in its neighborhood at even greater distances—up to about 9–10 meters. The waves reflected from the conducting surface reinforce the advancing waves at certain points.■ At other points again the two sets of waves weaken one another. In front of the plane wall one can recognize with the rectilinear conductor very distinct maxima and minima, and with the circular conductor the characteristic interference phenomena of stationary waves which I have described in an earlier paper. I was able to distinguish four nodal points,■ which were situated at the wall and at 33, 65, and 98 cm distance from it. We thus get 33

The opposite side of the oscillator, of course, where there would now be both the direct radiation and the radiation reflected from the screen.

By destructive interference.

The mirror served to concentrate or focus the radiation.

Forming *standing waves* very similar to the case of sound.

Minima.

c = f λ, where f is the frequency, λ the wavelength, and c the velocity, 3 × 10¹⁰ cm per sec in free space.

On wires the propagation speed depends upon the inductance and capacitance per unit length of wire.

Thus the second mirror served to collect and focus more of the radiation upon the detector, in effect increasing its sensitivity.

cm as a closer approximation to the half wavelength of the waves used, and 1.1 thousand-millionth of a second as their period of oscillation, assuming that they travel with the velocity of light.■ In wires the oscillation gave a wavelength of 29 cm. Hence it appears that these short waves also have a somewhat lower velocity in wires than in air; but the ratio of the two velocities comes very near to the theoretical value— unity—and does not differ from it so much as appeared to be probable from our experiments on longer waves.■ This remarkable phenomenon still needs elucidation. Inasmuch as the phenomena are only exhibited in the neighborhood of the optic axis of the mirror, we may speak of the result produced as an electric ray proceeding from the concave mirror.

I now constructed a second mirror, exactly similar to the first, and attached the rectilinear secondary conductor to it in such a way that the two wires of 50 cm length lay in the focal line, and the two wires connected to the spark gap passed directly through the wall of the mirror without touching it.■ The spark gap was thus situated directly behind the mirror, and the observer could adjust and examine it without obstructing the course of the waves. I expected to find that, on intercepting the ray with this apparatus, I should be able to observe it at even greater distances; and the event proved that I was not mistaken. In the rooms at my disposal I could now perceive the sparks from one end to the other. The greatest distance to which I was able, by availing myself of a doorway, to follow the ray was 16 meters; but according to the results of the reflection experiments (to be presently described), there can be no doubt that sparks could be obtained at any rate up to 20 meters in open spaces. For the remaining experiments such great distances are not necessary, and it is convenient that the sparking in the secondary conductor should not be too feeble; for most of the experiments a distance of 6–10 meters is most suitable. We shall not describe the simple phenomena which can be exhibited with the ray without difficulty. When the contrary is not expressly stated, it is to be assumed that the focal lines of both mirrors are vertical.

RECTILINEAR PROPAGATION

If a screen of sheet zinc 2 meters high and 1 meter broad is placed on the straight line joining both mirrors, and at right angles to the direction of the ray, the secondary sparks disappear completely. An equally complete shadow is thrown by a screen of tin foil or gold paper. If an assistant walks across the path of the ray, the secondary spark gap becomes dark as soon as he intercepts the ray, and again lights up when he leaves the path clear. Insulators do not stop the ray—it passes right through a wooden partition or door; and it is not without astonishment that one sees the sparks appear inside a closed room. If two conducting screens, 2 meters high and 1 meter broad, are set up symmetrically on the right and left of the ray, and perpendicular to it, they do not interfere at all with the secondary spark so long as the width of the opening between them is not less than the aperture of the mirrors, viz., 1.2 meters. If the opening is made narrower the sparks become weaker, and disappear when the width of the opening is reduced below 0.5 meter. The sparks also disappear if the opening is left with a breadth of 1.2 meters, but is shifted to one side of the straight line joining the mirrors.■ If the optical axis of the mirror containing the oscillator is rotated to the right or left about 10° out of the proper position, the secondary sparks become weak, and a rotation through 15° causes them to disappear.

There is no sharp geometrical limit to either the ray or the shadows; it is easy to produce phenomena corresponding to diffraction. As yet, however, I have not succeeded in observing maxima and minima at the edge of the shadows.■

It will be noted that the propagation was essentially along optical paths, which is characteristic of the short wavelengths employed.

A more sensitive means of detection would have been required.

POLARIZATION

From the mode in which our ray was produced we can have no doubt whatever that it consists of transverse vibrations and is plane polarized in the optical sense. We can also prove by experiment that this is the case. If the receiving mirror be rotated about the ray as axis until its focal line, and therefore the secondary conductor also, lies in a horizontal plane,■ the secondary sparks become more and more feeble, and when the two focal lines are at right angles, no sparks whatever are obtained even if the mirrors are moved close up to one another. The two mirrors behave like the polarizer and analyzer of a polarization apparatus.

While the radiating antenna lies in a vertical plane.

I next had made an octagonal frame, 2 meters high and 2 meters broad; across this were stretched copper wires 1 mm thick, the wires being parallel to each other and 3 cm apart. If the two mirrors were now set up with their focal lines parallel, and the wire screen was interposed perpendicularly to the ray and so that the direction of the wires was perpendicular to the direction of the focal lines, the screen practically did not interfere at all with the secondary sparks. But if the screen was set up in such a way that its wires were parallel to the focal lines, it stopped the ray completely. With regard, then, to transmitted energy the screen behaves towards our ray just as a tourmaline plate behaves towards a plane-polarized ray of light.■ The receiving mirror

Prior to the development of Polaroid screens, tourmaline was commonly used as a polarizer. It is a doubly refracting crystal, that is, dichroic—it absorbs one of the polarized components strongly, while transmitting the other.

was now placed once more so that its focal line was horizontal; under these circumstances, as already mentioned, no sparks appeared. Nor were any sparks produced when the screen was interposed in the path of the ray, so long as the wires in the screen were either horizontal or vertical. But if the frame was set up in such a position that the wires were inclined at 45° to the horizontal on either side, then the interposition of the screen immediately produced sparks in the secondary spark gap. Clearly the screen resolves the advancing oscillation into two components and transmits only that component which is perpendicular to the direction of its wires. This component is inclined at 45° to the focal line of the second mirror, and may thus, after being again resolved by the mirror, act upon the secondary conductor. The phenomenon is exactly analogous to the brightening of the dark field of two crossed Nicols by the interposition of a crystalline plate in a suitable position.■

With regard to the polarization it may be further observed that, with the means employed in the present investigation, we are only able to recognize the electric force. When the primary oscillator is in a vertical position the oscillations of this force undoubtedly take place in the vertical plane through the ray, and are absent in the horizontal plane. But the results of experiments with slowly alternating currents leave no room for doubt that the electric oscillations are accompanied by oscillations of magnetic force which take place in the horizontal plane through the ray and are zero in the vertical plane.■ Hence the polarization of the ray does not so much consist in the occurrence of oscillations in the vertical plane, but rather in the fact that the oscillations in the vertical plane are of an electrical nature, while those in the horizontal plane are of a magnetic nature. Obviously, then, the question, in which of the two planes the oscillation in our ray occurs, cannot be answered unless one specifies whether the question relates to the electric or the magnetic oscillation. It was Herr Kolaček* who first pointed out clearly that this consideration is the reason why an old optical dispute has never been decided.

Nicol prisms invented by William Nicol (1768-1851), were generally constructed of Iceland spar or calcite. They act as polarizers by reflecting one component out of the path of the beam. A crystalline plate interposed between the Nicols would rotate the plane of polarization.

According to Maxwell's theory, the radiation had both electric and magnetic properties; hence the term, *electromagnetic radiation.*

REFLECTION

We have already proved the reflection of the waves from conducting surfaces by the interference between the reflected and the advancing waves, and have also made use of the reflection in the construction of our concave mirrors. But now we are able to go further and to separate

* F. Kolaček, *Wied. Ann.,* 1888, 34, p. 676.

the two systems of waves from one another. I first placed both mirrors in a large room side by side, with their apertures facing in the same direction, and their axes converging to a point about 3 meters off. The spark gap of the receiving mirror naturally remained dark. I next set up a plane vertical wall made of thin sheet zinc, 2 meters high and 2 meters broad, at the point of intersection of the axes, and adjusted it so that it was equally inclined to both. I obtained a vigorous stream of sparks arising from the reflection of the ray by the wall. The sparking ceased as soon as the wall was rotated around a vertical axis through about 15° on either side of the correct position; from this it follows that the reflection is regular, not diffuse.■ When the wall was moved away from the mirrors, the axes of the latter being still kept converging towards the wall, the sparking diminished very slowly. I could still recognize sparks when the wall was 10 meters away from the mirrors, i.e., when the waves had to traverse a distance of 20 meters. This arrangement might be adopted with advantage for the purpose of comparing the rate of propagation through air with other and slower rates of propagation, e.g., through cables.

Such reflection, as we have seen, is typical of short-wave radiation. This would not be the case for wavelengths long compared with the reflector dimensions.

In order to produce reflection of the ray at angles of incidence greater than zero, I allowed the ray to pass parallel to the wall of the room in which there was a doorway. In the neighboring room to which this door led I set up the receiving mirror so that its optic axis passed centrally through the door and intersected the direction of the ray at right angles. If the plane conducting surface was now set up vertically at the point of intersection, and adjusted so as to make angles of 45° with the ray and also with the axis of the receiving mirror, there appeared in the secondary conductor a stream of sparks which was not interrupted by closing the door. When I turned the reflecting surface about 10° out of the correct position the sparks disappeared. Thus the reflection is regular, and the angles of incidence and reflection are equal. That the action proceeded from the source of disturbance to the plane mirror, and hence to the secondary conductor, could also be shown by placing shadow-giving screens at different points of this path. The secondary sparks then always ceased immediately; whereas no effect was produced when the screen was placed anywhere else in the room. With the aid of the circular secondary conductor it is possible to determine the position of the wave front in the ray; this was found to be at right angles to the ray before and after reflection, so that in the reflection it was turned through 90°.

The account of an experiment in which both mirrors were oriented with their focal lines horizontal and of another showing reflection from a grid of parallel wires are omitted.

■....................................

REFRACTION

In order to find out whether any refraction of the ray takes place in passing from air into another insulating medium, I had a large prism made of so-called hard pitch, a material like asphalt. The base was an isosceles triangle 1.2 meters in the side, and with a refracting angle of nearly 30°. The refracting edge was placed vertical, and the height of the whole prism was 1.5 meters. But since the prism weighed about 12 cwt,■ and would have been too heavy to move as a whole, it was built up of three pieces, each 0.5 meter high, placed one above the other. The material was cast in wooden boxes which were left around it, as they did not appear to interfere with its use. The prism was mounted on a support of such height that the middle of its refracting edge was at the same height as the primary and secondary spark gaps. When I was satisfied that refraction did take place, and had obtained some idea of its amount, I arranged the experiment in the following manner: The producing mirror was set up at a distance of 2.6 meters from the prism and facing one of the refracting surfaces, so that the axis of the beam was directed as nearly as possible towards the center of mass of the prism, and met the refracting surface at an angle of incidence of 25° (on the side of the normal towards the base). Near the refracting edge and also at the opposite side of the prism were placed two conducting screens which prevented the ray from passing by any other path than that through the prism. On the side of the emerging ray there was marked upon the floor a circle of 2.5 meters radius, having as its center the center of mass of the lower end of the prism. Along this the receiving mirror was now moved about, its aperture being always directed towards the center of the circle. No sparks were obtained when the mirror was placed in the direction of the incident ray produced; in this direction the prism threw a complete shadow. But sparks appeared when the mirror was moved towards the base of the prism, beginning when the angular deviation from the first position was about 11°. The sparking increased in intensity until the deviation amounted to about 22°, and then again decreased. The last sparks were observed with a deviation of about 34°. When the mirror was placed in a position of maximum effect, and then moved away from the prism along the radius of the circle, the sparks could be traced up to a distance of 5–6 meters. When an assistant stood either in front of the prism or behind it the sparking invariably ceased, which shows that the action reaches the secondary conductor through the prism and not in any other way.■ The experiments were repeated after placing both mirrors with their focal lines horizontal, but without alter-

cwt = hundred-weight, which is 100 pounds.

Refraction of the rays meant that they had a smaller velocity in the pitch than in the air.

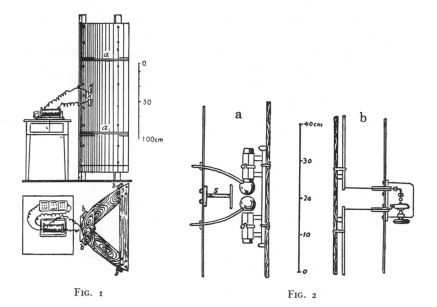

FIG. 1   FIG. 2

ing the position of the prism. This made no difference in the phenomena observed. A refracting angle of 30° and a deviation of 22° in the neighborhood of the minimum deviation corresponds to a refractive index of 1.69. The refractive index of pitchlike materials for light is given as being between 1.5 and 1.6. We must not attribute any importance to the magnitude or even the sense of this difference, seeing that our method was not an accurate one, and that the material used was impure.

We have applied the term rays of electric force to the phenomena which we have investigated. We may perhaps further designate them as rays of light of very great wavelength. The experiments described appear to me, at any rate, eminently adapted to remove any doubt as to the identity of light, radiant heat, and electromagnetic wave motion.■ I believe that from now on we shall have greater confidence in making use of the advantages which this identity enables us to derive both in the study of optics and of electricity.

Apart from interference effects, which Hertz was unable to observe with his apparatus, the rays exhibited all the properties usually associated with light.

Explanation of the Figures

In order to facilitate the repetition and extension of these experiments, I append in the accompanying Figures 1, 2a, and 2b, illustrations of the apparatus which I used, although these were constructed simply for the purpose of experimenting at the time and without any regard to durability. Figure 1 shows in plan and elevation (section) the producing mirror. It will be seen that the framework of it consists of two horizontal frames (*a*, *a*) of parabolic form, and four vertical supports (*b*, *b*) which are screwed to each of the frames so as to support and connect them. The sheet metal reflector is clamped between the frames and the supports, and fastened to both by numerous screws. The supports project above and below beyond the sheet metal so that they can be used as handles in handling the mirror. Figure 2a represents the primary conductor on a somewhat larger scale. The two metal parts slide with friction in two sleeves of strong paper which are held together by indiarubber bands.■ The sleeves themselves are fastened by four rods of sealing wax to a board which again is tied by indiarubber bands to a strip of wood forming part of the frame which can be seen in Figure 1. The two leading wires (covered with gutta-percha) terminate in two holes bored in the knobs of the primary conductor. This arrangement allows of all necessary motion and adjustments of the various parts of the conductor; it can be taken to pieces and put together again in a few minutes, and this is essential in order that the knobs may be frequently repolished.■ Just at the points where the leading wires pass through the mirror, they are surrounded during the discharge by a bluish light.■ The smooth wooden screen *s* is introduced for the purpose of shielding the spark gap from this light, which otherwise would interfere seriously with the production of the oscillations. Lastly, Fig. 2b represents the secondary spark gap. Both parts of the secondary conductor are again attached by sealing-wax rods and indiarubber bands to a slip forming part of the wooden framework. From the inner ends of these parts the leading wires, surrounded by glass tubes, can be seen proceeding through the mirror and bending towards one another. The upper wire carries at its pole a small brass knob. To the lower wire is soldered a piece of watch spring which carries the second pole, consisting of a fine copper point. The point is intentionally chosen of softer metal than the knob; unless this precaution is taken the point easily penetrates into the knob, and the minute sparks disappear from sight in the small hole thus produced. The figure shows how the point is adjusted by a screw which presses against the spring that is insulated from it by a glass plate. The spring is bent in a particular way in order to secure finer motion of the point than would be possible if the screw alone were used.

No doubt the apparatus here described can be considerably modified without interfering with the success of the experiments. Acting upon friendly advice, I have tried to replace the spark gap in the secondary

---

Or, more commonly, India rubber.

Oxide layers on the knobs interfere with the formation of sparks, probably by increasing the work function of the surface.

Corona discharge in the air, evidently caused by the high electric field between the wire lead and the mirror.

conductor by a frog's leg prepared for detecting currents; but this arrangement which is so delicate under other conditions does not seem to be adapted for these purposes.■ This may have been because of the alternating nature of the current at such high frequencies.

SUPPLEMENTARY READING

Appleyard, R., *Pioneers of Electrical Communication* (New York: Macmillan, 1930).

Crew, H., "Portraits of Famous Physicists," *Scripta Mathematica,* Pictorial Mathematics, Portfolio No. 4, 1942.

Dampier, W. C., *A History of Science* (New York: Cambridge University, 1946), pp. 244.

Lenard, P., *Great Men of Science* (New York: British Book Centre, 1934), pp. 358ff.

Magie, W. F., *A Source Book in Physics* (New York: McGraw-Hill, 1935), pp. 549ff.

14

Wilhelm K. Roentgen

1845 - 1923

X-Rays

THE LAST DECADE of the nineteenth century witnessed the discovery of two phenomena so totally different from the wide range of experience then known that it could best be characterized as the start of a new era in physics. These were the discovery of x-rays by Roentgen in 1895, and of natural radioactivity by Henri Becquerel (1852–1908) in the following year (see pp. 210 ff.). Both were found *accidentally,* although Becquerel was inspired to some extent by the discovery of x-rays. Purely accidental discoveries of a basic nature are exceedingly rare in physics. In a sense, every unexpected result could be considered accidental, but a careful experimenter recognizes the unusual, provided it falls within his range of imagination. Roentgen's discovery, while bound to occur eventually, involved a conceptual level far beyond the most fanciful thinking. Yet it could not have occurred much earlier, nor was it likely to have escaped notice for many more years. It came about as a result of experiments on cathode rays involving electric discharges in gases, a late nineteenth-century development.

Wilhelm Konrad Roentgen was born in the small town of Lennep, Prussia, on March 27, 1845, the son of a well-to-do textile merchant. He grew up in an age fertile with ideas, not only in physics but in almost all areas of intellectual activity; Darwin's theory of natural selection, Pasteur's germ theory of disease, Pavlov and Freud in psychology, the philosophy of Nietzsche, William James, and John Dewey, the literature of Mark Twain, Dostoevski, and Shaw, the art of Renoir, Cézanne, and Rodin, the music of Tchaikovsky and Brahms—all these men were among his contemporaries in the last half of the nineteenth century. Small wonder that his tastes for the arts and for intellectual pursuits generally, sharpened by his early home environment, grew with him during his career.

When Roentgen was three, his family moved to Apeldoorn, Holland, where he attended the public schools and, for a short time, a private boarding school. The records of his early schooling are incomplete, but apparently he made no great impression on his teachers. Following brief periods at the Utrecht Technical School and the University of Utrecht, Roentgen was admitted to the Polytechnical School in Zurich, where he received a diploma as a mechanical engineer in 1868. He was more intrigued by the pure sciences than by engineering and decided to continue his studies along these lines. Accordingly, he studied more mathematics and physics, choosing to make the latter his career after working for a time in the laboratory of August Kundt (1839–1894). Following a year with Kundt he was awarded the Ph.D. degree from the University of Zurich[1] for his thesis, *Studies on Gases*. During his years in Zurich, he developed a keen interest in mountain climbing, a hobby that he followed throughout his life.

After receiving his doctorate Roentgen remained at Zurich as assistant to Kundt, and went with him when the latter was called to the University of Würzburg. In 1872 the pair went to the (Kaiser-Wilhelm) University of Strasbourg, where, two years later, Roentgen was appointed a *privat-dozent*, or lecturer. Thus began an academic career which established him as an outstanding teacher and experimenter. He made important contributions in a number of fields; his diverse interests and universal competence were partly the result of an extensive knowledge of the literature which he gained by persistent study of current scientific journals. His most widely recognized achievement was a proof, in 1888, that magnetic effects are produced in a dielectric when it is moved in an electric field. Its great importance was its relationship to the electromagnetic concepts of Faraday and Maxwell. That same year he accepted the post of professor of physics and director of the Physical Institute at the University of Würzburg. It was here that Roentgen found his greatest inspiration, and he remained there for the next twelve years. The years at Würzburg were highly productive; Roentgen worked almost without pause, giving his energy completely to his researches and academic duties. In recognition of his academic abilities he was elected rector of the University in 1894. Late the following year, while working with a Hittorf-Crookes tube with which he was investigating some of the effects of cathode rays, he discovered x-rays.

Roentgen's discovery excited immediate interest in many quarters, not only for its obvious scientific import, but because of its value in diagnostic medicine. Within weeks of his first announcement, and even while his investigations were in the preliminary stage, the medical profession began to

---

[1] The university and the Polytechnical School shared the same building and facilities.

experiment with the new tool. In an incredibly short time, it was in routine use for diagnosis in hospitals the world over. The international acclaim that fell to Roentgen knew no bounds; he received countless honors and awards. The most impressive of these, from the point of view of pure science, was the award, in 1901, of the first Nobel prize in physics.

In 1900, at the special request of the Bavarian government, Roentgen accepted the directorship of the Physical Institute at Munich, where he finished out his career. While he did not find the conditions at the university entirely to his liking, Roentgen could not resist the many cultural attractions of Munich. It was largely because of the opportunity it afforded him to indulge his love for art and music that he remained there until his death in 1923. He continued his researches at Munich until his retirement in 1920, at the age of seventy-five, but the many demands made on his time as a result of his fame cut seriously into his scholarly work. Nevertheless, he trained many students and guided the affairs of his department with considerable skill, encouraging others in their researches. The latter part of his life was marred somewhat by a controversy over the originality of his discovery. Professor Philipp Lenard (1862–1947), at Heidelberg implied that his own investigations of cathode rays, which Roentgen acknowledged, had played a more significant role in the discovery than was generally credited to him. It appears, however, that there was no sound basis for these allegations, although from time to time friends of Lenard sought to credit him with a greater role than can be supported by the historical evidence.

Apart from the great benefits to humanity that stemmed from the medical aspects of Roentgen's discovery, its role in the basic scheme of things cannot be overestimated. It was the first significant development of the *modern era*[2] in physics—the key that opened the doors to a whole sequence of discoveries upon which are based our modern views of the nature of matter.

The following extracts are taken from Roentgen's first two communications on the subject. Entitled *On a New Kind of Rays, a Preliminary Communication*, they were published in the *Sitzungsberichte der Wurzburger Physikalischen—Medicinischen Gesellschaft*, December 1895 and March 1896, and in the *Annalen der Physik und Chemie*, 64, 1898. The translation is by Barker.[3]

---

[2] Generally known as *microphysics* or *quantum physics*, as distinguished from *classical physics*.

[3] G. F. Barker (ed.), *Roentgen Rays* (New York: Harper, 1899). Another translation, by Otto Glasser, differs slightly from this: *Dr. W. C. Roentgen* (Springfield, Ill.: Thomas, 1945). Still another translation will be found in reprint No. 22 of the Alembic Club: *X-rays and the Electric Conductivity of Gases* (Edinburgh: Livingstone, 1958).

Roentgen's Experiment

FIRST COMMUNICATION

1. If the discharge of a fairly large induction coil be made to pass through a Hittorf vacuum tube, or through a Lenard tube, a Crookes tube, or other similar apparatus,■ which has been sufficiently exhausted, the tube being covered with thin, black cardboard which fits it with tolerable closeness, and if the whole apparatus be placed in a completely darkened room, there is observed at each discharge a bright illumination of a paper screen covered with barium platinocyanide,■ placed in the vicinity of the induction coil, the fluorescence thus produced being entirely independent of the fact whether the coated or the plain surface is turned towards the discharge tube. This fluorescence is visible even when the paper screen is at a distance of two meters from the apparatus.

These tubes differed only in detail, such as in size, shape, or location of electrodes.

Which was used in studies of fluorescence or to show the presence of ultraviolet light.

It is easy to prove that the cause of the fluorescence proceeds from the discharge apparatus, and not from any other point in the conducting circuit.

2. The most striking feature of this phenomenon is the fact that an active agent here passes through a black cardboard envelope, which is opaque to the visible and the ultraviolet rays of the sun or of the electric arc; an agent, too, which has the power of producing active fluorescence. Hence we may first investigate the question whether other bodies also possess this property.

We soon discover that all bodies are transparent to this agent, though in very different degrees. I proceed to give a few examples: Paper is very transparent;* behind a bound book of about one thousand pages I saw the fluorescent screen light up brightly, the printers' ink offering scarcely a noticeable hinderance. In the same way the fluorescence appeared behind a double pack of cards; a single card held between the apparatus and the screen being almost unnoticeable to the eye. A single sheet of tin foil is also scarcely perceptible; it is only after several layers have been placed over one another that their shadow is distinctly seen on the screen. Thick blocks of wood are also transparent, pine boards two or three centimeters thick absorbing only slightly. A plate of aluminium about fifteen millimeters thick, though it enfeebled the action

---

* By "transparency" of a body I denote the relative brightness of a fluorescent screen placed close behind the body, referred to the brightness which the screen shows under the same circumstances, though without the interposition of the body.

seriously, did not cause the fluorescence to disappear entirely. Sheets of hard rubber several centimeters thick still permit the rays to pass through them.* Glass plates of equal thickness behave quite differently, according as they contain lead (flint glass) or not; the former are much less transparent than the latter. If the hand be held between the discharge tube and the screen, the darker shadow of the bones is seen within the slightly dark shadow image of the hand itself. Water, carbon disulphide, and various other liquids, when they are examined in mica vessels, seem also to be transparent. That hydrogen is to any considerable degree more transparent than air I have not been able to discover.■ Behind plates of copper, silver, lead, gold, and platinum the fluorescence may still be recognized, though only if the thickness of the plates is not too great. Platinum of a thickness of 0.2 millimeter is still transparent; the silver and copper plates may even be thicker.■ Lead of a thickness of 1.5 millimeters is practically opaque; and on account of this property this metal is frequently most useful. A rod of wood with a square cross-section (20 × 20 millimeters), one of whose sides is painted white with lead paint,■ behaves differently according as to how it is held between the apparatus and the screen. It is almost entirely without action when the x-rays pass through it parallel to the painted side; whereas the stick throws a dark shadow when the rays are made to traverse it perpendicular to the painted side. In a series similar to that of the metals themselves their salts can be arranged with reference to their transparency, either in the solid form or in solution.

3. The experimental results which have now been given, as well as others, lead to the conclusion that the transparency of different substances, assumed to be of equal thickness, is essentially conditioned upon their density: no other property makes itself felt like this, certainly to so high a degree.

The following experiments show, however, that the density is not the only cause acting. I have examined, with reference to their transparency, plates of glass, aluminium, calcite;■ and quartz, of nearly the same thickness; and while these substances are almost equal in density, yet it was quite evident that the calcite was sensibly less transparent than the other substances, which appeared almost exactly alike. No particularly strong fluorescence of calcite, especially by comparison with glass, has been noticed.

4. All substances with increase in thickness become less transparent. In order to find a possible relation between transparency and thickness,

This would have been extremely difficult to observe, inasmuch as the absorption in either gas is so slight.

The absorption of x-rays increases rapidly with the atomic number of the absorber. Hence the effectiveness of lead as an absorber.

Paint which contains basic lead carbonate.

Calcite is calcium carbonate. Calcium has atomic number 20, compared with 13 for aluminum and 14 for the silicon in glass.

Others called them *Roentgen rays*, to which Roentgen objected strongly.

---

* For brevity's sake I shall use the expression "rays"; and to distinguish them from others of this name I shall call them "x-rays."■

I have made photographs in which portions of the photographic plate were covered with layers of tin foil, varying in the number of sheets superposed.∎ Photometric measurements of these will be made when I am in possession of a suitable photometer.

5. Sheets of platinum, lead, zinc, and aluminium were rolled of such thickness that all appeared nearly equally transparent. The following table contains the absolute thickness of these sheets measured in millimeters, the relative thickness referred to that of the platinum sheet, and their densities:

Thickness		Relative Thickness	Density
Pt	0.018 mm	1	21.5
Pb	0.05 mm	3	11.3
Zn	0.10 mm	6	7.1
Al	3.5 mm	200	2.6

We may conclude from these values that different metals possess transparencies which are by no means equal, even when the product of thickness and density are the same. The transparency increases much more rapidly than this product decreases.

6. The fluorescence of barium platinocyanide is not the only recognizable effect of the x-rays. It should be mentioned that other bodies also fluoresce; such, for instance, as the phosphorescent calcium compounds, the uranium glass, ordinary glass, calcite, rock salt, and so on.

∎..

I have not yet been able to prove experimentally that the x-rays are able also to produce a heating action; yet we may well assume that this effect is present, since the capability of the x-rays to be transformed is proved by means of the observed fluorescence phenomena.∎ It is certain, therefore, that all the x-rays which fall upon a substance do not leave it again as such.

The retina of the eye is not sensitive to these rays. Even if the eye is brought close to the discharge tube, it observes nothing, although, as experiment has proved, the media contained in the eye must be sufficiently transparent to transmit the rays.

7. After I had recognized the transparency of various substances of relatively considerable thickness, I hastened to see how the x-rays behaved on passing through a prism, and to find whether they were thereby deviated or not.

Experiments with water and with carbon disulphide enclosed in mica prisms of about 30° refracting angle showed no deviation, either with the fluorescent screen or on the photographic plate. For the purposes of

A technique that is still used to measure the absorption of materials for x-rays.

Actually, the lead (atomic number 82) should have appeared relatively more opaque than the platinum (atomic number 78). Probably the observations were made in the vicinity of an absorption peak of platinum (a region of energy in which a substance exhibits marked absorption properties).

A discussion of the action of x-rays on photographic plates is omitted.

The heating effect is small; the energy absorbed per gram of material is about 83 ergs per roentgen (for materials of low atomic number).

comparison the deviation of rays of ordinary light under the same conditions was observed; and it was noted that in this case the deviated images fell on the plate about 10 or 20 millimeters distant from the direct image. By means of prisms made of hard rubber and of aluminium, also of about 30° refracting angle, I have obtained images on the photographic plate in which some small deviation may perhaps be recognized. However, the fact is quite uncertain; the deviation, if it does exist, being so small that in any case the refractive index of the x-rays in the substances named cannot be more than 1.05 at the most.■ With a fluorescent screen I was also unable to observe any deviation.

The index of refraction of x-rays in material media actually is less than, but very close to, unity.

Up to the present time experiments with prisms of denser metals have given no definite results, owing to their feeble transparency and the consequently diminished intensity of the transmitted rays.

With reference to the general conditions here involved on the one hand, and on the other to the importance of the question whether the x-rays can be refracted or not on passing from one medium into another, it is most fortunate that this subject may be investigated in still another way than with the aid of prisms. Finely divided bodies in sufficiently thick layers scatter the incident light and allow only a little of it to pass, owing to reflection and refraction; so that if powders are as transparent to x-rays as the same substances are in mass—equal amounts of material being presupposed—it follows at once that neither refraction nor regular reflection takes place to any sensible degree. Experiments were tried with finely powdered rock salt, with fine electrolytic silver powder, and with zinc dust, such as is used in chemical investigations. In all these cases no difference was detected between the transparency of the powder and that of the substance in mass, either by observation with the fluorescent screen or with the photographic plate.■

Diffraction effects, which are readily obtained with powders, would be too small to have been observed unless sought for particularly.

From what has now been said it is obvious that the x-rays cannot be concentrated by lenses; neither a large lens of hard rubber nor a glass lens having any influence upon them. The shadow picture of a round rod is darker in the middle than at the edge; while the image of a tube which is filled with a substance more transparent than its own material is lighter at the middle than at the edge.■

Because of the difference in absorption.

■ . . . .

Some general observations on the reflection (not observed) and penetration of x-rays through air are omitted.

Other substances behave in general like air; they are more transparent to x-rays than to cathode rays.

11. A further difference, and a most important one, between the behavior of cathode rays and of x-rays lies in the fact that I have not succeeded, in spite of many attempts, in obtaining a deflection of the x-rays by a magnet, even in very intense fields.

The possibility of deflection by a magnet has, up to the present time, served as a characteristic property of the cathode rays; although it was observed by Hertz and Lenard that there are different sorts of cathode rays, "which are distinguished from each other by their production of phosphorescence, by the amount of their absorption, and by the extent of their deflection by a magnet."■ A considerable deflection, however, was noted in all of the cases investigated by them; so that I do not think that this characteristic will be given up except for stringent reasons.

These were simply of different energy.

12. According to experiments especially designed to test the question, it is certain that the spot on the wall of the discharge tube which fluoresces the strongest is to be considered as the main center from which the x-rays radiate in all directions. The x-rays proceed from that spot where, according to the data obtained by different investigators, the cathode rays strike the glass wall. If the cathode rays within the discharge-apparatus are deflected by means of a magnet, it is observed that the x-rays proceed from another spot—namely, from that which is the new terminus of the cathode rays.

For this reason, therefore, the x-rays, which it is impossible to deflect, cannot be cathode rays simply transmitted or reflected without change by the glass wall. The greater density of the gas outside of the discharge tube certainly cannot account for the great difference in the deflection, according to Lenard.■

Lenard held at first that x-rays were a manifestation of cathode rays, which he believed to be a wave phenomenon.

I therefore reach the conclusion that the x-rays are not identical with the cathode rays, but that they are produced by the cathode rays at the glass wall of the discharge apparatus.

13. This production does not take place in glass alone, but, as I have been able to observe in an apparatus closed by a plate of aluminium 2 millimeters thick, in this metal also. Other substances are to be examined later.

14. The justification for calling by the name "rays" the agent which proceeds from the wall of the discharge apparatus I derive in part from the entirely regular formation of shadows, which are seen when more or less transparent bodies are brought between the apparatus and the fluorescent screen (or the photographic plate).

I have observed, and in part photographed, many shadow pictures of this kind, the production of which has a particular charm. I possess, for instance, photographs of the shadow of the profile of a door which separates the rooms in which, on one side, the discharge apparatus was placed, on the other the photographic plate; the shadow of the bones of the hand; the shadow of a covered wire wrapped on a wooden spool; of a set of weights enclosed in a box; of a galvanometer in which the mag-

A pinhole camera consists of a small aperture and a photographic plate.

Largely because of their short wavelength. Interference can be observed between the incident and diffracted beams in suitable crystals.

They do not, of course, except that in very intense electric fields, such as would be experienced by a photon passing close to a nucleus, pair production can occur.

netic needle is entirely enclosed by metal; of a piece of metal whose lack of homogeneity becomes noticeable by means of the x-rays, etc.

Another conclusive proof of the rectilinear propagation of the x-rays is a pinhole photograph which I was able to make of the discharge apparatus while it was enveloped in black paper; the picture is weak but unmistakably correct.■

15. I have tried in many ways to detect interference phenomena of the x-rays; but unfortunately, without success, perhaps only because of their feeble intensity.■

16. Experiments have been begun, but are not yet finished, to ascertain whether electrostatic forces affect the x-rays in any way.■

17. In considering the question what are the x-rays—which, as we have seen, cannot be cathode rays—we may perhaps at first be led to think of them as ultraviolet light, owing to their active fluorescence and their chemical actions. But in so doing we find ourselves opposed by the most weighty considerations. If the x-rays are ultraviolet light, this light must have the following properties:

(a) On passing from air into water, carbon disulphide, aluminium, rock salt, glass, zinc, etc., it suffers no noticeable refraction.

(b) By none of the bodies named can it be regularly reflected to any appreciable extent.

(c) It cannot be polarized by any of the ordinary methods.

(d) Its absorption is influenced by no other property of substances so much as by their density.

That is to say, we must assume that these ultraviolet rays behave entirely differently from the ultrared, visible, and ultraviolet rays which have been known up to this time.

I have been unable to come to this conclusion, and so have sought for another explanation.

There seems to exist some kind of relationship between the new rays and light rays; at least this is indicated by the formation of shadows, the fluorescence, and the chemical action produced by them both. Now, we have known for a long time that there can be in ether longitudinal vibrations besides the transverse light vibrations; and, according to the views of different physicists, these vibrations must exist. Their existence, it is true, has not been proved up to the present, and consequently their properties have not been investigated by experiment.

Ought not, therefore, the new rays to be ascribed to longitudinal vibrations in the ether?

I must confess that in the course of the investigation I have become more and more confident of the correctness of this idea, and so, therefore, permit myself to announce this conjecture, although I am perfectly aware that the explanation given still needs further confirmation.■

The idea was, of course, incorrect, but it serves well to illustrate how puzzled Roentgen was by his new radiation.

SECOND COMMUNICATION■

Since my work must be interrupted for several weeks, I take the opportunity of presenting in the following paper some new phenomena which I have observed.

Delivered about three months after the first.

18. It was known to me at the time of my first publication that x-rays can discharge electrified bodies; and I conjecture that in Lenard's experiments it was the x-rays, and not the cathode rays, which had passed unchanged through the aluminium window of his apparatus, which produced the action described by him upon electrified bodies at a distance.■ I have, however, delayed the publication of my experiments until I could contribute results which are free from criticism.

Roentgen was undoubtedly correct in this, for Lenard's aluminium window would not have permitted the cathode rays to pass through.

These results can be obtained only when the observations are made in a space which is protected completely, not only from the electrostatic forces proceeding from the vacuum tube, from the conducting wires, from the induction apparatus, etc., but is also closed against air which comes from the neighborhood of the discharge apparatus.■

Since the air contains ions produced by the discharge apparatus.

To secure these conditions I had a chamber made of zinc plates soldered together, which was large enough to contain myself and the necessary apparatus, which could be closed airtight, and which was provided with an opening which could be closed by a zinc door. The wall opposite the door was for the most part covered with lead. At a place near the discharge apparatus, which was set up outside the case, the zinc wall, together with the lining of sheet lead, was cut out for a width of 4 centimeters; and the opening was covered again airtight with a thin sheet of aluminium. The x-rays penetrated through this window into the observation space.

I observed the following phenomena:

(a) Electrified bodies in air, charged either positively or negatively, are discharged if x-rays fall upon them; and this process goes on the more rapidly the more intense the rays are.■ The intensity of the rays was estimated by their action on a fluorescent screen or a photographic plate.

It is immaterial in general whether the electrified bodies are conductors or insulators. Up to the present I have not found any specific difference in the behavior of different bodies with reference to the rate of dis-

They were discharged primarily because the air surrounding them was ionized, rather than by any direct effect. See (d), below.

charge; nor as to the behavior of positive and negative electricity. Yet it is not impossible that small differences may exist.

(b) If the electrified conductor be surrounded not by air but by a solid insulator, e.g., paraffin, the radiation has the same action as would result from exposure of the insulating envelope to a flame connected to the earth.∎

That is, it provided a conducting path to the earth.

(c) If this insulating envelope be surrounded by a close-fitting conductor which is connected to the earth, and which, like the insulator, is transparent to x-rays, the radiation produces on the inner electrified conductor no action which can be detected by my apparatus.

(d) The observations noted under (a), (b), (c) indicate that air through which x-rays have passed possesses the power of discharging electrified bodies with which it comes in contact.

(e) If this is really the case, and if, further, the air retains this property for some time after it has been exposed to the x-rays, then it must be possible to discharge electrified bodies which have not been themselves exposed to the rays, by conducting to them air which has thus been exposed.

We may convince ourselves in various ways that this conclusion is correct. One method of experiment, although perhaps not the simplest, I shall describe.

I used a brass tube 3 centimeters wide and 45 centimeters long; at a distance of some centimeters from one end a part of the wall of the tube was cut away and replaced by a thin aluminium plate; at the other end, through an airtight cap, a brass ball fastened to a metal rod was introduced into the tube in such a manner as to be insulated. Between the ball and the closed end of the tube there was soldered a side tube which could be connected with an exhaust apparatus; so that when this is in action the brass ball is subjected to a stream of air which on its way through the tube has passed by the aluminium window. The distance from the window to the ball was over 20 centimeters.

I arranged this tube inside the zinc chamber in such a position that the x-rays could enter through the aluminium window of the tube perpendicular to its axis. The insulated ball lay then in the shadow, out of the range of the action of these rays. The tube and the zinc case were connected by a conductor, the ball was joined to a Hankel electroscope.

It was now observed that a charge (either positive or negative) given to the ball was not influenced by the x-rays so long as the air remained at rest in the tube, but that the charge instantly decreased considerably if by exhaustion the air which had been subjected to the rays was drawn past the ball. If by means of storage cells the ball was maintained at a constant potential, and if the modified air was drawn continuously through the tube, an electric current arose just as if the ball were connected to the wall of the tube by a poor conductor.

(f) The question arises, How does the air lose the property which is given it by the x-rays? It is not yet settled whether it loses this property gradually of itself—i.e., without coming in contact with other bodies.■ On the other hand, it is certain that a brief contact with a body of large surface, which does not need to be electrified, can make the air inactive. For instance, if a thick enough stopper of wadding is pushed into the tube so far that the modified air must pass through it before it reaches the electrified ball, the charge on the ball remains unaffected even while the exhaustion is taking place.

■..................................

The ions recombine in a short time, particularly upon collision with a surface.

Omitted are some observations on the effects of x-rays on other gases and on the efficiency of different cathode materials for the production of the rays. Roentgen concluded that a platinum cathode was the best.

SUPPLEMENTARY READING

Dampier, W. C., *A History of Science* (New York: Cambridge University, 1946), pp. 369ff.

Glasser, O., *Wilhelm Conrad Roentgen and the Early History of the Roentgen Rays* (Springfield, Ill.: Thomas, 1934).

———, *Dr. W. C. Roentgen* (Springfield, Ill.: Thomas, 1945).

Magie, W. F., *A Source Book in Physics* (New York: McGraw-Hill, 1935), pp. 600ff.

15

Henri Becquerel
1852 - 1908

Natural Radioactivity

WITHIN a few months after Roentgen announced his discovery of x-rays Becquerel made his equally startling discovery of radioactivity. Here was the first evidence that atoms possessed some structure, even though it was not apparent at the time. In fact, Becquerel's discovery seems to have made comparatively little impression at the outset, coming as it did on the heels of the more exciting discovery of Roentgen. He was led to his discovery, as we have seen, by Roentgen's observation that the x-rays emanated from the glass walls of the tube where these were bombarded by cathode rays. Furthermore, the glass showed a visible fluorescence at these points. Thinking that a connection might exist between fluorescence and the emission of x-rays, Becquerel proceeded to study various materials known to exhibit fluorescence,[1] including certain uranium salts. Following a number of experiments and a fortuitous circumstance forced upon him by a spell of bad weather, Becquerel concluded that the particular uranium compounds which he had investigated gave off a radiation that could penetrate various thicknesses of matter.

That this might be a form of x-rays was considered, of course. There were a number of similarities: the penetrating power of the radiation, the fact that it could expose photographic plates, and by no means the least, the radiation ionized the air through which it passed. The latter was of particular significance because it provided a simple means to detect and measure the radiation. Shortly after Becquerel's initial observations it was found that thorium exhibited similar properties, and several years later, while making a systematic search for these effects in other substances, Pierre and Marie Curie discovered radium.

---

[1] Or phosphorescence, which means that the material continues to give off light after the source of excitation is removed.

Radioactivity was discovered early in 1896, but it remained for Ernest Rutherford (1871–1942), three years later, to identify the radiation. He found that it consisted of two parts, one easily absorbed by very small thicknesses, the other a good deal more penetrating. To these he gave the names *alpha* and *beta* rays, respectively. Later, the more penetrating *gamma* radiation was found generally to accompany these radiations. Becquerel showed that the beta rays were readily deflected in a magnetic field, and not long afterward it was established that they were identical with cathode rays.

The discovery that some elements were naturally unstable and decayed ultimately to a stable form by the emission of one or more forms of radiation was of inestimable value to the subsequent development of nuclear physics. Not only did it provide clues to the structure of nuclei, it also furnished an important probe (the alpha particle) for the earliest experiments on the nuclear atom.

Antoine Henri Becquerel was born in Paris on December 15, 1852, the third in a line of distinguished scientists. Both his father and grandfather were well-known physicists; each, in turn occupied the chair of physics at the Musée d'Histoire Naturelle, and probably each had some part in young Becquerel's choice of career. His contemporaries were Hertz, Roentgen, and J. J. Thomson, all of whom played important roles in the early stages of the *modern age* of quantum physics, destined to become the most productive period in the entire history of that science. Following conventional early training Becquerel attended the École Polytechnique, where he earned the degree of doctor of science. In 1875 he entered government service as an engineer in the department of roads and bridges, becoming chief engineer in 1894. Meanwhile, he taught physics at the museum where his father and grandfather had taught, and in 1892, following the death of his father, he succeeded to the same chair each of them had occupied. He was elected to membership in L'Institut de France in 1889 and in 1895 was appointed professor of physics at the École Polytechnique.

His early work was concerned largely with various properties of light, including polarization, absorption in crystals, and phosphorescence. He discovered radioactivity, originally called Becquerel rays, in 1896, after which he continued his studies in that field until his death in Brittany in 1908. He won the Rumford Medal of the Royal Society, and in 1903 shared the Nobel prize in physics with Pierre and Marie Curie.

The extracts that follow are all taken from the *Comptes Rendus*, vol. 122 (1896); the first starts on page 420, the second on page 501, and the third on page 1086. Other papers by Becquerel will be found in the same and succeeding years, but these contain the essence of his discovery.

Becquerel's Experiment

Comptes Rendus,
122 (1896), page
420.

I

At a previous meeting, *M*. Ch. Henry announced that phosphorescent zinc sulphide interposed in the path of the rays emitted from a Crooke's tube increased the intensity of the radiations which passed through Aluminum.

Further *M*. Niewenglowski found that commercial phosphorescent calcium sulphide emits radiations which traverse opaque bodies.

This fact can be extended to different phosphorescent bodies and in particular to uranium salts, where the phosphorescence lasts only for a very short time.

The double
sulphate.

By using bisulfate∎ of uranium and potasium, of which I have crystals in the form of a thin transparent crust, I have been able to perform the following experiment:

An optical photo-
graphic plate,
sensitive to
visible light.

A *Lumière* photographic plate∎ having bromide emulsion was wrapped with two sheets of thick black paper, so thick that the plate was not clouded by exposure to the sun for a whole day. Externally, over the paper sheet, was placed a piece of the phosphorescent substance, and all were exposed to the sun for many hours. Upon developing the photographic plate I recognized the silhouette of the phosphorescent substance in black on the negative. If a coin or a metallic screen with an open-work design were placed between the phosphorescent substance and the paper, the image of these objects appeared on the negative.

The same experiment can be repeated by interposing between the phosphorescent substance and the paper a thin glass plate; this excludes the possibility of a chemical action resulting from vapors that could be emitted by the substance upon being heated by the sun's rays.

From these experiments it may be concluded that the phosphorescent substance emits radiations which penetrate paper that is opaque to light and reduce the silver salts in a photographic plate.

Comptes Rendus,
122 (1896),
page 501.

II

At the last meeting I summarized some experiments which I have conducted to show that radiations are emitted by phosphorescent bodies, and that these radiations can penetrate different bodies opaque to light.

The plates or
crusts apparently
were obtained by
crystallization
from solution.

The experiments which I will report today were made with radiations emitted by crystalline plates of uranyl and potassium bisulphates.∎

$$[SO_4(UO)K + H_2O]$$

The phosphorescence of this substance is very strong, but its persistence is less than $\frac{1}{100}$ of a second. The character of the luminous radiations emitted by this substance was studied at another time by my father, and I shall have the opportunity later of mentioning some interesting particulars regarding these radiations.

It can be verified very easily that the radiation emitted by this substance when exposed to the sun or to diffuse daylight traverse, not only black paper sheets, but also different metals, for instance and aluminum plate and a thin sheet of copper.■ I have performed the following experiment, among others:

A *Lumière* plate with silver bromide emulsion was enclosed in an opaque housing of black cloth, closed at one side by an aluminum plate. When this arrangement was exposed to a bright sun, even for an entire day, the plate was not clouded. But when the uranium salt was placed on the outside, over the aluminum plate, and again exposed to the sun for many hours, it was observed in developing the plate by the usual procedure that the image of the crystalline material appeared black on the photographic plate.■ When the aluminum plate was a little thicker, the magnitude of the effect was less than that found after traversing two black paper sheets.

If, between the uranium salt and the aluminum plate or the black paper, a screen made of a copper sheet about 0.10 mm thick were interposed, for instance in the shape of a cross, it was observed on the negative that the image of the cross was lighter, but of a sort indicating that the radiation had traversed the copper plate. In another experiment a thinner copper plate (0.04 mm) weakened the active radiations very much less.

Phosphorescence produced, not by direct sunlight, but by the solar radiation reflected from the metallic mirror of a heliostat,■ and then refracted by a prism and a quartz lens, showed the same phenomena.

I insist particularly on the following fact, which seems to me very important and not in accordance with the phenomena that one would expect to observe: the same crystalline plates, placed with respect to the photographic plates under the same conditions and traversing the same screens, but sheltered from the incident radiations and kept in darkness, still produced the same photographic effect.■ I might point out how I was led to this observation:

Some of the preceding experiments were prepared during Wednesday the 26th and Thursday the 27th of February,■ and since on those days the sun appeared only intermittently, I stopped all experiments and left them in readiness by placing the wrapped plates in the drawer of a

Becquerel thought initially that the activity was caused by exposure to sunlight.

It became clear later that the sun played no role whatever in this experiment.

An instrument for tracking the sun by means of a clockwork in order to reflect its light onto a given spot.

Here was the realization that the activity was not connected with sunlight.

1896.

cabinet, leaving in place the uranium salts. The sun did not appear on the following days and I developed the plates on March 1st, expecting to find only very faint images. The silhouettes appeared, on the contrary, with great intensity. I thought therefore that the action must continue in the dark and arranged the following experiment:

At the bottom of an opaque cardboard box I placed a photographic plate, then on the sensitive face I placed a crust of uranium salt which was convex, and touched the emulsion only at a few points; then, alongside, I placed another crust of the same salt separated from the emulsion by a thin glass plate; this operation was executed in a darkroom, the box was closed and then placed in another cardboard box, which I put into a drawer.

I did the same thing with a holder closed by an aluminum plate, inside which I placed a photographic plate, and then laid on it a piece of uranium salt. Everything was enclosed in an opaque box and placed in a drawer. After about five hours I developed the plates, and the images of the crystals appeared black, the same as in the preceding experiment, and as if they had been made phosphorescent by means of light. For the salt placed directly over the emulsion, there was scarcely any difference between the action at the points in contact and at other parts of the salt about one millimeter away from the emulsion;■ the difference may be attributed to the different distances of the sources of active radiation. The action of the salt placed over the glass plate was slightly weakened but the shape of the source was very well reproduced.■ Finally, in passing through the aluminum plate the effect was considerably weakened, but was nevertheless very clear.

It is important to note that this phenomenon does not seem to be due to luminous radiation emitted by phosphorescence, since the latter become very weak after about $\frac{1}{100}$ of a second and are barely perceptible.

A hypothesis presents itself very naturally to explain these radiations (by noting that the effects bear a great resemblance to the effects caused by the radiations studied by *M*. Lenard and *M*. Roentgen■), as invisible radiation emitted by phosphorescence, but whose persistence is infinitely greater than the duration of the luminous radiations emitted by these bodies. The present experiments, although not contrary to this hypothesis, do not lead to any formulae.■ The experiments which I intend to do from now on, I hope, will clarify this new kind of phenomena.

The uranium salt was close enough so that even the alpha particles would have darkened the emulsion.

The glass plate would have absorbed the alpha particles, leaving the betas and gammas to expose the plate.

Becquerel was impressed by the similarities between x-rays and his own radiations.

That is, to any definite conclusion

III

Comptes Rendus,
122 (1896), page
1086.

Several months ago I showed that uranium salts emit radiations hitherto unknown, and that the radiations exhibit remarkable properties, some of which are not unlike the effects studied by *M.* Roentgen. The radiations are emitted not only when the salts are exposed to light, but also when they are kept in the dark, and for two months the same salts continued to emit, without noticeable decrease in amount, these new radiations. From March 3d to May 3d the salts were kept in a closed opaque box. Since May 3d they have been in a lead-walled box, which was kept in the darkroom. By means of a simple arrangement, a photographic plate can be placed under a black paper at the bottom of the box, on which rest the salts under test, without exposing them to any radiation not passing through the lead.

Under these circumstances the salts continued to emit active radiation.

■..................................

Omitted are descriptions of a number of similar experiments on different uranium compounds.

All the uranium salts I have studied, whether phosphorescent or not under light, whether in crystal form or in solution, gave me corresponding results. I have thus been led to the conclusion that the effect is due to the presence of the element uranium in these salts, and that the metal should give more noticeable effects than its compounds.■ An experiment performed several weeks ago confirmed this belief; the effect on photographic plates is much greater for the element than that produced by one of the salts, particularly by the bisulphate of uranium and potassium.

■..................................

It will be noted that Becquerel was at last on the right track.

A description of the ionizing power of the radiation is omitted.

SUPPLEMENTARY READING

Dampier, W. C., *A History of Science* (New York: Cambridge University, 1946), pp. 371ff.

Holton, G., and Roller, D. H. D., *Foundations of Modern Physical Science* (Cambridge, Mass.: Addison-Wesley, 1958), Ch. 36.

Jauncey, G. E. M., "The early years of radioactivity," *American Journal of Physics,* vol. 14 (1946), pp. 226–241.

Magie, W. F., *A Source Book in Physics* (New York: McGraw-Hill, 1935), pp. 610ff.

Thomson, J. J., *Recollections and Reflections* (London: Macmillan, 1937), pp. 411ff.

16

J. J. Thomson
1856-1940

The Electron

NEW DISCOVERIES in science often lead to developments in the most unexpected quarters. We have seen how Roentgen's discovery of x-rays led Becquerel to the investigations from which he chanced upon his discovery of radioactivity. The discovery of x-rays led as well to increased activity in the field of gas discharges. The atomic theory of matter was, by that time, firmly established on the strength of chemical evidence and the kinetic theory of gases. The electrical nature of matter was also readily apparent. But the actual connection between electrical and atomic properties was not clearly recognized. No doubt those who investigated the passage of electricity through gases believed that their studies might provide the essential link between these phenomena. At least this was the goal that induced J. J. Thomson to undertake research in gas discharges during the last decade of the nineteenth century. There were others who preceded him, notably J. W. Hittorf (1824–1914), Sir William Crookes (1832–1919), Eugen Goldstein (1850–1930), and Jean Baptiste Perrin (1870–1942), but their investigations added chiefly to the qualitative aspects of gas discharge phenomena.

The technique was generally the same in each case; a glass tube with platinum electrodes was gradually evacuated while a (large) potential difference was applied between the electrodes. The visible discharge was then studied as a function of pressure, electric field, and nature of the gas. There were several significant advances prior to the discovery of x-rays. It was known, for example, that in these discharges a radiation emanated from the cathode or negative electrode which exhibited certain interesting properties. Perhaps the most significant of these was the green fluorescence that was produced in the glass walls wherever the radiation fell upon them. In 1869 Hittorf showed the rectilinear propagation of the radiation by placing obstacles between the cathode and glass walls and showing that "optical" shadows were formed. This was confirmed several years later by Goldstein, who introduced the name *cathode rays,* and who held, in common with many of his colleagues, that the radiation was a wave phenomenon, similar to light. On the other hand, Crookes, who showed that the radiation could be deflected by a magnetic field, and Perrin, who found that an insulated conductor assumed a negative charge when cathode rays fell upon it, believed the radiation to be corpuscular. A spirited controversy continued for many years until Thomson proved conclusively that the radiation consisted of material particles originating in the vicinity of the cathode.

The discovery of x-rays contributed indirectly but significantly to Thomson's solution of the problem. When Roentgen showed that x-rays rendered the air conducting, Thomson saw in this a means of investigating the mechanism of conduction in gases, from which he found that the currents were carried by positive and negative *ions,* such as would be expected if the x-rays disrupted the molecules of gas. Proceeding from these observations Thomson had the necessary insight to devise telling experiments on the electric discharges in rarefied gases, and from these to identify the electron as the unit of electric charge.

Joseph J. Thomson was born near Manchester, England, on December 18, 1856, the son of a publisher and bookseller. As he recollected when writing his autobiography,

> . . . both time and place were fortunate, for the period between now and then has been one of the most eventful in the history of the world. From the beginning to the end, and especially in the latter half, there has been a quick succession of one stupendous event after another. Monarchies have fallen, and have been replaced by Republics and Dictatorships. Free trade, which as a Manchester man I naturally regarded for long as essential to the prosperity of the country, has gone too. . . . When I was a boy there were no bicycles, no motor cars, no aeroplanes, no electric light, no telephones, no wireless, no gramophones, no electrical engineering, no x-ray photographs, no cinemas, and no germs, at least none recognized by the doctors.[1]

Among his contemporaries were Hertz, Roentgen, and Becquerel, each of whom made substantial contributions to the *modern age of physics,* and who, like Thomson, were awarded Nobel prizes for their extraordinary grasp and elucidation of physical phenomena.

Thomson was educated at a private school until he was fourteen, when he entered Owens College, a small school in Manchester which had a great

---

[1] J. J. Thomson, *Recollections and Reflections* (London: Macmillan, 1937), p. 1.

influence on his choice of career. He had intended to be an engineer, which in those days required an apprenticeship, and while waiting for an opening at the particular company to which he had applied, it was decided that he should attend the local college. There he developed an interest in pure science, and in 1876 entered Trinity College, Cambridge, on a small scholarship. He achieved an outstanding record at Trinity, where, on the basis of a thesis on the transformation of energy, he was elected a fellow in 1880. In the following year Thomson published the first of many papers that were to show his deep insight into physical problems. This was a theoretical paper (*Phil. Mag.*, vol. xi) on the inertia of electric charge, a study that proved of great value to him in his subsequent experimental work.

In 1882 Thomson was appointed a lecturer in mathematics at Trinity, and in the following year became a university lecturer. In 1894 he succeeded Lord Rayleigh as Cavendish professor of experimental physics. For some time prior to this he had been investigating the properties of electrical discharges in gases, but it was not until 1895 that he found himself on the right track and in 1897 he discovered the electron. His discovery immediately opened wide avenues for further study; the electron had to be fitted into the scheme of things. The atomic theory of matter, as well as older branches of physics such as physical optics, electricity, and magnetism had to be re-examined in the light of the new discovery. In all this Thomson played an active and major role. He pioneered the field of mass spectroscopy and discovered *isotopes*. He calculated the scattering of x-rays by the electrons bound to atoms, from which it appeared that the number of electrons in a heavy atom was roughly one half its atomic weight. All this contributed greatly to the gradual evolution of our modern theories of atomic and nuclear structure.

Thomson was president of the Royal Society from 1915 to 1920, during the war years, and was heavily engaged at the same time in defense activities for various government agencies. In 1918 be became master of Trinity College, but the Cavendish Laboratory remained his primary interest throughout his long and active life. He had been instrumental in building there one of the greatest research laboratories in the world, and to it he returned at every opportunity. He was appointed professor of natural philosophy at the Royal Institution in 1905, received the Nobel prize in 1906, and was knighted two years later. He resigned his chair at the Cavendish Laboratory and the Royal Institution in 1919 and 1920 respectively, but until his death in 1940 continued his duties as master of Trinity. The honors that came to him are too numerous to mention. He received all that one might imagine could be awarded a scientist; even so, they could not begin to measure his contributions to science.

The extract which follows is taken from his paper on cathode rays (*Philosophical Magazine,* vol. 44, Series 5, 1897, page 293), containing his account of the discovery of the electron.

Thomson's Experiment

The experiments* discussed in this paper were undertaken in the hope of gaining some information as to the nature of the cathode rays. The most diverse opinions are held as to these rays; according to the almost unanimous opinion of German physicists they are due to some process in the ether to which—inasmuch as in a uniform magnetic field their course is circular and not rectilinear—no phenomenon hitherto observed is analogous:■ another view of these rays is that, so far from being wholly ethereal, they are in fact wholly material, and that they mark the paths of particles of matter charged with negative electricity. It would seem at first sight that it ought not to be difficult to discriminate between views so different, yet experience shows that this is not the case, as amongst the physicists who have most deeply studied the subject can be found supporters of either theory.

The electrified-particle theory has for purposes of research a great advantage over the ethereal theory, since it is definite and its consequences can be predicted; with the etherial theory it is impossible to predict what will happen under any given circumstances, as on this theory we are dealing with hitherto unobserved phenomena in the ether, of whose laws we are ignorant.

The following experiments were made to test some of the consequences of the electrified-particle theory.

It appears that opinion on the nature of cathode rays was divided almost along national lines, most of those holding the view that it was a wave phenomenon being from the German school, while the adherents to the corpuscular view were, like Thomson, English. The division stemmed from traditions in science in these two countries rather than differences in political thought.

CHARGE CARRIED BY THE CATHODE RAYS

If these rays are negatively electrified particles, then when they enter an enclosure they ought to carry into it a charge of negative electricity. This has been proved to be the case by Perrin,■ who placed in front of a plane cathode two coaxial metallic cylinders which were insulated from

Comptes Rendus, vol. 121 (1895), page 1130.

---

* Some of these experiments have already been described in a paper read before the Cambridge Philosophical Society (*Proceedings,* vol. ix, 1897); and in a Friday Evening Discourse at the Royal Institution ("Electrician," May 21, 1897).

each other: the outer of these cylinders was connected with the earth, the inner with a gold-leaf electroscope. These cylinders were closed except for two small holes, one in each cylinder, placed so that the cathode rays could pass through them into the inside of the inner cylinder. Perrin found that when the rays passed into the inner cylinder the electroscope received a charge of negative electricity, while no charge went to the electroscope when the rays were deflected by a magnet so as no longer to pass through the hole.

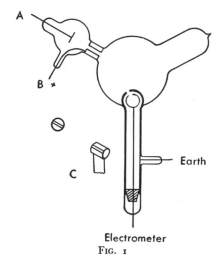

Electrometer

FIG. 1

This experiment proves that something charged with negative electricity is shot off from the cathode, traveling at right angles to it, and that this something is deflected by a magnet; it is open, however, to the objection that it does not prove that the cause of the electrification in the electroscope has anything to do with the cathode rays. Now the supporters of the ethereal theory do not deny that electrified particles are shot off from the cathode; they deny, however, that these charged particles have any more to do with the cathode rays than a rifle ball has with the flash when a rifle is fired. I have therefore repeated Perrin's experiment in a form which is not open to this objection. The arrangement used was as follows: Two coaxial cylinders (Fig. 1) with slits in them are placed in a bulb connected with the discharge tube; the cathode rays from the cathode A pass into the bulb through a slit in a metal plug fitted into the neck of the tube; this plug is connected with the anode and is put to earth. The cathode rays thus do not fall upon the cylinders unless they are deflected by a magnet.■ The outer cylinder is connected with the earth, the inner with the electrometer. When the

The objection to Perrin's experiment stemmed from the fact that the anode of his discharge tube served also as the outer cylinder. Thus, deflecting the *charged particles* away proved nothing, according to the supporters of the wave picture.

cathode rays (whose path was traced by the phosphorescence on the glass) did not fall on the slit, the electrical charge sent to the electrometer when the induction coil producing the rays was set in action was small and irregular;■ when, however, the rays were bent by a magnet so as to fall on the slit there was a large charge of negative electricity sent to the electrometer. I was surprised at the magnitude of the charge; on some occasions enough negative electricity went through the narrow slit into the inner cylinder in one second to alter the potential of a capacity of 1.5 microfarads by 20 volts.■ If the rays were so much bent by the magnet that they overshot the slits in the cylinder, the charge passing into the cylinder fell again to a very small fraction of its value when the aim was true. Thus this experiment shows that however we twist and deflect the cathode rays by magnetic forces, the negative electrification follows the same path as the rays, and that this negative electrification is indissolubly connected with the cathode rays.

When the rays are turned by the magnet so as to pass through the slit into the inner cylinder, the deflection of the electrometer connected with this cylinder increases up to a certain value, and then remains stationary although the rays continue to pour into the cylinder. This is due to the fact that the gas in the bulb becomes a conductor of electricity when the cathode rays pass through it, and thus, though the inner cylinder is perfectly insulated when the rays are not passing, yet as soon as the rays pass through the bulb the air between the inner cylinder and the outer one becomes a conductor, and the electricity escapes from the inner cylinder to the earth. Thus the charge within the inner cylinder does not go on continually increasing; the cylinder settles down into a state of equilibrium in which the rate at which it gains negative electricity from the rays is equal to the rate at which it loses it by conduction through the air. If the inner cylinder has initially a positive charge it rapidly loses that charge and acquires a negative one; while if the initial charge is a negative one, the cylinder will leak if the initial negative potential is numerically greater than the equilibrium value.

> Generally, several thousand volts were required to initiate such discharges; hence the use of an induction coil.
>
> This corresponds to a current of 30 microamperes.

DEFLECTION OF THE CATHODE RAYS BY AN ELECTROSTATIC FIELD

An objection very generally urged against the view that the cathode rays are negatively electrified particles, is that hitherto no deflection of the rays has been observed under a small electrostatic force, and though the rays are deflected when they pass near electrodes connected with sources of large differences of potential, such as induction coils or electrical machines, the deflection in this case is regarded by the supporters of the etherial theory as due to the discharge passing between the electrodes, and not primarily to the electrostatic field. Hertz[■] made the rays travel between two parallel plates of metal placed inside the discharge tube, but found that they were not deflected when the plates were connected with a battery of storage cells; on repeating this experiment I at first got the same result, but subsequent experiments showed that the absence of deflection is due to the conductivity conferred on the rarefied gas by the cathode rays. On measuring this conductivity it was found that it diminished very rapidly as the exhaustion increased; it seemed then that on trying Hertz's experiment at very high exhaustions there might be a chance of detecting the deflection of the cathode rays by an electrostatic force.

The apparatus used is represented in Figure 2.

Hertz was a supporter of the etherial view—perhaps understandably so in view of his main interests.

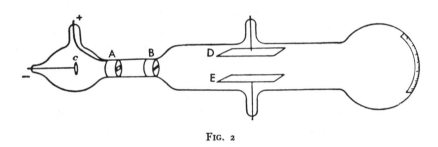

FIG. 2

The rays from the cathode C pass through a slit in the anode A, which is a metal plug fitting tightly into the tube and connected with the earth; after passing through a second slit[■] in another earth-connected metal plug B, they travel between two parallel aluminium plates about 5 cm long by 2 broad and at a distance of 1.5 cm apart; they then fall on the end of the tube and produce a narrow well-defined phosphorescent patch. A scale pasted on the outside of the tube serves to measure the deflection of this patch. At high exhaustions the rays were deflected when the two aluminum plates were connected with the terminals of a battery of small storage cells; the rays were depressed when the upper plate was connected with the negative pole of the battery, the lower with the positive, and raised when the upper plate was connected with the positive, the lower with the negative pole. The deflection was proportional to the difference of potential between the plates, and I could detect the deflection when the potential-difference was as small as two volts. It was only when the vacuum was a good one that the deflection

To define the beam more precisely.

took place, but that the absence of deflection is due to the conductivity of the medium is shown by what takes place when the vacuum has just arrived at the stage at which the deflection begins. At this stage there is a deflection of the rays when the plates are first connected with the terminals of the battery, but if this connection is maintained the patch of phosphorescence gradually creeps back to its undeflected position. This is just what would happen if the space between the plates were a conductor, though a very bad one, for then the positive and negative ions between the plates would slowly diffuse, until the positive plate became coated with negative ions, the negative plate with positive ones; thus the electric intensity between the plates would vanish and the cathode rays be free from electrostatic force. Another illustration of this is afforded by what happens when the pressure is low enough to show the deflection and a large difference of potential, say 200 volts, is established between the plates; under these circumstances there is a large deflection of the cathode rays, but the medium under the large electromotive force breaks down every now and then and a bright discharge passes between the plates;■ when this occurs the phosphorescent patch produced by the cathode rays jumps back to its undeflected position. When the cathode rays are deflected by the electrostatic field, the phosphorescent band breaks up into several bright bands separated by comparatively dark spaces; the phenomena are exactly analogous to those observed by Birkeland when the cathode rays are deflected by a magnet, and called by him the magnetic spectrum.

> That is, the residual air between the plates breaks down under the high electric field.

A series of measurements of the deflection of the rays by the electrostatic force under various circumstances will be found later on in the part of the paper which deals with the velocity of the rays and the ratio of the mass of the electrified particles to the charge carried by them.■ It may, however, be mentioned here that the deflection gets smaller as the pressure diminishes,■ and when in consequence the potential-difference in the tube in the neighborhood of the cathode increases.

> See page 228. Since the velocity of the cathode rays increases with increase in mean free path.

> Probably the current decreased at the same time because less gas was available for ionization.

■...

As the cathode rays carry a charge of negative electricity, are deflected by an electrostatic force as if they were negatively electrified, and are acted on by a magnetic force in just the way in which this force would act on a negatively electrified body moving along the path of these rays, I can see no escape from the conclusion that they are charges of negative electricity carried by particles of matter. The question next arises, what are these particles? are they atoms, or molecules, or matter in a still finer state of subdivision? To throw some light on this point, I have made a series of measurements of the ratio of the mass of these

> Some measurements on the conductivity of gases and on the behavior of the cathode rays in magnetic fields are omitted.

particles to the charge carried by it. To determine this quantity, I have used two independent methods. The first of these is as follows: Suppose we consider a bundle of homogeneous cathode rays. Let m be the mass of each of the particles, e the charge carried by it. Let N be the number of particles passing across any section of the beam in a given time; then Q the quantity of electricity carried by these particles is given by the equation.

$$Ne = Q$$

As in Thomson's first experiment.

We can measure Q if we receive the cathode rays in the inside of a vessel connected with an electrometer.■ When these rays strike against a solid body, the temperature of the body is raised; the kinetic energy of the moving particles being converted into heat; if we suppose that all this energy is converted into heat, then if we measure the increase in the temperature of a body of known thermal capacity caused by the impact of these rays, we can determine W, the kinetic energy of the particles, and if v is the velocity of the particles,

$$(\tfrac{1}{2})Nmv^2 = W$$

If ρ is the radius of curvature of the path of these rays in a uniform magnetic field H, then

$$\frac{mv}{e} = H\rho = I$$

where I is written for $H\rho$ for the sake of brevity. From these equations we get

Replacing N by $\frac{Q}{e}$

$$\frac{m}{2e}\, v^2 = \frac{W}{Q}$$

$$v = \frac{2W}{QI}$$

$$\frac{m}{e} = \frac{I^2 Q}{2W}$$

Thus, if we know the values of Q, W, and I, we can deduce the values of v and $\frac{m}{e}$.

To measure these quantities, I have used tubes of three different types. The first I tried is like that represented in Figure 2, except that the plates E and D are absent, and two coaxial cylinders are fastened to the end of the tube. The rays from the cathode C fall on the metal plug B, which is connected with the earth, and serves for the anode; a horizontal slit is cut in this plug. The cathode rays pass through this slit, and then strike against the two coaxial cylinders at the end of the tube; slits are cut in these cylinders, so that the cathode rays pass into the inside of the inner cylinder. The outer cylinder is connected with the earth, the inner cylinder, which is insulated from the outer one, is connected with an electrometer, the deflection of which measures Q, the quantity of electricity brought into the inner cylinder by the rays. A thermoelectric couple is placed behind the slit in the inner cylinder; this couple is made of very thin strips of iron and copper fastened to very fine iron and copper wires. These wires passed through the cylinders, being insulated from them, and through the glass to the outside of the tube, where they were connected with a low-resistance galvanometer, the deflection of which gave data for calculating the rise of temperature of the junction produced by the impact against it of the cathode rays. The strips of iron and copper were large enough to ensure that every cathode ray which entered the inner cylinder struck against the junction. In some of the tubes the strips of iron and copper were placed end to end, so that some of the rays struck against the iron, and others against the copper; in others, the strip of one metal was placed in front of the other; no difference, however, could be detected between the results got with these two arrangements. The strips of iron and copper were weighed, and the thermal capacity of the junction calculated. In one set of junctions this capacity was 5×10^{-3}, in another 3×10^{-3}.■ If we assume that the cathode rays which strike against the junction give their energy up to it, the deflection of the galvanometer gives us W or $\frac{1}{2}Nmv^2$.

In calories per degree centigrade

The value of I, i.e., $H\rho$, where ρ is the curvature of the path of the rays in a magnetic field of strength H was found as follows: The tube was fixed between two large circular coils■ placed parallel to each other, and separated by a distance equal to the radius of either; these coils produce a uniform magnetic field, the strength of which is got by measuring with an ammeter the strength of the current passing through them. The cathode rays are thus in a uniform field, so that their path is circular. Suppose that the rays, when deflected by a magnet, strike against the glass of the tube at E (Fig. 3), then, if ρ is the radius of the circular path of the rays,

Known as Helmholtz coils.

$$2\rho = \frac{CE^2}{AC} + AC$$

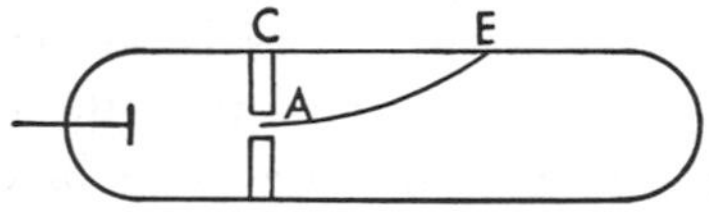

FIG. 3

From the expression relating the chord radius, and height of a segment of a circle.

thus, if we measure CE and AC we have the means of determining the radius of curvature of the path of the rays.

The determination of ρ is rendered to some extent uncertain, in consequence of the pencil of rays spreading out under the action of the magnetic field, so that the phosphorescent patch at E is several millimeters long;■ thus values of ρ differing appreciably from each other will be got by taking E at different points of this phosphorescent patch. Part

This would indicate a spread in velocity of the cathode rays.

of this patch was, however, generally considerably brighter than the rest; when this was the case, E was taken as the brightest point; when such a point of maximum brightness did not exist, the middle of the patch was taken for E. The uncertainty in the value of ρ thus introduced amounted sometimes to about 20 percent; by this I mean that if we took E first at one extremity of the patch and then at the other, we should get values of ρ differing by this amount.

The measurement of Q, the quantity of electricity which enters the inner cylinder, is complicated by the cathode rays making the gas through which they pass a conductor, so that though the insulation of the inner cylinder was perfect when the rays were off, it was not so when they were passing through the space between the cylinders; this caused some of the charge communicated to the inner cylinder to leak away so that the actual charge given to the cylinder by the cathode rays was larger than that indicated by the electrometer. To make the error from this cause as small as possible, the inner cylinder was connected to the largest capacity available, 1.5 microfarad, and the rays were only kept on for a short time, about 1 or 2 seconds, so that the alteration in potential of the inner cylinder was not large, ranging, in the various experiments from about .5 to 5 volts. Another reason why it is necessary to limit the duration of the rays to as short a time as possible, is to avoid the correction for the loss of heat from the thermoelectric junction by conduction along the wires;∎ the rise in temperature of the junction was of the order $2°$ C; a series of experiments showed that with the same tube and the same gaseous pressure Q and W were proportional to each other when the rays were not kept on too long.

Tubes of this kind gave satisfactory results, the chief drawback being that sometimes in consequence of the charging up of the glass of the tube, a secondary discharge started from the cylinder to the walls of the tube, and the cylinders were surrounded by glow; when this glow appeared, the readings were very irregular; the glow could, however, be got rid of by pumping and letting the tube rest for some time. The results got with this tube are given in the table∎ under the heading Tube 1.

The second type of tube was like that used for photographing the path of the rays;∎ double cylinders with a thermoelectric junction like those used in the previous tube were placed in the line of fire of the rays, the inside of the bell jar was lined with copper gauze connected with the earth. This tube gave very satisfactory results; we were never troubled with any glow round the cylinders, and the readings were most concordant; the only drawback was that as some of the connections had

The heat conduction would be proportional to the temperature difference.

See page 227.

This was a bell jar with a projecting tube in which the discharge was produced. The rays caused visible paths in passing from the tube into the rarefied gas in the bell jar.

to be made with sealing wax,■ it was not possible to get the highest exhaustions with this tube, so that the range of pressure for this tube is less than that for Tube 1. The results got with this tube are given in the table under the heading Tube 2.

The third type of tube was similar to the first, except that the openings in the two cylinders were made very much smaller; in this tube the slits in the cylinders were replaced by small holes, about 1.5 mm in diameter. In consequence of the smallness of the openings, the magnitude of the effect was very much reduced; in order to get measurable results it was necessary to reduce the capacity of the condenser in connection with the inner cylinder to .15 microfarad,■ and to make the galvanometer exceedingly sensitive, as the rise in temperature of the thermoelectric junction was in these experiments only about .5° C on the average. The results obtained in this tube are given in the table under the heading Tube 3.

The results of a series of measurements with these tubes are given in the following table.

Gas	Value of W/Q	I	m/e	v
		Tube 1.		
Air.....................	$4 \cdot 6 \times 10^{11}$	230	$\cdot 57 \times 10^{-7}$	4×10^9
Air.....................	$1 \cdot 8 \times 10^{12}$	350	$\cdot 34 \times 10^{-7}$	1×10^{10}
Air..............	$6 \cdot 1 \times 10^{11}$	230	$\cdot 43 \times 10^{-7}$	$5 \cdot 4 \times 10^9$
Air.....................	$2 \cdot 5 \times 10^{12}$	400	$\cdot 32 \times 10^{-7}$	$1 \cdot 2 \times 10^{10}$
Air.....................	$5 \cdot 5 \times 10^{11}$	230	$\cdot 48 \times 10^{-7}$	$4 \cdot 8 \times 10^9$
Air.....................	1×10^{12}	285	$\cdot 4 \times 10^{-7}$	7×10^9
Air.....................	1×10^{12}	285	$\cdot 4 \times 10^{-7}$	7×10^9
Hydrogen...............	6×10^{12}	205	$\cdot 35 \times 10^{-7}$	6×10^9
Hydrogen...............	$2 \cdot 1 \times 10^{12}$	460	$\cdot 5 \times 10^{-7}$	$9 \cdot 2 \times 10^9$
Carbonic acid■.........	$8 \cdot 4 \times 11^{11}$	260	$\cdot 4 \times 10^{-7}$	$7 \cdot 5 \times 10^9$
Carbonic acid...........	$1 \cdot 47 \times 10^{12}$	340	$\cdot 4 \times 10^{-7}$	$8 \cdot 5 \times 10^9$
Carbonic acid...........	$3 \cdot 0 \times 10^{12}$	480	$\cdot 39 \times 10^{-7}$	$1 \cdot 3 \times 10^{10}$
		Tube 2.		
Air.....................	$2 \cdot 8 \times 10^{11}$	175	$\cdot 53 \times 10^{-7}$	$3 \cdot 3 \times 10^9$
Air.....................	$4 \cdot 4 \times 10^{11}$	195	$\cdot 47 \times 10^{-7}$	$4 \cdot 1 \times 10^9$
Air.....................	$3 \cdot 5 \times 10^{11}$	181	$\cdot 47 \times 10^{-7}$	$3 \cdot 8 \times 10^9$
Hydrogen...............	$2 \cdot 8 \times 10^{11}$	175	$\cdot 53 \times 10^{-7}$	$3 \cdot 3 \times 10^9$
Air.....................	$2 \cdot 5 \times 10^{11}$	160	$\cdot 51 \times 10^{-7}$	$3 \cdot 1 \times 10^9$
Carbonic acid...........	2×10^{11}	148	$\cdot 54 \times 10^{-7}$	$2 \cdot 5 \times 10^9$
Air.....................	$1 \cdot 8 \times 10^{11}$	151	$\cdot 63 \times 10^{-7}$	$2 \cdot 3 \times 10^9$
Hydrogen...............	$2 \cdot 8 \times 10^{11}$	175	$\cdot 53 \times 10^{-7}$	$3 \cdot 3 \times 10^9$
Hydrogen...............	$4 \cdot 4 \times 10^{11}$	201	$\cdot 46 \times 10^{-7}$	$4 \cdot 4 \times 16^9$
Air.....................	$2 \cdot 5 \times 10^{11}$	176	$\cdot 61 \times 10^{-7}$	$2 \cdot 8 \times 10^9$
Air.....................	$4 \cdot 2 \times 10^{11}$	200	$\cdot 48 \times 10^{-7}$	$4 \cdot 1 \times 10^9$
		Tube 3.		
Air.....................	$2 \cdot 5 \times 10^{11}$	220	$\cdot 9 \times 10^{-7}$	$2 \cdot 4 \times 10^9$
Air.....................	$3 \cdot 5 \times 10^{11}$	225	$\cdot 7 \times 10^{-7}$	$3 \cdot 2 \times 10^9$
Hydrogen...............	3×10^{11}	250	$1 \cdot 0 \times 10^{-7}$	$2 \cdot 5 \times 10^9$

In the early days of vacuum practice, the use of sealing wax to fasten electrodes into tubes was fairly common. In fact, wax is still used to some extent for such purposes in research laboratories.

In order to increase the potential resulting from the reduced charge.

Carbon dioxide.

It will be noticed that the value of $\frac{m}{e}$ is considerably greater for Tube 3, where the opening is a small hole, than for Tubes 1 and 2, where the opening is a slit of much greater area. I am of opinion that the values of $\frac{m}{e}$ got from Tubes 1 and 2 are too small,■ in consequence of the leakage from the inner cylinder to the outer by the gas being rendered a conductor by the passage of the cathode rays.

Not so; see below.

It will be seen from these tables that the value of $\frac{m}{e}$ is independent of the nature of the gas.■ Thus, for the first tube the mean for air is .40 × 10⁻⁷, for hydrogen .42 × 10⁻⁷, and for carbonic acid gas .4 × 10⁻⁷; for the second tube the mean for air is .52 × 10⁻⁷, for hydrogen .50 × 10⁻⁷, and for carbonic acid gas .54 × 10⁻⁷.

The value of $\frac{m}{e}$ for electrons is 0.569×10^{-7} emu per gram.

Experiments were tried with electrodes made of iron instead of aluminium; this altered the appearance of the discharge and the value of v at the same pressure, the values of $\frac{m}{e}$ were, however, the same in the two tubes; the effect produced by different metals on the appearance of the discharge will be described later on.

In all the preceding experiments, the cathode rays were first deflected from the cylinder by a magnet, and it was then found that there was no deflection either of the electrometer or the galvanometer, so that the deflections observed were entirely due to the cathode rays; when the glow mentioned previously surrounded the cylinders there was a deflection of the electrometer even when the cathode rays were deflected from the cylinder.■

Owing to the collection of charge from the ionized gas.

Before proceeding to discuss the results of these measurements I shall describe another method of measuring the quantities $\frac{m}{e}$ and v of an entirely different kind from the preceding; this method is based upon the deflection of the cathode rays in an electrostatic field. If we measure the deflection experienced by the rays when traversing a given length under a uniform electric intensity, and the deflection of the rays when they traverse a given distance under a uniform magnetic field, we can find the values of $\frac{m}{e}$ and v in the following way:

This is the well-known Thomson experiment.

■Let the space passed over by the rays under a uniform electric intensity F be 1, the time taken for the rays to traverse this space is $\frac{1}{v}$, the velocity in the direction of F is therefore

$$\frac{Fe}{m}\frac{1}{v}$$

so that θ, the angle through which the rays are deflected when they leave the electric field and enter a region free from electric force, is given by the equation

$$\theta = \frac{Fe}{m}\frac{1}{v^2}$$

If, instead of the electric intensity, the rays are acted on by a magnetic force H at right angles to the rays, and extending across the distance 1, the velocity at right angles to the original path of the rays is

$$\frac{Hev}{m}\frac{1}{v}$$

so that ϕ, the angle through which the rays are deflected when they leave the magnetic field, is given by the equation

$$\phi = \frac{He}{m}\frac{1}{v}$$

From these equations we get

$$v = \frac{\phi}{\theta}\frac{F}{H}$$

and

$$\frac{m}{e} = \frac{H^2\theta}{F\phi^2}$$

In actual experiments H was adjusted so that $\phi = \theta$; in this case the equations become

$$v = \frac{F}{H},$$

$$\frac{m}{e} = \frac{H^2}{F\theta}$$

The apparatus used to measure v and $\frac{m}{e}$ by this means is that represented in Figure 2. The electric field was produced by connecting the two aluminium plates to the terminals of a battery of storage cells. The phosphorescent patch at the end of the tube was deflected, and the deflection measured by a scale pasted to the end of the tube. As it was necessary to darken the room to see the phosphorescent patch, a needle coated with luminous paint was placed so that by a screw it could be moved up and down the scale; this needle could be seen when the room was darkened, and it was moved until it coincided with the phosphorescent patch. Thus, when light was admitted, the deflection of the phosphorescent patch could be measured.

The magnetic field was produced by placing outside the tube two coils whose diameter was equal to the length of the plates;■ the coils were placed so that they covered the space occupied by the plates, the distance between the coils was equal to the radius of either. The mean

So that the magnetic field acted over the same distance as the electric field. A Helmholtz pair was employed.

value of the magnetic force over the length 1 was determined in the following way: a narrow coil C whose length was 1, connected with a ballistic galvanometer, was placed between the coils; the plane of the windings of C was parallel to the planes of the coils; the cross section of the coil was a rectangle 5 cm by 1 cm. A given current was sent through the outer coils and the kick α of the galvanometer observed when this current was reversed. The coil C was then placed at the center of two very large coils, so as to be in a field of uniform magnetic force: the current through the large coils was reversed and the kick β of the galvanometer again observed; by comparing α and β we can get the mean value of the magnetic force over a length 1; this was found to be

$$60 \times i$$

where i is the current flowing through the coils.

> Probably in emu (electromagnetic units).

A series of experiments was made to see if the electrostatic deflection was proportional to the electric intensity between the plates; this was found to be the case. In the following experiments the current through the coils was adjusted so that the electrostatic deflection was the same as the magnetic:

Gas	θ	H	F	l	m/e	v
Air.............	8/110	5·5	$1·5 \times 10^{10}$	5	$1·3 \times 10^{-7}$	$2·8 \times 10^{9}$
Air.............	9·5/110	5·4	$1·5 \times 10^{10}$	5	$1·1 \times 10^{-7}$	$2·8 \times 10^{9}$
Air.............	13/110	6·6	$1·5 \times 10^{10}$	5	$1·2 \times 10^{-7}$	$2·3 \times 10^{9}$
Hydrogen.......	9/110	6·3	$1·5 \times 10^{10}$	5	$1·5 \times 10^{-7}$	$2·5 \times 10^{9}$
Carbonic acid....	11/110	6·9	$1·5 \times 10^{10}$	5	$1·5 \times 10^{-7}$	$2·2 \times 10^{9}$
Air.............	6/110	5	$1·8 \times 10^{10}$	5	$1·3 \times 10^{-7}$	$3·6 \times 10^{9}$
Air.............	7/110	3·6	1×10^{10}	5	$1·1 \times 10^{-7}$	$2·8 \times 10^{9}$

> It appears that the earlier measurements gave results in better agreement with the known value for the electron.

The cathode in the first five experiments was aluminium, in the last two experiments it was made of platinum; in the last experiment Sir William Crookes's method of getting rid of the mercury vapor by inserting tubes of pounded sulphur, sulphur iodide, and copper filings between the bulb and the pump was adopted.■ In the calculation of $\frac{m}{e}$ and v no allowance has been made for the magnetic force due to the coil in the region outside the plates; in this region the magnetic force will be in the opposite direction to that between the plates, and will tend to bend the cathode rays in the opposite direction: thus the effective value of H will be smaller than the value used in the equations, so that the values of $\frac{m}{e}$ are larger, and those of v less than they would be if this correction were applied. This method of determining the values of $\frac{m}{e}$ and v is much less laborious and probably more accurate than the former method; it cannot, however, be used over so wide a range of pressures.

> Apparently Thomson used a mercury pump, probably of the Sprengel or Toepler type. Mercury vapor reacts with the substances listed, resulting in compounds having lower vapor pressures.

From these determinations we see that the value of $\frac{m}{e}$ is independent of the nature of the gas, and that its value 10^{-7} is very small compared with the value 10^{-4}, which is the smallest value of this quantity previously known, and which is the value for the hydrogen ion in electrolysis.■

Thus for the carriers of the electricity in the cathode rays $\frac{m}{e}$ is very small compared with its value in electrolysis.■ The smallness of $\frac{m}{e}$ may be due to the smallness of m or the largeness of e, or to a combination of these two. That the carriers of the charges in the cathode rays are small compared with ordinary molecules is shown, I think, by Lenard's results as to the rate at which the brightness of the phosphorescence produced by these rays diminishes with the length of path traveled by the ray. If we regard this phosphorescence as due to the impact of the charged particles, the distance through which the rays must travel before the phosphorescence fades to a given fraction (say $\frac{1}{e}$, where $e = 2.71$) of its original intensity, will be some moderate multiple of the mean free path. Now Lenard found that this distance depends solely upon the density of the medium, and not upon its chemical nature or physical state.■ In air at atmospheric pressure the distance was about half a centimeter, and this must be comparable with the mean free path of the carriers through air at atmospheric pressure. But the mean free path of the molecules of air is a quantity of quite a different order.■ The carrier, then, must be small compared with ordinary molecules.

The two fundamental points about these carriers seem to me to be (1) that these carriers are the same whatever the gas through which the discharge passes, (2) that the mean free paths depend upon nothing but the density of the medium traversed by these rays.

> The mass of the hydrogen ion (proton) is approximately 1830 times the electron mass.

> Where the carriers are ions.

> It is essentially the mass of gas traversed that determines the absorption of the cathode rays (electrons).

> It is much smaller.

SUPPLEMENTARY READING

Cambridge Readings in the Literature of Science (New York: Cambridge University, 1924), pp. 132ff.

Dampier, W. C., *A History of Science* (New York: Cambridge University, 1946).

Magie, W. F., *A Source Book in Physics* (New York: McGraw-Hill, 1935), pp. 583ff.

Thomson, J. J., *Recollections and Reflections* (London: Macmillan, 1937).

17

Albert Einstein
1879-1955

The Photoelectric Effect

EARLY in the twentieth century the scientific world had the good fortune to witness the achievements of one of its most remarkable minds. Only when viewed in full historical perspective will the contributions of Albert Einstein assume their total significance in the evolution of scientific thought. Nonetheless, for the great reach of his imagination and the power of his conceptual schemes, he had no equal among contemporary scientists. His scientific stature was enormous, both in the eyes of his colleagues and of the public at large. Rarely in the history of physics has a scientist been accorded so much public acclaim. Not since Newton, in fact, was there a physicist who attracted as much universal attention. Indeed, the two are frequently compared on the basis of the great influence each had on the development of physics. In their methods, however, they differed in one major respect: Newton was primarily an experimenter, although his law of gravitation was largely the result of theoretical speculation. Einstein, on the other hand, was a theorist who did no experimental work whatever in the course of his investigations. They shared the distinction of unifying great blocks of knowledge having universal application.

Einstein is perhaps best known for his theories of relativity,* yet he worked in many fields, particularly in the kinetic theory of matter and the quantum theory of light. It was in connection with the latter, for his explanation of the photoelectric effect,[1] that he won the Nobel prize in 1921. The photoelectric effect was discovered experimentally by Hertz in 1887, in

---

* See Appendix, page 313.

[1] Photoelectricity generally includes three distinct phenomena, the *photovoltaic, photoconductive,* and *photoemissive* effects. The last, with which we are concerned here, is usually called the photoelectric effect.

the course of his researches on electric waves. He noticed that the sparks produced in the gap of his secondary or *detector* circuit were influenced by the light falling upon the gap from the sparks in the primary or *transmitting* circuit. Upon further investigation, Hertz concluded that it was the ultraviolet portion of the light that was responsible for the phenomenon, and that the effect was greatest when the light was incident upon the negative terminal (cathode) of the gap. Being concerned mainly with other problems, Hertz did not carry these studies very far, leaving to others the more detailed investigations. Many were attracted to the problem, but the most significant contributions were made by Wilhelm Hallwachs (1859–1922), who showed that the emission consisted of negative electricity, and by Lenard, who measured the $\frac{e}{m}$ of the photoelectric carriers and found it to be the same as that determined by Thomson for cathode rays.

By the early part of the twentieth century two empirical laws had been firmly established. First, the photoelectric current, or number of electrons emitted per unit time, was proportional to the intensity of the incident light. Second, the maximum energy of the emitted electrons was proportional to the frequency of the light, not to its intensity. It was at this point, in 1905, that Einstein showed how Planck's new quantum theory of radiation could be used to account for the photoelectric effect. His solution was notable for its simplicity, yet it accounted fully for the observed facts. In that same year he made his first discoveries in the field of relativity.

Albert Einstein was born on March 14, 1879, at Ulm, Württemberg, in South Germany, where his father was a small businessman. Germany was then in a period of rapid economic growth following the Franco-Prussian War and the formation of the empire several years earlier. Bismarck had forged an empire that could not be contained in its own territory; its enormous industrial expansion, with the consequent need for new markets, coupled with notions of racial and cultural superiority, led to two world wars. Einstein was involved in both; in the first, although a Swiss neutral, he supported the pacifist movement. In the second, while he found the thought of war no less deplorable, he could no longer justify his own total pacifism and therefore became instrumental in persuading the United States government to attempt the development of atomic weapons.

His early years were spent in Munich, where his family had settled shortly after he was born. There he attended elementary school and the gymnasium,[2] showing no particular aptitude nor interest in his studies. He enjoyed learning by himself, however, and devoted much of his time to reading in mathematics and science. When he was fifteen, his family moved to

---

[2] Roughly the equivalent of high school plus junior college.

Milan to seek better economic conditions and young Einstein left the gymnasium without taking the examinations required for admission to a university. The following year he attended the Cantonal School at Aarau, Switzerland, where he won his certificate, and then entered the Zurich Polytechnic. There, strangely enough, his chief interest was in experimental physics. He spent a great deal of time in the laboratories, but continued to read all that he could of the current ideas in physics, studying on his own, for the most part.

Einstein graduated from Zurich in 1900, then did some private tutoring while he acquired Swiss citizenship, and in 1902 was appointed an examiner in the Swiss Patent Office. By his own account his years there were very pleasant; there he found the time to develop his major ideas. During the next two years he published five papers, chiefly on kinetic theory and thermodynamics. These showed marked ability, but in 1905 he achieved true greatness with his papers on the quantum theory of light and on special relativity. That same year he completed a dissertation[3] for his doctorate. By then he was reasonably well known; he began to lecture at Bern University, meanwhile retaining his position in the Patent Office. In 1909 he became assistant professor at the Zurich Polytechnic, and two years later accepted an appointment as full professor at the German University in Prague. He returned to Zurich shortly afterward as a full professor, remaining there until 1914, when he was invited to the Kaiser Wilhelm Institute in Berlin, then one of the leading research centers in the world. While there he published his theory of general relativity in 1916, the predictions of which were confirmed by astronomical observations three years later.

As the Nazi mentality began to creep over Germany it became clear that Einstein could no longer remain. German culture was being sacrificed to distorted ambitions, and even scientific truths were suspected of racial taint. Einstein, being a Jew, realized that his own future, both as a scientist and as an individual, was no longer secure in Hitler's Germany and reluctantly decided to leave. In 1933 Einstein joined the Institute for Advanced Studies at Princeton, where he remained for the rest of his life. His presence there enriched the Institute, as well as the entire scientific world. He continued to work quietly, seeking to complete his unified field theory, lending his support to various movements in search of world peace, and condemning ignorance and bigotry wherever he found it. He died on April 18, 1955, shortly before the fiftieth anniversary of his most important discovery, the theory of relativity.

The following extract, taken from the *Annalen der Physik,* vol. 17 (1905), page 144, contains his explanation of the photoelectric effect.

---

[3] On a new method of determining the size of molecules.

Einstein's "Experiment"

The common conception that the energy of the light is distributed evenly over the space through which it is propagated, encounters especially great difficulties in the attempt to explain the photoelectric effects which have already been shown in the pioneering work due to Mr. Lenard.*

According to the theory, that the incident light is composed of quanta of energy $\left(\dfrac{R}{N}\right)\beta\nu$, the origin of the cathode rays may be interpreted in the following manner.■ The quanta of energy penetrate the surface of the material and their respective energies are at least in part changed into the kinetic energy of electrons. The simplest process conceivable is that a quantum of light gives up all its energy to a single electron. We shall assume that this happens, but at the same time not exclude the possibility that the electron absorbs only a fraction of the incident energy. Upon reaching the surface, an electron originally inside the body will have lost a part of its kinetic energy. Furthermore one may assume that each electron in leaving the body does an amount of work P, which is characteristic of the material. Those electrons that are ejected normal to and from the immediate surface will have the greatest velocities.■ The kinetic energy of these electrons is

$$\frac{R}{N}\,\beta\nu - P$$

If the body is charged to the positive potential π and surrounded by conductors at zero potential, and if this voltage is just sufficient to prevent the loss of charge from the surface then it must be that

$$\pi\,\epsilon = \frac{R}{N}\,\beta\nu - P$$

where ϵ is the electronic charge, or

$$\pi E = R\beta\nu - P'$$

where E is the charge of a gram equivalent of the univalent ion and P' the potential of this amount of negative charge with respect to the body.■

If E is set equal to 9.6×10^3 then $\pi \times 10^{-8}$ is the potential in volts which the body will assume when irradiated in a vacuum.

---

* P. Lenard, *Ann. d. Phys. 8* (1902), pp. 169-170.

Margin notes

$\dfrac{R}{N} = k$, the Boltzman constant, where R is the Molar gas constant and N the Avogadro number. Thus, $\beta = \dfrac{h}{k}$, where h is Planck's constant, and the quantum of energy is then $h\nu$.

The electron must overcome the attractive forces holding it to the material. P is known as the photoelectric *work function* of the material.

Usually written as ½ $mv^2 = h\nu - W = \epsilon V$ where V is the retarding potential required to stop, the fastest photoelectrons.

E is the Faraday, equal to 9650 emu per gram-equivalent, or 96,500 coulombs per gram equivalent.

R = 8.31 × 10⁷
erg/mole K°.

To see whether the relation derived agrees with observations in order of magnitude we take $P' = O$, $\nu = 1.03 \times 10^{15}$ (which corresponds to the limit of the sun's spectrum in the direction of the ultraviolet), and $\beta = 4.866 \times 10^{-11}$. We thus obtain $\pi \times 10^{-8} = 4.3$ volts, which in order of magnitude agrees quite well with the results of Mr. Lenard.*■

If the formula derived is correct, it would follow that π, if plotted in cartesian coordinates as a function of the frequency of the exciting photons, would yield a straight line whose slope is independent of the material under investigation.■

From which Planck's constant is determined. The experimental verification of Einstein's photoelectric equation, and the determination of the Planck constant, were accomplished by R. A. Millikan in 1916.

As far as I can see our ideas are not contrary to Mr. Lenard's observations on the photoelectric effect. If each quantum of light were to give its energy to the electrons independently of all the others then the velocity distribution, i.e., the quality of the cathode rays produced, will be independent of the intensity of the exciting radiation; on the other hand the numbers of electrons leaving the body under equal conditions will be directly proportional to the intensity of the incident radiation.†

In the preceding it has been assumed that the energy of at least some of the quanta of the incident light had been given completely to individual electrons. If this plausible assumption had not been made, then one would have obtained, instead of the above equation, the following inequality

$$\pi E + P' \leqq R\beta\nu$$

The emission of light from a surface when bombarded by electrons.

For the case of cathode luminescence,■ which is the inverse of the process just considered, one obtains by similar reasoning:

$$\pi E + P' \geqq R\beta\nu$$

For the materials investigated by Mr. Lenard, πE is always considerably greater than $R\beta\nu$, since the voltage through which the electrons must have fallen in order to produce visible light is in some cases hundreds and in others, thousands of volt.** It must therefore be assumed that the kinetic energy of an electron is employed in the production of a great number of quanta of light.

We shall have to assume, that in the ionization of a gas by means of ultraviolet light each absorbed quantum of light energy is used up in the ionization of a single gas molecule. It follows therefrom that the work necessary theoretically for ionization of the molecule cannot be greater

---

* *Ibid.*, p. 165 and p. 184.
† *Ibid.*, p. 150 and pp. 166–168.
** P. Lenard, *Ann. d. Phys. 12* (1903), p. 469.

than the energy of the absorbed effective quantum of light energy. If J represents the theoretical ionization energy per gram equivalent then

$$R\beta v = J$$

According to Lenard's measurements the greatest effective wavelength for air■ is about 1.9×10^{-5} cm and therefore

$$R\beta v = 6.4 \cdot 10^{12} \text{ erg} \gtrless J$$

That is, the longest wavelength of light that can ionize the air.

An upper limit for the ionization potential may be obtained from the ionization voltages in rarefied gases. According to J. Stark* the smallest measured ionization voltage in air is approximately 10 volts (for Pt anodes)■ and therefore the upper limit for J is 9.6×10^{12}, which is rather close to the value just found. There is one other consequence whose experimental verification seems to me of great importance. If each absorbed quantum ionizes a molecule, then there has to exist a relation between the absorbed amount of light L and the number j of gram molecular weights, thereby ionized, namely

The nature of the electrodes, in principle, should not affect the ionization potential of the gas.

$$j = \frac{L}{R\beta v}$$

This relation must, if our concept corresponds to reality, be valid for every gas which (at the frequency considered) does not show absorption without corresponding ionization.■

At potentials below the ionization potential the gas could absorb by excitation.

---

* J. Stark, *Die Elektrizität in Gasen* (Leipzig, 1902), p. 57.

SUPPLEMENTARY READING

Crowther, J. G., *Six Great Scientists* (London: Hamilton, 1955), pp. 223ff.

Einstein, A., *Out of My Later Years* (New York: Philosophical Library, 1950).

Infeld, L., *Albert Einstein* (New York: Scribner, 1950).

Knedler, J. W. (ed.), *Masterworks of Science* (New York: Doubleday, 1949), pp. 599ff.

Zworykin, V. K., and Ramberg, E. G., *Photoelectricity* (New York: Wiley, 1949).

18

Robert A. Millikan
1865 - 1953

The Elementary Electric Charge

FOLLOWING J. J. Thomson's identification of the electron in 1897, it was natural that attempts would be made to determine precisely its properties. Thomson had measured the ratio of charge to mass of this elementary particle, thereby showing the ratio, at least, to be unique. The next obvious step was the determination of its mass or charge separately. A little reflection will show that the first cannot be measured independently of the second. Any experiment designed to reveal its inertial properties must take advantage of the charge on the electron, since in practice it can be accelerated only by electric or magnetic fields. The charge, on the other hand, can be determined independently, assuming it to be always the same. Prior to Thomson's work some estimates had been made of the magnitude of the basic unit of electric charge; G. J. Stoney, in 1881, proposed the value 3 times 10^{-11} esu, which was roughly a factor of ten smaller than the true value. It was Stoney, by the way, who in 1891 suggested the name *electron* for the "natural" unit of charge. A better estimate was made on the basis of kinetic theory by O. E. Meyer, namely 3 times 10^{-10} esu. J. S. E. Townsend, working in Thomson's laboratory, found a value roughly in agreement with Meyer's, but his observations, which involved the settling rate of ionized clouds of water vapor, were incapable of high precision. In 1903, both J. J. Thomson and H. A. Wilson attempted similar measurements, with modified equipment, but without marked improvement in accuracy. The chief difficulty lay in reproducing conditions in successive cloud formations. Thus, when Millikan became interested in the problem in 1907, the value of this highly important constant was known only within fairly wide limits. He devised a method of studying the motion of *single* drops of water vapor[1] under the action of electric and gravitational fields. This method,

---

[1] Because of evaporation of the water droplets during observation, Millikan was unable to achieve great precision until he went to the use of oil droplets.

which has come to be known as the *oil-drop experiment,* was a great improvement over earlier methods, yielding reliable and reproducible values for the electronic charge. Millikan's convincing demonstration of the discreteness of charge helped considerably to establish finally the atomic theory of matter, for even in the first decade of the twentieth century there were some who held that a *continuous* theory of electricity and matter could account equally well for the known facts.

Robert A. Millikan was born March 22, 1868, at Morrison, Illinois, the son of a small-town minister. He lived and worked in a period characterized by some of the most revolutionary developments in physics, a period in which were established many of the concepts that serve as the bases for our present views of the physical universe. New concepts of matter, of radiation, even of space and time, were associated with the last part of the nineteenth and early part of the twentieth century. These were not simply ideas that came about in the course of normal progress in science, if indeed there be such a thing. Instead they were radical departures from conventional lines of thought and were not necessarily confined to science. As Millikan pointed out in his autobiography,[2] " . . . Indeed, I suspect that the changes that have taken place during the last century in the average man's fundamental beliefs, in his philosophy, in his conception of religion, in his whole world outlook, are greater than the changes that occurred during the preceding four thousand years all put together." In answer to the question of why these great changes took place during the period in question, Millikan went on to say, " . . . Unquestionably because of the growth since the middle of the nineteenth century in man's knowledge and control of nature —that is, *because of science and its applications to human life,* for these have bloomed in my time as no one in history had ever dreamed could be possible."

Millikan attended public schools in Iowa, where his family had moved when he was five, graduating from the Maquoketa High School in 1885. Following a brief experience as a court reporter, he entered Oberlin College, where he acquired an interest in physics when asked to teach the subject during his junior year. There was not yet a strong tradition in the physical sciences in American universities. Johns Hopkins University, where such a tradition first developed, had been established but recently, and the American Physical Society was yet to be founded. The first American scientific journal in physics, the *Physical Review,* began publication while Millikan was a graduate fellow at Columbia University, where he went in 1893 after receiving a master's degree at Oberlin. A measure of the great change that has taken place in American science during the past half cen-

---

[2] R. A. Millikan, *Autobiography* (Englewood Cliffs, N. J.: Prentice-Hall, 1950), page xii.

tury may be taken from the fact that Millikan was for a time the sole graduate student in physics at Columbia.

After obtaining his doctorate in 1895, Millikan spent a year in residence at the Universities of Berlin and Göttingen. He then accepted an appointment as assistant in physics at the newly estabished Ryerson Laboratory of the University of Chicago. There he devoted himself to teaching and research, achieving recognition in both areas. He began his studies of the electron charge in 1907, and obtained, several years later, remarkably accurate results. In 1910 he was appointed a full professor at Chicago.

With his reputation firmly established as a result of his oil-drop experiment, Millikan turned to other research problems. In 1916 he confirmed experimentally Einstein's photoelectric equation, presenting thereby convincing proof of the photon concept and determining directly the value of Planck's constant.

During World War I he gave his time almost completely to government work in various capacities, following which, in 1921, he left Chicago to become director of the Norman Bridge Laboratory of Physics at Pasadena, then newly established. At the same time he became chairman of the Executive Council of the California Institute of Technology, a post held by him until his retirement in 1945. For his work on the elementary electric charge, as well as the photoelectric effect, Millikan was awarded the Nobel prize in 1923.

Until his retirement he remained active in research, particularly in the field of cosmic rays, to which he contributed much of the early empirical data. He died in San Marino, California, in 1953, having played a major role in moving American physics into the front ranks of science.

The following extract, containing the bulk of Millikan's oil-drop experiment, is taken from *Physical Review,* 32 (1911), page 349.

Millikan's Experiment

1. INTRODUCTION

In a preceding paper* a method of measuring the elementary electrical charge was presented which differed essentially from methods which had been used by earlier observers only in that all of the measurements from which the charge was deduced were made upon one individual charged carrier. This modification eliminated the chief sources of

---

* Millikan, *Phys. Rev.,* December 1909, and *Phil. Mag.,* 19, p. 209.

uncertainty which inhered in preceding determinations by similar methods such as those made by Sir Joseph Thomson,† H. A. Wilson,‡ Ehrenhaft§ and Broglie,¶ all of whom had deduced the elementary charge from the average behavior in electrical and gravitational fields of swarms of charged particles.■

The method used in the former work consisted essentially in catching ions by C. T. R. Wilson's method of droplets of water or alcohol, in then isolating by a suitable arrangement a single one of these droplets, and measuring its speed first in a vertical electrical and gravitational field combined, then in a gravitational field alone.*

The modification consisted in making observations upon single isolated droplets carrying multiple charges, rather than upon the surface of a cloud.

The sources of error or uncertainty which still inhered in the method arose from: (1) the lack of complete stagnancy in the air through which the drop moved; (2) the lack of perfect uniformity in the electrical field used; (3) the gradual evaporation of the drops, rendering it impossible to hold a given drop under observation for more than a minute, or to time the drop as it fell under gravity alone through a period of more than five or six seconds; (4) the assumption of the exact validity of Stokes's law for the drops used.■ The present modification of the method is not only entirely free from all of these limitations, but it constitutes an entirely new way of studying ionization and one which seems to be capable of yielding important results in a considerable number of directions.

Stokes's law relates the limiting velocity of falling objects to the properties of the medium.

With its aid it has already been found possible:

1. To catch upon a minute droplet of oil and to hold under observation for an indefinite length of time one single atmospheric ion or any desired number of such ions between 1 and 150.■

2. To present direct and tangible demonstration, through the study of the behavior in electrical and gravitational fields of this oil drop, carrying its captured ions, of the correctness of the view advanced many years ago and supported by evidence from many sources that all electrical charges, however produced, are exact multiples of one definite, elementary, electrical charge, or in other words, that an electrical charge instead of being spread uniformly over the charged surface has a definite granular structure, consisting, in fact, of an exact number of specks, or atoms of electricity, all precisely alike, peppered over the surface of the charged body.

These were ions formed of the gas molecules in the air, such as nitrogen or oxygen.

---

† Thomson, *Phil. Mag.*, 46, 1898, p. 528; 48, 1899, p. 547; 5, 1903, p. 346.
‡ H. A. Wilson, *Phil. Mag.*, 5, 1903, p. 429.
§ Ehrenhaft, *Phys. Zeit.*, Mai 1909.
¶ Broglie, *Le Radium*, Juillet 1909.
* In work reported since this paper was first presented, Ehrenhaft (*Phys. Zeit.*, July 1910) has adopted this vertical-field arrangement so that he also now finds it possible to make all his measurements upon individual charged particles.

3. To make an exact determination of the value of the elementary electrical charge which is free from all questionable theoretical assumptions and is limited in accuracy only by that attainable in the measurement of the coefficient of viscosity of air.

4. To observe directly the order of magnitude of the kinetic energy of agitation of a molecule, and thus to bring forward new direct and most convincing evidence of the correctness of the kinetic theory of matter.

5. To demonstrate that the great majority, if not all, of the ions of ionized air, of both positive and negative sign, carry the elementary electrical charge.

6. To show that Stokes's law for the motion of a small sphere through a resisting medium breaks down as the diameter of the sphere becomes comparable with the mean free path of the molecules of the medium, and to determine the exact way in which it breaks down.

2.   THE METHOD

The only essential modification in the method consists in replacing the droplet of water or alcohol by one of oil, mercury or some other nonvolatile substance, and in introducing it into the observing space in a new way.

Figure 1 shows the apparatus used in the following experiments. By means of a commercial "atomizer" A^* a cloud of fine droplets of oil is blown with the aid of dust-free air into the dust-free chamber C. One or more of the droplets of this cloud is allowed to fall through a pinhole p into the space between the plates M, N of a horizontal air condenser and the pinhole is then closed by means of an electromagnetically operated cover not shown in the diagram. If the pinhole is left open air currents are likely to pass through it and produce irregularities. The plates M, N are heavy, circular, ribbed brass castings 22 cm in diameter having surfaces which are ground so nearly to true planes that the error is nowhere more than .02 mm. These planes are held exactly 16 mm apart by means of three small ebonite posts, held firmly in place by ebonite screws. A

At the University
of Chicago.

* The atomizer method of producing very minute but accurately spherical drops for the purpose of studying their behavior in fluid media, was first conceived and successfully carried out in January 1908 at the Ryerson Laboratory■ by Mr. J. Y. Lee, while he was engaged in a quantitative investigation of Brownian movements. His spheres were blown from Wood's metal, wax, and other like substances which solidify at ordinary temperatures. Since then the method has been almost continuously in use here, upon this and a number of other problems, and elsewhere upon similar problems.

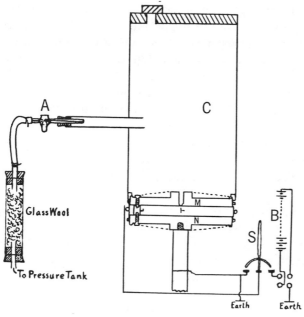

Fig. 1

strip of thin-sheet ebonite *C* passes entirely around the plates, thus forming a completely enclosed air space. Three glass windows, 1.5 cm square,
are placed in this ebonite strip at the angular positions 0°, 165°, and
180°. A narrow parallel beam of light from an arc lamp enters the condenser through the first window and emerges through the last. The other
window serves for observing, with the aid of a short focus telescope
placed about 2 feet distant, the illuminated oil droplet as it floats in the
air between the plates. The appearance of this drop is that of a brilliant
star on a black background. It falls,■ of course under the action of
gravity, toward the lower plate; but before it reaches it, an electrical
field of strength between 3,000 volts and 8,000 volts per centimeter is
created between the plates by means of the battery *B*, and, if the droplet
had received a frictional charge of the proper sign and strength as it was
blown out through the atomizer, it is pulled up by this field against
gravity, toward the upper plate. Before it strikes it the plates are short-
circuited by means of the switch *S* and the time required by the drop to
fall under gravity the distance corresponding to the space between the
cross hairs of the observing telescope is accurately determined. Then the
rate at which the droplet moves up under the influence of the field is
measured by timing it through the same distance when the field is on.■
This operation is repeated and the speeds checked an indefinite number

Owing to the
resistance of the
air for so small a
drop, it falls at
constant velocity.

This velocity is
also constant,
because of the **air**
resistance.

of times, or until the droplet catches an ion from among those which exist normally in air, or which have been produced in the space between the plates by any of the usual ionizing agents like radium or x-rays. The fact that an ion has been caught and the exact instant at which the event happened is signaled to the observer by the change in the speed of the droplet under the influence of the field. From the sign and magnitude of this change in speed, taken in connection with the constant speed under gravity, the sign and the exact value of the charge carried by the captured ion are determined.■ The error in a single observation need not exceed one third of one percent. It is from the values of the speeds observed that all of the conclusions above mentioned are directly and simply deduced.

The experiment is particularly striking when, as often happens, the droplet carries but one elementary charge and then by the capture of an ion of opposite sign is completely neutralized so that its speed is altogether unaffected by the field. In this case the computed charge is itself the charge on the captured ion.

The measurement of the distance between the cross hairs, correct to about .01 mm, is made by means of a standard scale placed vertically at exactly the same distance from the telescope as the pinhole p.

The mass of an ion is negligible in comparison with that of the oil drop; hence, only the charge was altered when an ion was captured.

3.   THE DEDUCTION OF THE RELATIVE VALUES OF THE CHARGES CARRIED BY A GIVEN DROPLET

The relations between the apparent mass* m of a drop, the charge e_n, which it carries, its speed, v_1 under gravity, and its speed v_2 under the influence of an electrical field of strength F, are given by the simple equation

$$\frac{v_1}{v_2} = \frac{mg}{Fe_n - mg} \text{ or } e_n = \frac{mg}{F}\left(\frac{v_1 + v_2}{v_1}\right). \tag{1}$$

This equation involves no assumption whatever save that the speed of the drop is proportional to the force acting upon it, an assumption which is fully and accurately tested experimentally in the following work. Furthermore, equation (1) is sufficient not only for the correct determination of the relative values of all of the charges which a given drop may have through the capture of a larger or smaller number of ions, but it is also sufficient for the establishment of all of the assertions made above, except 3, 4, and 6. However, for the sake of obtaining a pro-

---

* The term "apparent mass" is used to denote the difference between the actual mass and the buoyancy of the air.

visional estimate of the value of m in equation (1), and therefore of making at once a provisional determination of the absolute values of the charges carried by the drop, Stokes's law will for the present be assumed to be correct, but it is to be distinctly borne in mind that the conclusions just now under consideration are not at all dependent upon the validity of this assumption.

This law in its simplest form states that if μ is the coefficient of viscosity of a medium, x the force acting upon a spherical drop of radius a in that medium, and v the velocity with which the drop moves under the influence of the force, then

$$x = 6\pi\mu av \qquad (2)$$

It can be seen that the velocity is proportional to the force in Stokes's law.

The substitution in this equation of the resulting gravitational force acting on a spherical drop of density σ in a medium of density ρ gives the usual expression for the rate of fall, according to Stokes, of a drop under gravity, viz.,

$$v_1 = \frac{2}{9\mu} ga^2 (\sigma - \rho) \qquad (3)$$

The elimination of m from (1) by means of (3), and the further relation $m = \frac{4}{3}\pi a^3 (\sigma - \rho)$ gives the charge e_n in the form■

Note that m is the difference between the actual mass of the drop and the mass of the displaced air.

$$e_n = \frac{4}{3}\pi \left(\frac{9\mu}{2}\right)^{\frac{3}{2}} \left(\frac{1}{g(\sigma - \rho)}\right)^{\frac{1}{2}} \frac{(v_1 + v_2)v_1^{\frac{1}{2}}}{F} \qquad (4)$$

It is from this equation that the values of e_n in tables I–XII are obtained.

4. PRELIMINARY OBSERVATIONS UPON THE CATCHING OF IONS BY OIL DROPS

Table I presents the record of the observations taken upon a drop which was watched through a period of four and one half hours as it was alternately moved up and down between the cross hairs of the observing telescope under the influence of the field F and gravity G. How completely the errors arising from evaporation, convection currents, or any sort of disturbances in the air were eliminated is shown by the constancy during all this time in the value of the velocity under gravity. This constancy was not attained without a considerable amount of experimenting which will be described in section II. It is sufficient here to state that the heating effects■ of the illuminating arc were eliminated, first by filtering the light through about two feet of water, and second, by

Which might have produced convection currents.

TABLE 1
Negative Drop

Distance between cross-hairs = 1.010 cm
Distance between plates   = 1.600 cm
Temperature   = 24.6° C
Density of oil at 25° C   = .8960
Viscosity of air at 25.2° C = .0001836

	G sec	F sec	n	$e_n \times 10^{10}$	$e_1 \times 10^{10}$
	22.8	29.0	7	34.47	4.923
	22.0	21.8	8	39.45	4.931
	22.3	17.2			
$G = 22.28$	22.4	—	9	44.42	4.936
$V = 7950$	22.0	17.3			
	22.0	17.3			
	22.0	14.2	10	49.41	4.941
	22.7	21.5	8	39.45	
	22.9	11.0	12	59.12	4.927
	22.4	17.4	9	44.42	
	22.8	14.3	10	49.41	
$V = 7920$	22.8	12.2	11	53.92	4.902
$G = 22.80$	22.8	12.3			
	23.0	—			
	22.8	14.2			
$F = 14.17$	—	—	10	49.41	4.941
	22.8	14.0			
	22.8	17.0			
$F = 17.13$	—	17.2	9	44.42	4.936
	22.9	17.2			
	22.8	10.9			
$F = 10.73$	22.8	10.9	12	59.12	4.927
	22.8	10.6			
	22.8	12.2	11	53.92	4.902
$V = 7900$	22.8	8.7	14	68.65	4.904
$G = 22.82$	22.7	6.8	17	83.22	4.894
$F = 6.7$	22.9	6.6			
	22.8	7.2			
	—	7.2			
	—	7.3			
$F = 7.25$	—	7.2	16	78.34	4.897
	23.0	7.4			
	—	7.3			
	—	7.2			
$F = 8.65$	22.8	8.6	14	68.65	4.904
	23.1	8.7			
	23.2	9.8	13	63.68	4.900
	—	9.8			
	23.5	10.7	12	59.12	4.927
$F = 10.63$	23.4	10.6			
	23.2	9.6			
	23.0	9.6			
	23.0	9.6			
	23.2	9.5			
$V = 7820$	23.0	9.6	13	63.68	4.900
$G = 23.14$	—	9.4			
$F = 9.57$	22.9	9.6			
	—	9.6			
	22.9	9.6			
	—	10.6	12	59.12	4.927
$F = 8.65$	—	8.7	14	68.65	4.904
	23.4	8.6			

TABLE 1—*Continued.*

$F = 12.25$	23.0	12.3	11	53.92	4.902
	23.3	12.2			
	—	12.1			
	23.2	12.4			

Change forced with radium.

$F = 72.10$	23.4	72.4	5	24.60	4.920
	22.9	72.4			
	23.2	72.2			
	23.5	71.8			
	23.0	71.7			
$V = 7800$	23.0	39.2	6		
$G = 23.22$	23.2	39.2			
	—	27.4	7	34.47	
	—	20.7	8	39.38	4.922
	—	26.9	7	34.47	4.923
	—	27.2			
$F = 39.20$	23.3	39.5	6	29.62	4.937
	23.3	39.2			
	23.4	39.0			
	23.3	39.1			
	23.2	71.8	5	24.60	4.920
	23.4	382.5	4		
	23.2	374.0			
	23.4	71.0	5	24.60	4.920
	23.8	70.6			
$V = 7760$	23.4	38.5	6		
$G = 23.43$	23.1	39.2			
	23.5	70.3			
	23.4	70.5			
	23.6	71.2	5	24.60	4.920
	23.4	71.4			
	23.6	71.0			
	23.4	71.4			
	23.5	380.6			
	23.4	384.6			
	23.2	380.0			
$F = 379.6$	23.4	375.4	4	19.66	4.915
	23.6	380.4			
	23.3	374.0			
	23.4	383.6			
$F = 39.18$	—	39.2	6	29.62	4.937
$V = 7730$	23.5	39.2			
$G = 23.46$	23.5	39.0			
	23.4	39.6			
$F = 70.65$	—	70.8	5	24.60	4.920
	—	70.4			
	—	70.6			
	23.6	378.0	4	19.66	

Saw it, here, at end of 305 sec, pick up two negatives.

	23.6	39.4	6	29.62	4.937
	23.6	70.8	5	24.60	4.920

Mean of all e_1's $= 4.917$

Differences

24.60 − 19.66 = 4.94

29.62 − 24.60 = 5.02

34.47 − 29.62 = 4.85

39.38 − 34.47 = 4.91

Mean dif. $= 4.93$

shutting off the light from the arc altogether except at occasional instants, when the shutter was opened to see that the star was in place, or
to make an observation of the instant of its transit across a cross hair.
Further evidence of the complete stagnancy of the air is furnished by the
fact that for an hour or more at a time the drop would not drift more
than two or three millimeters to one side or the other of the point at
which it entered the field.

The observations in Table I are far less accurate than many of those
which follow, the timing being done in this case with a stopwatch, while
many of the later timings were taken with a chronograph. Nevertheless
this series is presented because of the unusual length of time over which
the drop was observed, and because of the rather unusual variety of
phenomena which it presents.

The column headed G shows the successive times in seconds taken
by the droplet to fall, under gravity, the distance between the cross hairs.
It will be seen that, in the course of the four and one half hours, the
value of this time increases very slightly, thereby showing that the drop
is very slowly evaporating. Furthermore, there are rather marked fluc-
tuations recorded in the first ten observations which are probably due to
the fact that, in this part of the observation, the shutter was open so
much as to produce very slight convection currents.

The column headed F is the time of ascent of the drop between the
cross hairs under the action of the field. The column headed e_n is the
value of the charge carried by the drop as computed from (4). The
column headed n gives the number■ by which the values of the preceding column must be divided to obtain the numbers in the last column.
The numbers in the e_n column are in general averages of all the observations of the table which are designated by the same numeral in the n
column. If a given observation is not included in the average in the e_n
column, a blank appears opposite that observation in the last two
columns. On account of the slow change in the value of G, the observations are arranged in groups and the average value of G for each group
is placed opposite that group in the first column. The reading of the
voltmeter, taken at the mean time corresponding to each group, is
labelled V and placed just below or just above the mean G corresponding to that group. The volts were in this case read with a ten-thousand-
volt Braun electrometer■ which had been previously calibrated, but
which may in these readings be in error by as much as one percent,
though the error in the relative values of the volts will be exceedingly
slight. The PD was applied by means of a storage battery. It will be

Note that this number was always an integer.

A type of electro-static voltmeter.

seen from the readings that the potential fell somewhat during the time of observation, the rate of fall being more rapid at first than it was later on.

5. MULTIPLE RELATIONS SHOWN BY THE CHARGES ON A GIVEN DROP

Since the original drop in this case was negative, it is evident that a sudden increase in the speed due to the field, that is, a decrease in the time given in column F, means that the drop has caught a negative ion from the air, while a decrease in the speed means that it has caught a positive ion.■

If attention be directed, first, to the latter part of the table, where the observations are most accurate, it will be seen that, beginning with the group for which $G = 23.43$, the time of the drop in the field changed suddenly from 71 sec to 380 sec, then back to 71, then down to 39, then up again to 71, and then up again to 380. These numbers show conclusively that the positive ion caught in the first change, i.e., from 71 to 380, carried exactly the same charge as the negative ion caught in the change from 380 to 71. Or again, that the negative ion caught in the change from 71 to 39, had exactly the same charge as the positive ion caught in the change from 39 to 71.

Furthermore, the exact value of the charge caught in each of the above cases is obtained in terms of mg from the difference in the values of e_n, given by equation (1), and if it be assumed that the value of m is approximately known through Stokes's law, then the approximately correct value of the charge on the captured ion is given by the difference between the values of e_n obtained through equation (4). The mean value of this difference obtained from all the changes in the latter half of Table 1 (see Differences) is 4.93×10^{-10}.■

■..

Or lost a negative ion in a collision.

It was found later that a small correction in the viscosity of air was required. The present best value for the charge is 4.8025×10^{-10} esu.

Omitted are detailed studies of some of the finer points of the experiment, and a good deal of additional data. The essentials, however, are contained in the preceding extract.

SUPPLEMENTARY READING

Jones, G. O., Rotblat, J., and Whitrow, G. J., *Atoms and The Universe* (New York: Scribner, 1956), pp. 26ff.

Millikan, R. A., *The Electron*, 2d ed., (Chicago: University of Chicago, 1924).

———, *Autobiography* (Englewood Cliffs, N. J.: Prentice-Hall, 1950).

19

Ernest Rutherford

1871 - 1937

Induced Transmutation

THE FIRST HALF of the twentieth century, a period of immense progress in all the natural sciences, was notable particularly for the new insight gained into the structure of matter. It was the age of *microphysics,* as distinct from large-scale or *macrophysical* phenomena; the age of atomic and nuclear physics. It began essentially with the discovery of radioactivity by Becquerel in 1896. If atoms could transform from one to another with the emission of charged particles, it was clear that they could not be the ultimate, indivisible *building blocks* of nature. But the interpretation of radioactivity in terms of atomic structure was not given by Becquerel. This remained for the rare imagination and experimental skill of Ernest Rutherford, who contributed to the development of this branch of physics in a most remarkable fashion. Rutherford found that the radiation from uranium consisted of at least two types, which he termed *alpha* and *beta* radiation, and he later identified the alpha particle as the nucleus of the helium atom. For this work he was awarded the Nobel prize in chemistry for 1908.

By no means does this complete the record of Rutherford's accomplishments. With F. Soddy he established, in 1902, the laws governing the transformation of radioactive substances. He proposed the then revolutionary theory that radioactivity (the emission of radiation) was a phenomenon that accompanied the spontaneous transformation of the atoms of radioactive substances into other kinds of atoms. He presented impressive experimental evidence for the new theory, but it was several years before his contemporaries generally accepted this interpretation, so radically did it depart from the customary views of the indestructibility of matter.

From his observations on the scattering of α particles, Rutherford suggested a model of the atom in 1911, the so-called *nuclear atom,* which

N. Bohr later combined with the quantum theory of light to form the basis of his famous theory of the hydrogen atom. As Bohr described it in his celebrated paper,[1]

> In order to explain the results of experiments on scattering of α rays by matter Prof. Rutherford has given a theory of the structure of atoms. According to this theory, the atoms consist of a positively charged nucleus surrounded by a system of electrons kept together by attractive forces from the nucleus; the total negative charge of the electrons is equal to the positive charge of the nucleus. Further, the nucleus is assumed to be the seat of the essential part of the mass of the atom, and to have linear dimensions exceedingly small compared with the linear dimensions of the whole atom. The number of electrons in an atom is deduced to be approximately equal to half the atomic weight. Great interest is to be attributed to this atom model; for, as Rutherford has shown, the assumption of the existence of nuclei, as those in question, seems to be necessary in order to account for the results of the experiments on large angle scattering of the α rays.

Nor was this the climax of Rutherford's remarkable career. In 1919 he discovered that the nucleus itself had structure, when, by bombarding nitrogen with α particles, he produced the first artificial nuclear transformation. Thus began the age of nuclear physics.

Ernest Rutherford was born at Brightwater, near Nelson, New Zealand, on August 30, 1871, the son of a farmer who specialized in the production of flax. He attended state schools at Foxhill and Havelock until the age of fifteen, when he won a scholarship and entered Nelson College. A bright but not precocious youngster, he distinguished himself in college by winning all the important scholastic prizes and honors. In 1890 he transferred to Canterbury College, Christchurch, again with a scholarship, where he received the B.A. degree in 1892 and, in the following year, the M.A. degree, with first class honors in mathematics and physics. In 1894 he graduated with the B.Sc., following which he was admitted to Trinity College, Cambridge, to work under J. J. Thomson. He was in residence in Cambridge at an opportune time. Shortly after his arrival Roentgen announced his discovery of x-rays; in rapid succession came Becquerel's discovery of radioactivity and Thomson's proof of the existence of the electron.

His early work at Cambridge was on the detection of electric or "wireless" waves, but after the announcements of Roentgen and Becquerel he turned his attention to the more timely subject of electrical conduction in gases. In this field he quickly gained recognition for his extraordinary ability.

---

[1] *Philosophical Magazine,* vol. 26, July 1913, p. 1.

In 1898, then twenty-seven, Rutherford was appointed to the Macdonald research professorship of physics at McGill University, Montreal. There he spent the next nine years, completing, with his colleague F. Soddy, the ingenious experiments that led to their explanation of the cause of radio-activity. Largely as a result of this work, Rutherford was elected a fellow of the Royal Society in 1903, and was awarded the Rumford Medal of that society in the following year. He left McGill in 1907 to accept the Lang-worthy chair of physics at Manchester University, where he carried on his investigations on the nature of the α particle. Using the newly invented · Geiger counter he found the charge on the α particle; later with T. Royds, he provided the most convincing proof that α particles were helium nuclei by collecting the gas that resulted when α particles passed through the walls of an evacuated chamber and showing by spectroscopic means that the gas was helium.

He won the Nobel prize for chemistry in 1908, an indication of how closely related the two disciplines were in the area of atomic structure. It was at Manchester that he developed his theory of the nuclear atom, follow-ing studies on the scattering of α particles, and it was there that Bohr applied Planck's quantum theory to the Rutherford atom, showing thereby how the hydrogen spectrum could be predicted. During World War I Rutherford devoted his time to various government projects, but in 1917, with the war drawing to a close, he returned to his studies of the collision of α particles with atoms of gas. In the course of these investigations he observed an anomalous effect in air, which was traced to nitrogen, and in 1919 he an-nounced the artificial transmutation of matter. That same year he suc-ceeded J. J. Thomson as Cavendish professor of experimental physics in the University of Cambridge, and was elected to fellowship at Trinity Col-lege. So ended his major discoveries. He devoted his time from that point on to organizing research activities and guiding his students in their research problems. His prestige was enormous. He became president of the Royal Society in 1925. That same year he was appointed to the Order of Merit, and six years later was vested with the title of Baron Rutherford of Nelson, taken from the name of the town where he attended college. Until his death in 1937 he remained actively in the forefront of physics, writing, ad-vising colleagues and students, seeking support for scientific research. His influence on contemporary physics cannot easily be assessed; he was truly one of the giants of modern science. Before his death he had the satisfaction of witnessing the developments of the early 1930's, when a number of major breakthroughs occurred in the field of nuclear physics.

The following extract, which contains his discovery of artificial transmu-tation, is taken from the *Philosophical Magazine,* vol. 37 (1919), page 537.

Rutherford's Experiment

COLLISION OF α PARTICLES WITH LIGHT ATOMS

I. Hydrogen

1. On the nucleus theory of atomic structure, it is to be anticipated that the nuclei of light atoms should be set in swift motion by intimate collisions with α particles. From consideration of impact, it can be simply shown that as a result of a head-on collision, an atom of hydrogen should acquire a velocity 1.6 times that of the α particle before impact, and should possess .64 of the energy of the incident α particle.■ Such high-speed "H" atoms should be readily detected by the scintillation method. This was shown to be the case by Marsden,* who found that the passage of α particles through hydrogen gave rise to numerous faint scintillations on a zinc sulphide screen placed far beyond the range of the α particles.■ The maximum range of the H particles, set in motion by the α particles from radium C was over 100 cm in hydrogen or about four times the range of the colliding α particles in that gas. This range agreed well with the value calculated by Darwin† from Bohr's** theory of the absorption of α particles by matter.

In most of the experiments of Marsden, a thin glass α-ray tube, containing purified radium emanation, was used as an intense source of rays.■ This was placed in a closed vessel at a suitable distance from a zinc sulphide screen, and the space between filled with compressed hydrogen. It was found that the number of H scintillations fell off approximately according to an exponential law when absorbing screens of matter were interposed,■ and the relative absorption of metal foils was in good accord with the square root law observed by Bragg for α particles.■

In a second paper, Marsden†† showed that the α-ray tube itself gave rise to a number of scintillations like those from hydrogen. Similar results were observed with an α-ray tube made from quartz instead of glass,■ and also with a nickel plate coated with radium C. The number of H scintillations observed in all cases appeared to be too large to be accounted for by the possible presence of hydrogen in the material, and Marsden concluded that there was strong evidence that hydrogen arose

By equating momenta and energies before and after impact, keeping in mind that the α-particle mass is very nearly 4 times the hydrogen mass.

Swift, charged particles, upon striking a zinc sulphide screen (or other similar phosphor), have a part of their energy transformed to light, seen as minute flashes or *scintillations*. In Rutherford's time these were observed visually. The more modern method is to observe them with a photoelectric detector.

The decay product of radium is a gas, radon, originally known as *radium emanation*.

Probably the result of different energy groups and straggling.

Bragg found that for most elements the atomic stopping power varied directly as the square root of the atomic weight.

The glass in these tubes was thin enough for the α particles to penetrate.

---

* Marsden, *Phil. Mag.*, xxvii (1914), p. 814.
† Darwin, *Phil. Mag.*, xxvii (1914), p. 499.
** Bohr, *Phil. Mag.*, xxv (1913), p. 10.
†† Marsden, *Phil. Mag.*, xxx (1915), p. 240.

On the theory that the hydrogen was ejected by collision with the α particles.

from the radioactive matter itself.■ Further experiments were interrupted by the departure of Mr. Marsden to New Zealand early in 1915 to fill the professorship of physics in Victoria College, Wellington. The quantity of radium available there was too small to continue observations, while the possibility of further work was precluded by the return of Professor Marsden to Europe on Active Service.■

This was during World War I.

That is, that protons might be emitted in radioactive decay.

We have seen that Marsden in his second paper had some indications that the radioactive matter itself gave rise to swift H atoms.■ This, if correct, was a very important result, for previously the presence of no light element except helium had been observed in radioactive transformations.

It was thought desirable to continue these experiments in more detail, and during the past four years I have made a number of experiments on this point and on other interesting problems that have arisen during the progress of the work. The experiments recorded in this and subsequent papers have been carried out at very irregular intervals, as the pressure of routine and war work permitted, and in some cases experiments have been entirely dropped for long intervals.

2. Source of the Scintillations from Active Matter

Marsden had observed that the number of H scintillations from a nickel plate, coated with radium C, was considerably greater than for a corresponding quantity of emanation—measured by γ rays—from an α-ray tube. It thus seemed possible that H atoms might arise from the disintegration of radium C, for it is well known that this product is transformed in an anomalous manner.■ In order to test this point, observations were made on the variations of the number of H scintillations from an α-ray tube immediately after it was filled with emanation. It is well known that the amount of radium C in such a tube increases at first very slowly. For example, after filling a tube with emanation, the fraction of the final amount of radium C present after 10 minutes is only 2 percent,■ but reaches 9 percent after 20 minutes.* Consequently, observations made on the number of scintillations within 10 minutes after filling should decide definitely whether the scintillations arise from radium C alone and not from the other α-ray products present, viz., the emanation and radium A. In the latter case, the number of scintillations after 10 minutes should be only 2 percent of the final number reached about three hours later when radium C is in transient equilibrium with the emanation.

Radium C decays in two ways. Some of the atoms in a given sample transform by α emission, the rest by β decay.

As the radon decays it ultimately forms radium C, which in turn decays. Hence, the amount of radium C increases until equilibrium is established, when the radium C decays at the same rate as it is formed.

---

* "Radioactive Substances and their Radiations," Rutherford, p. 499.

A number of α-ray tubes were kindly made and filled for me by Mr. N. Tunstall, B.Sc. The whole process of filling and removal for testing was done as rapidly as possible, and the counting of scintillations was usually begun within four minutes after filling. The α-ray tube was placed between the poles of a strong electromagnet in order to reduce the luminosity due to β rays on the zinc sulphide screen, placed 2 centimeters beyond the range of the α rays.■ After every precaution had been taken to avoid radioactive contamination, the number of scintillations observed between 4 and 10 minutes was greatly in excess of the number to be expected if they had their origin in the transformation of radium C alone. The actual ratio of the maximum number varied with the thickness of the α-ray tube, but the fraction observed initially was from 20 to 40 percent of the maximum reached three hours later.

These results showed conclusively that, if the H atoms from a glass α-ray tube were a product of radioactive disintegration, they arose not only from radium C but also from radium A or the emanation or both. It is hoped to discuss in a later paper the results of a number of experiments to test whether hydrogen is a product of radioactive change. It is not easy to give a decisive answer to this important problem on account of the numerous factors involved. It will be seen later that the number of scintillations from hydrogen is much greater than is to be expected on the simple theory, and it is difficult to be sure of the absence of hydrogen as a contamination in the source and absorbers of the radiation. In addition, both nitrogen and oxygen atoms are set in such swift motion by collision with α particles that they cause scintillations outside the range of the α particles. It seems probable that the large number of scintillations observed by Marsden *(loc. cit.)* from a nickel plate coated with radium C were mainly due, not to H atoms, but to high-velocity N and O atoms produced from the air between the source and the screen.

■·····································

4. Counting Scintillations

As the systematic counting of H scintillations under varied conditions is a rather difficult and trying task, it may be of some value to mention the general arrangements found most suitable and convenient in practice. Using the excellent zinc sulphide screens, specially prepared by Mr. Glew, the scintillation due to a high-speed H atom appears as a fine brilliant star or point of light, very similar in appearance and intensity to that produced by an alpha particle about 3 mm from the end of its

The β rays were deflected away from the screen by the magnetic field.

Omitted is a section describing the preparation of the α-particle sources.

The ionization produced per-unit path of air by α particles depends upon their residual range. Hence, it was customary to identify the α particles in terms of residual range.

Having different energies.

range.■ Near the end of the range of the H atom, the scintillation becomes very feeble, and can only be observed on a dark background. Consequently, in a heterogeneous beam of H atoms,■ the actual number counted per minute is to some extent dependent on the luminosity of the background seen in the microscope. It is important to adjust and keep the luminosity of the screen to the right amount throughout the whole interval of an experiment. This is most simply done by means of a small "pea" lamp fixed in a metal tube in which the current is varied. While weak scintillations are readily counted on a dark background, it is difficult under such conditions to keep the eye focused on the microscope image and the eye rapidly becomes fatigued and counting becomes erratic. The microscope employed had a magnification of about 40 and covered a field of 2 mm diameter. This in practice was found to be a very convenient magnification. In later experiments, special zinc sulphide screens were prepared in which the smaller crystals were sifted through a fine gauze on to a glass plate covered with a thin layer of adhesive material. These fine crystals completely covered the plate several crystals deep. With such a screen, the H scintillations appeared larger and more diffuse, probably due to the scattering of the light in passing through the thick layer of crystals, and were more easily counted, while weak scintillations could be counted on a brighter background than with the ordinary screen. At the same time, the layer of crystals was so uniform, that each incident H atom produced a scintillation.

In these experiments, two workers are required, one to remove the source of radiation and to make experimental adjustments, and the other to do the counting.■ Before beginning to count, the observer rests his eyes for half an hour in a dark room and should not expose his eyes to any but a weak light during the whole time of counting.■ The experiments were made in a large darkened room with a small dark chamber attached, to which the observer retired when it was necessary to turn on the light for experimental adjustments. It was found convenient in practice to count for 1 minute and then rest for an equal interval, the times and data being recorded by the assistant. As a rule, the eye becomes fatigued after an hour's counting and the results become erratic and unreliable. It is not desirable to count for more than 1 hour per day, and preferably only a few times per week.

Under good conditions, counting experiments are quite reliable from day to day. Those obtained by my assistant Mr. W. Kay and myself were always in excellent accord under the most varied conditions. It was usually arranged that the number of scintillations to be counted varied between 15 and 40 per minute.■

Note the elaborate procedure required for counting the scintillations.

Known as *dark adaption*.

By varying the source strength, or distance, or so on.

5. Experimental Arrangement

For experiments with hydrogen and other gases, the active disk *D* (Fig. 1) was mounted at a convenient height parallel to the screen on a metal bar *B* which slid into a rectangular brass box *A*, 18 cm long, 6 cm deep, and 2 cm wide, with metal flanges at both ends fitting between the rectangular poles of a large electromagnet. One end was closed by a ground-glass plate *C*, and the other by a waxed brass plate *E*, in the center of which was cut a rectangular opening 1 cm long and 3 mm wide. This opening was covered by a thin plate of metals of silver, aluminium, or iron, whose stopping power for α particles lay between

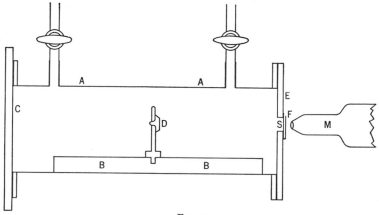

FIG. I

4 and 6 cm of air.■ The zinc sulphide screen *F* was fixed opposite the opening and distant 1 or 2 mm from the metal covering. By means of two stopcocks, the vessel was filled with the gas to be examined either by exhaustion or displacement.■ It is a great advantage to have the zinc sulphide screen outside the apparatus, in order to avoid contamination due to volatilized active matter, and for the easy introduction of absorbing material between the end plate and the screen.

That is, whose thicknesses were the equivalent of 4 to 6 cm of air in stopping α particles.

Either by evacuating the box, then filling, or by flowing the gas through to displace the air.

In practice, the source was introduced into the brass vessel at a convenient distance from the screen, and the air exhausted. The α rays after traversing the end plate fell on the screen, and the marked luminosity due to them was a guide in fixing the microscope *M* in the center of the opening. The diameter of the field of view (2 mm) was less than the width of the opening (3 mm).

Since the number of H atoms observed under ordinary conditions is less than one in a hundred thousand of the number of α particles, H atoms, projected in the direction of the α particles, can only be detected when the α rays are stopped by the absorbing screens. It was not found possible to bring an intense source closer than 3 cm from the screen on account of the luminosity excited in it by the γ rays and swift β rays,■ which prevented counting of weak scintillations. A strong magnetic field was necessary to bend away the β rays which caused a very marked luminosity on the screen. A field of 6000 gauss was generally employed for this purpose.

6.   Scintillations Due to Source and Absorbing Screens

When the containing vessel was exhausted of air, scintillations were always observed on the screen proportional in number to the activity of

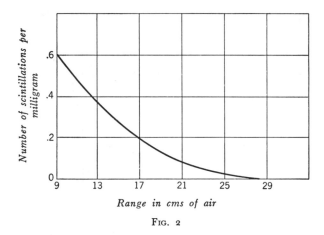

Fig. 2

the source. The number fell off rapidly between 7 and 12 cm air absorption and then more slowly, but a few could be observed nearly to 28 cm. The variation of number with amount of absorption in terms of cms of air is shown in Figure 2. This refers to a heated brass source, 3.3 cm from the screen, with a heated silver plate of stopping power 6 cm of air just before the screen.■

These scintillations appear to be due mainly to H atoms excited partly in the source and partly in the absorbing screens. Thin foils of aluminium, for example, placed close to the source increase the number of scintillations. This is due to the occlusion of hydrogen, which can be removed by heating the aluminium in an exhausted furnace just below the melting point. Similar effects were observed with silver but not with gold. In practice, all screens to be used in the path of the α rays were heated to drive off occluded gases as far as possible. This is very necessary when small numbers of scintillations have to be counted. Usually a silver plate was used to absorb the α rays. Gold was found to be very free from hydrogen, but it could not be used in place of silver close to the screen on account of the marked luminosity set up on the screen well beyond the range of the α particles. This peculiarity of gold had been previously noted by Marsden, but I was surprised to observe the magnitude of the effect with strong sources of radiation. A fuller account of the nature and cause of this luminosity will be postponed till a later paper. In a similar way, mica was found to cause a good deal of luminosity, apparently due to gamma rays.■ In addition, as is to be expected, mica gives rise to numerous H atoms and swift oxygen atoms. For these reasons, mica is unsuitable for an absorbing screen for α particles in this type of experiment.

■...

A material such as mica would be expected to fluoresce somewhat under bombardment by radiation.

Omitted is a detailed discussion of the theory of collision of α particles with light atoms.

COLLISION OF α PARTICLES WITH LIGHT ATOMS

IV. An Anomalous Effect in Nitrogen

It has been shown in Paper I that a metal source, coated with a deposit of radium C, always gives rise to a number of scintillations on a zinc sulphide screen far beyond the range of the α particles. The swift atoms causing these scintillations carry a positive charge and are deflected by a magnetic field, and have about the same range and energy as the swift H atoms produced by the passage of α particles through hydrogen. These "natural" scintillations are believed to be due mainly to swift H atoms from the radioactive source, but it is difficult to decide whether they are expelled from the radioactive source itself or are due to the action of α particles on occluded hydrogen.

The apparatus employed to study these "natural" scintillations is the same as that described in Paper I. The intense source of radium C was placed inside a metal box about 3 cm from the end, and an opening in the end of the box was covered with a silver plate of stopping power equal to about 6 cm of air.■ The zinc sulphide screen was mounted outside, about 1 mm distant from the silver plate, to admit of the introduction of absorbing foils between them. The whole apparatus was placed in a strong magnetic field to deflect the β rays. The variation in the number of these "natural" scintillations with absorption in terms of cms of air is shown in Figure 3, curve A. In this case, the air in the box was exhausted and absorbing foils of aluminium were used. When dried

Which, together with the 3 cm of air, should have been sufficient to stop the α particles completely.

oxygen or carbon dioxide was admitted into the vessel, the number of scintillations diminished to about the amount to be expected from the stopping power of the column of gas.

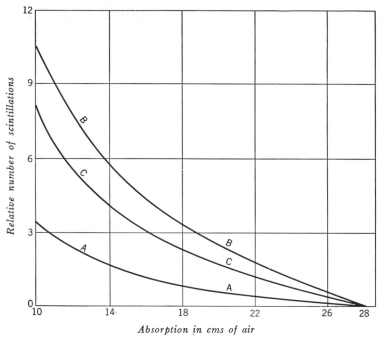

FIG. 3

A surprising effect was noticed, however, when dried air was introduced. Instead of diminishing, the number of scintillations was increased, and for an absorption corresponding to about 19 cm of air the number was about twice that observed when the air was exhausted. It was clear from this experiment that the α particles in their passage through air gave rise to long-range scintillations which appeared to the eye to be about equal in brightness to H scintillations. A systematic series of observations was undertaken to account for the origin of these scintillations. In the first place we have seen that the passage of α particles through nitrogen and oxygen gives rise to numerous bright scintillations which have a range of about 9 cm in air. These scintillations have about the range to be expected if they are due to swift N or O atoms, carrying unit charge, produced by collision with α particles. All experiments have consequently been made with an absorption greater than 9 cm of air, so that these atoms are completely stopped before reaching the zinc sulphide screen.

It was found that these long-range scintillations could not be due to the presence of water vapor in the air; for the number was only slightly reduced by thoroughly drying the air. This is to be expected, since on the average the number of the additional scintillations due to air was equivalent to the number of H atoms produced by the mixture of hydrogen at 6-cm pressure with oxygen.■ Since on the average the vapor pressure of water in air was not more than 1 cm, the effects of complete drying would not reduce the number by more than one sixth. Even when oxygen and carbon dioxide saturated with water vapor at 20° C were introduced in place of dry air, the number of scintillations was much less than with dry air.

That is, if produced by collision with hydrogen.

It is well known that the amount of hydrogen or gases containing hydrogen is normally very small in atmospheric air.■ No difference was observed whether the air was taken directly from the room or from outside the laboratory or was stored for some days over water.

Much less than 1 percent.

There was the possibility that the effect in air might be due to liberation of H atoms from the dust nuclei in the air. No appreciable difference, however, was observed when the dried air was filtered through long plugs of cotton wool, or by storage over water for some days to remove dust nuclei.

Since the anomalous effect was observed in air, but not in oxygen, or carbon dioxide, it must be due either to nitrogen or to one of the other gases present in atmospheric air. The latter possibility was excluded by comparing the effects produced in air and in chemically prepared nitrogen. The nitrogen was obtained by the well-known method of adding ammonium chloride to sodium nitrite, and stored over water. It was carefully dried before admission to the apparatus. With pure nitrogen, the number of long-range scintillations under similar conditions was greater than in air. As a result of careful experiments, the ratio was found to be 1.25, the value to be expected if the scintillations are due to nitrogen.■

On the basis of the amount of nitrogen present compared with that contained in air; i.e., there was 25 percent more nitrogen in the chamber when pure nitrogen was admitted.

The results so far obtained show that the long-range scintillations obtained from air must be ascribed to nitrogen, but it is important, in addition, to show that they are due to collision of α particles with atoms of nitrogen through the volume of the gas. In the first place, it was found that the number of the scintillations varied with the pressure of the air in the way to be expected if they resulted from collision of α particles along the column of gas.■ In addition, when an absorbing screen of gold or aluminium was placed close to the source, the range of the scintillations was found to be reduced by the amount to be expected if the range of the expelled atom was proportional to the range of the

Rather than in the immediate vicinity of the source.

The energy of the
expelled atom
was therefore
proportional to
the α-particle
energy.

Normal
temperature and
pressure;
generally 0°C and
76 cm Hg.

That is, at a
pressure
equivalent to 12
cm absorption,
etc.

From the nitrogen
in the air.

Near the end of
their range the
particles had too
little energy to
produce bright
scintillations.

colliding α particles.■ These results show that the scintillations arise from the volume of the gas and are not due to some surface effect in the radioactive source.

In Figure 3, curve *A,* the results of a typical experiment are given showing the variation in the number of natural scintillations with the amount of absorbing matter in their path measured in terms of centimeters of air for α particles. In these experiments carbon dioxide was introduced at a pressure calculated to give the same absorption of the α rays as ordinary air. In curve *B* the corresponding curve is given when air at NTP■ is introduced in place of carbon dioxide. The difference curve *C* shows the corresponding variation of the number of scintillations arising from the nitrogen in the air. It was generally observed that the ratio of the nitrogen effect to the natural effect was somewhat greater for 19 cm than for 12 cm absorption.

In order to estimate the magnitude of the effect, the space between the source and screen was filled with carbon dioxide at diminished pressure and a known pressure of hydrogen was added. The pressure of the carbon dioxide and of hydrogen were adjusted so that the total absorption of α particles in the mixed gas should be equal to that of the air. In this way it was found that the curve of absorption of H atoms produced under these conditions was somewhat steeper than curve *C* of Figure 3. As a consequence, the amount of hydrogen mixed with carbon dioxide required to produce a number of scintillations equal to that of air, increased with the increase of absorption. For example, the effect in air was equal to about 4 cm of hydrogen at 12 cm absorption, and about 8 cm at 19 cm absorption.■ For a mean value of the absorption, the effect was equal to about 6 cm of hydrogen. This increased absorption of H atoms under similar conditions indicated either that (1) the swift atoms from air■ had a somewhat greater range than the H atoms, or (2) that the atoms from air were projected more in the line of flight of the α particles.

While the maximum range of the scintillations from air using radium C as a source of α rays appeared to be about the same, viz., 28 cm, as for H atoms produced from hydrogen, it was difficult to fix the end of the range with certainty on account of the smallness of the number and the weakness of the scintillations.■ Some special experiments were made to test whether, under favorable conditions, any scintillations due to nitrogen could be observed beyond 28 cm of air absorption. For this purpose a strong source (about 60 mg Ra activity) was brought within 2.5 cm of the zinc sulphide screen, the space between containing dry air. On still further reducing the distance, the screen became too bright to detect very feeble scintillations. No certain evidence of scintillations was found beyond a range of 28 cm. It would therefore appear that (2) above is the more probable explanation.

In a previous paper we have shown that the number of swift atoms of nitrogen or oxygen produced per-unit path by collision with α particles is about the same as the corresponding number of H atoms in hydrogen. Since the number of long-range scintillations in air is equivalent to that produced under similar conditions in a column of hydrogen at 6-cm pressure, we may consequently conclude that only one long-range atom is produced for every 12 close collisions giving rise to a swift nitrogen atom of maximum range 9 cm.■

> Since 6 cm is roughly $\frac{1}{12}$ of an atmosphere.

It is of interest to give data showing the number of long-range scintillations produced in nitrogen at atmospheric pressure under definite conditions. For a column of nitrogen 3.3 cm long, and for a total absorption of 19 cm of air from the source, the number due to nitrogen per milligram of activity is .6 per minute on a screen of 3.14 sq mm area.

Both as regards range and brightness of scintillations, the long-range atoms from nitrogen closely resemble H atoms, and in all probability are hydrogen atoms. In order, however, to settle this important point definitely, it is necessary to determine the deflection of these atoms in a magnetic field. Some preliminary experiments have been made by a method similar to that employed in measuring the velocity of the H atom. The main difficulty is to obtain a sufficiently large deflection of the stream of atoms and yet have a sufficient number of scintillations per minute for counting. The α rays from a strong source passed through dry air between two parallel horizontal plates 3 cm long and 1.6 mm apart, and the number of scintillations on the screen placed near the end of the plates was observed for different strengths of the magnetic field. Under these conditions, when the scintillations arise from the whole length of the column of air between the plates, the strongest magnetic field available reduced the number of scintillations by only 30 percent. When the air was replaced by a mixture of carbon dioxide and hydrogen of the same stopping power for α rays, about an equal reduction was noted. As far as the experiment goes, this is an indication that the scintillations are due to H atoms; but the actual number of scintillations and the amount of reduction was too small to place much reliance on the result.

> Omitted is Rutherford's speculation that the question might be settled definitely by the use of a solid nitrogen compound.

■..

DISCUSSION OF RESULTS

From the results so far obtained it is difficult to avoid the conclusion that the long-range atoms arising from collision of α particles with nitrogen are not nitrogen atoms but probably atoms of hydrogen, or atoms of mass 2. If this be the case, we must conclude that the nitrogen atom is disintegrated■ under the intense forces developed in a close collision with a swift α particle, and that the hydrogen atom which is liberated formed a constituent part of the nitrogen nucleus. We have drawn attention in another paper to the rather surprising observation that the range of the nitrogen atoms in air is about the same as the oxygen atoms, although we should expect a difference of about 19 percent.■ If in collisions which give rise to swift nitrogen atoms, the hydrogen is at the same time disrupted, such a difference might be accounted for, for the energy is then shared between two systems.■

It is of interest to note, that while the majority of the light atoms, as is well known, have atomic weights represented by $4n$ or $4n+3$ where n is a whole number, nitrogen is the only atom which is expressed by $4n+2$. We should anticipate from radioactive data that the nitrogen nucleus consists of three helium nuclei each of atomic mass 4 and either two hydrogen nuclei or one of mass 2.■ If the H nuclei were outriders of the main system of mass 12, the number of close collisions with the bound H nuclei would be less than if the latter were free, for the α particle in a collision comes under the combined field of the H nucleus and of the central mass. Under such conditions, it is to be expected that the α particle would only occasionally approach close enough to the H nucleus to give it the maximum velocity, although in many cases it may give it sufficient energy to break its bond with the central mass. Such a point of view would explain why the number of swift H atoms from nitrogen is less than the corresponding number in free hydrogen and less also than the number of swift nitrogen atoms. The general results indicate that the H nuclei, which are released, are distant about twice the diameter of the electron (7×10^{-13} cm)■ from the center of the main atom. Without a knowledge of the laws of force at such small distances, it is difficult to estimate the energy required to free the H nucleus■ or to calculate the maximum velocity that can be given to the escaping H atom. It is not to be expected, a priori, that the velocity or range of the H atom released from the nitrogen atom should be identical with that due to a collision in free hydrogen.

Taking into account the great energy of motion of the α particle expelled from radium C, the close collision of such an α particle with a light atom seems to be the most likely agency to promote the disruption

The nuclear equation is:
$_2He^4 + _7N^{14} \rightarrow _8O^{17} + _1H^1$

On the basis of the different masses.

Actually, the energy is shared between the hydrogen (proton) that is ejected and the product nucleus, oxygen.

It must be remembered that this was prior to the discovery of the neutron. In the current view, a nucleus would not be regarded as consisting of other nuclei.

The classical electron radius is
$$\frac{e^2}{2m_oc^2} = 1.4 \times 10^{-13}cm$$

The binding energy.

of the latter; for the forces on the nuclei arising from such collisions appear to be greater than can be produced by any other agency at present available. Considering the enormous intensity of the forces brought into play, it is not so much a matter of surprise that the nitrogen atom should suffer disintegration as that the α particle itself escapes disruption into its constituents. The results as a whole suggest that, if α particles—or similar projectiles—of still greater energy were available for experiment we might expect to break down the nucleus structure of many of the lighter atoms.■

I desire to express my thanks to Mr. William Kay for his invaluable assistance in counting scintillations.

A remarkable prophecy, which turned out to be of greater import than Rutherford or his colleagues had any reason to expect.

SUPPLEMENTARY READING

Dampier, W. C., *A History of Science* (New York: Cambridge University, 1946), pp. 377ff.

Eve, A. S., *Rutherford* (New York: Macmillan, 1939).

Holton, G., and Roller, D. H. D., *Foundations of Modern Physical Science* (Cambridge, Mass.: Addison-Wesley, 1958), Ch. 34.

Jones, G. O., Rotblat, J., and Whitrow, G. J., *Atoms and The Universe* (New York: Scribner, 1956), pp. 33ff.

Rutherford, E., *Radioactive Substances and their Radiations* (Cambridge, Eng.: Cambridge University, 1913).

———, Chadwick, J., and Ellis, C. D., *Radiations from Radioactive Substances* (Cambridge, Eng.: Cambridge University, 1930).

20

James Chadwick
1891-1974

The Neutron

THE BRANCH of physics dealing with the nucleus, which had its origin in Rutherford's classic demonstration of induced transmutation in 1919, advanced along a broad front during the 1930's. A number of separate developments accounted for the very substantial progress during that period. In 1932 J. Chadwick discovered the *neutron,* thereby clarifying the entire problem of nuclear structure, but at the same time rendering more complex the concept of *elementary particle.* Prior to the discovery of the neutron only the proton and electron were known, a scheme which had the advantage of simplicity; but the need to postulate electrons in the nucleus introduced problems more serious than the loss in "economy of thought" brought on by the discovery of the neutron.

That year Carl D. Anderson discovered the *positron,* or positive electron, and Harold C. Urey discovered the hydrogen isotope of mass 2, *deuterium.* Two years later the team of Irène (1897–1956) and Frédéric Joliot-Curie (1900–1958) found that radioactivity could be induced artificially in many elements. The first of the *mesons* (the μ meson) was observed by Anderson and Neddermeyer in 1936, and in 1939 Otto Hahn and Fritz Strassmann, following some of the earlier work of Enrico Fermi (1901–1954), discovered the fission of uranium. The same period witnessed impressive advances in the *machines* of nuclear physics, the accelerators that played so important a role in the development of this new branch of physics. Both the *Van de Graaff generator* and Ernest Lawrence's *cyclotron* were developed in 1931; the first experiment with high-energy protons had been accomplished the year before by John Cockcroft and E. Walton with their voltage-multiplying generator at Cambridge. By the start of World War II the experimental techniques of nuclear physics were well advanced, even though the understanding of nuclear phenomena left much to be desired.

Chadwick was led to his discovery of the neutron by the observation, made in 1930 by Bothe and Becker, that when the light elements, such as boron and beryllium, were bombarded by α particles they seemed to emit penetrating gamma rays; at least the radiation was uncharged. On the assumption that this was gamma radiation the Joliot-Curies measured its absorption by various materials, during which they found that the radiation was able to eject high-energy protons from hydrogen-containing materials such as paraffin. It was at this point that Chadwick conducted the well-known experiments from which he concluded that the radiation, instead of being gamma rays, consisted of particles having the same mass as the proton, but no charge. He called these particles *neutrons*.

James Chadwick was born in Manchester, England, on October 20, 1891. He received his secondary-school education in Manchester, after which he attended the universities of Manchester and Cambridge, as well as Charlottenburg Institution in Berlin, where he studied under Hans Geiger (1882–1945), the developer, with Rutherford and W. Müller, of the Geiger counter. In 1921 he became a fellow of Gonville and Caius College, Cambridge, and two years later was appointed assistant director of radioactive research at the Cavendish Laboratory. There he collaborated with Rutherford and C. D. Ellis in writing the most comprehensive book on nuclear phenomena then available, a book that remained the authority in the field for many years.[1]

In 1932 Chadwick announced his identification of the neutron, for which he was awarded the Nobel prize in 1935. He was knighted in 1945, and in 1948 became master of Gonville and Caius College. Chadwick died in Cambridge, England on July 24, 1974.

The extract that follows, containing Chadwick's account of the discovery of the neutron, is taken from the *Proc. Roy. Soc. Lond.,* vol. A136 (1932), page 692. It has been reprinted in *Foundations of Nuclear Physics* by R. T. Beyer (New York: Dover, 1949).

---

[1] E. Rutherford, J. Chadwick, and C. D. Ellis, *Radiations from Radioactive Substances* (Cambridge, Eng.: Cambridge University, 1930).

Chadwick's Experiment

It was shown by Bothe and Becker* that some light elements when bombarded by α particles of polonium emit radiations which appear to be of the γ-ray type. The element beryllium gave a particularly marked effect of this kind, and later observations by Bothe, by Mme. Curie-

---

* *Z. Physik,* vol. 66 (1930), p. 289.

Properly, Mme.
Joliot-Curie.

$I = I_o \varepsilon^{-ax}$, where I_o is the initial intensity, I the intensity after passing through the thickness x, and a is the absorption coefficient. This assumes, of course, that the absorption is of this exponential form.

Calculated from the difference in mass between the two nuclei. For example, assuming that $_4Be^9$ captured an α particle to form $_6C^{13}$, the difference in mass between these nuclei is less than the α-particle mass by 10.5×10^6 electron volts. The agreement was largely fortuitous.

expansion chamber: cloud chamber.

An initial velocity of this value.

Joliot,*[■] and by Webster[†] showed that the radiation excited in beryllium possessed a penetrating power distinctly greater than that of any γ radiation yet found from the radioactive elements. In Webster's experiments the intensity of the radiation was measured both by means of the Geiger-Müller tube counter and in a high-pressure ionization chamber. He found that the beryllium radiation had an absorption coefficient in lead of about 0.22 cm^{-1} as measured under his experimental conditions.[■] Making the necessary corrections for these conditions, and using the results of Gray and Tarrant to estimate the relative contributions of scattering, photoelectric absorption, and nuclear absorption in the absorption of such penetrating radiation, Webster concluded that the radiation had a quantum energy of about 7×10^6 electron volts. Similarly he found that the radiation from boron bombarded by α particles of polonium consisted in part of a radiation rather more penetrating than that from beryllium, and he estimated the quantum energy of this component as about 10×10^6 electron volts.[■] These conclusions agree quite well with the supposition that the radiations arise by the capture of the α particle into the beryllium (or boron) nucleus and the emission of the surplus energy as a quantum of radiation.

The radiations showed, however, certain peculiarities, and at my request the beryllium radiation was passed into an expansion chamber[■] and several photographs were taken. No unexpected phenomena were observed though, as will be seen later, similar experiments have now revealed some rather striking events. The failure of these early experiments was partly due to the weakness of the available source of polonium, and partly to the experimental arrangement, which, as it now appears, was not very suitable.

Quite recently, Mme. Curie-Joliot and M. Joliot** made the very striking observation that these radiations from beryllium and from boron were able to eject protons with considerable velocities from matter containing hydrogen. In their experiments the radiation from beryllium was passed through a thin window into an ionization vessel containing air at room pressure. When paraffin wax, or other matter containing hydrogen, was placed in front of the window, the ionization in the vessel was increased, in some cases as much as doubled. The effect appeared to be due to the ejection of protons, and from further experiment they showed that the protons had ranges in air up to about 26 cm, corresponding to a velocity of nearly 3×10^9 cm per second.[■] They suggested

---

* I. Curie, *C. R. Acad. Sci. Paris*, vol. 193 (1931), p. 1412.
† *Proc. Roy. Soc.*, A, vol. 136 (1932), p. 428.
** Curie and Joliot, *C. R. Acad. Sci. Paris*, vol. 194 (1932), p. 273.

that energy was transferred from the beryllium radiation to the proton by a process similar to the Compton effect■ with electrons, and they estimated that the beryllium radiation had a quantum energy of about 50×10^6 electron volts. The range of the protons ejected by the boron radiation was estimated to be about 8 cm in air, giving on a Compton process an energy of about 35×10^6 electron volts for the effective quantum.*

There are two grave difficulties in such an explanation of this phenomenon. Firstly, it is well established that the frequency of scattering of high energy quanta by electrons is given with fair accuracy by the Klein-Nishina formula, and this formula should also apply to the scattering of quanta by a proton.■ The observed frequency of the proton scattering is, however, many thousand times greater than that predicted by this formula. Secondly, it is difficult to account for the production of a quantum of 50×10^6 electron volts from the interaction of a beryllium nucleus and an α particle of kinetic energy of 5×10^6 electron volts. The process which will give the greatest amount of energy available for radiation is the capture of the α particle by the beryllium nucleus, Be⁹, and its incorporation in the nuclear structure to form a carbon nucleus C¹³. The mass defect■ of the C¹³ nucleus is known both from data supplied by measurements of the artificial disintegration of boron B¹⁰ and from observations of the band spectrum of carbon; it is about 10×10^6 electron volts. The mass defect of Be⁹ is not known,■ but the assumption that it is zero will give a maximum value for the possible change of energy in the reaction Be⁹ + α → C¹³ + quantum. On this assumption it follows that the energy of the quantum emitted in such a reaction cannot be greater than about 14×10^6 electron volts. It must, of course, be admitted that this argument from mass defects is based on the hypothesis that the nuclei are made as far as possible of α particles; that the Be⁹ nucleus consists of 2 α particles + 1 proton + 1 electron and the C¹³ nucleus of 3 α particles + 1 proton + 1 electron.■ So far as the lighter nuclei are concerned, this assumption is supported by the evidence from experiments on artificial disintegration, but there is no general proof.

Accordingly, I made further experiments to examine the properties of the radiation excited in beryllium. It was found that the radiation ejects particles not only from hydrogen but from all other light elements which were examined. The experimental results were very difficult to explain on the hypothesis that the beryllium radiation was a quantum radiation, but followed immediately if it were supposed that the radiation

The *Compton effect* relates to the interaction of a photon with an electron, wherein the photon transfers only a part of its energy and momentum to the electron. This differs from the photoelectric effect, in which all the energy of the photon is given to an electron, less that required to overcome the binding energy.

A theoretical calculation of the scattering, since the proton has the same charge as the electron, the number of interactions should be the same for similar proton and electron *targets*.

The difference between the mass of the nucleus and the sum of the masses of its constituents.

It is 58.0 Mev.

The Be⁹ nucleus consists, in the modern view, of 4 protons and 5 neutrons. Similar arguments apply, nonetheless.

---

* Many of the arguments of the subsequent discussion apply equally to both radiations, and the term "beryllium radiation" may often be taken to inlcude the boron radiation.

Chadwick applied the name *neutron*, thinking it to be the electron-proton combination proposed much earlier by Rutherford.

valve: electron tube. A *valve counter* refers to an electronic detector; in this case an ionization chamber connected to an electronic amplifier.

consisted of particles of mass nearly equal to that of a proton and with no net charge, or neutrons.■

2.   Observations of Recoil Atoms

The properties of the beryllium radiation were first examined by means of the valve counter used in the work* on the artificial disintegration by α particles and described fully there.■ Briefly, it consists of a small ionization chamber connected to a valve amplifier. The sudden production of ions in the chamber by the entry of an ionizing particle is

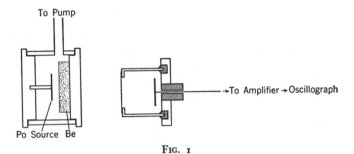

To Pump

To Amplifier → Oscillograph

Po Source Be

Fig. 1

Radium-beryllium mixtures later came into wide use as neutron sources.

That is, equivalent in stopping power to 4.5 cm of air.

The *natural*, or *background* effect arises primarily from radioactive contamination in the walls of the chamber and from cosmic rays.

Neutrons pass readily through materials of high atomic number, since the energy that can be transferred in a collision is greater the more nearly equal are the two masses.

detected by means of an oscillograph connected in the output circuit of the amplifier. The deflections of the oscillograph were recorded photographically on a film of bromide paper.

The source of polonium was prepared from a solution of radium (D+E+F)† by deposition on a disc of silver. The disc had a diameter of 1 cm and was placed close to a disc of pure beryllium■ of 2 cm diameter, and both were enclosed in a small vessel which could be evacuated (Fig. 1). The first ionization chamber used had an opening of 13 mm covered with aluminium foil of 4.5 cm air equivalent,■ and a depth of 15 mm. This chamber had a very low natural effect, giving on the average only about 7 deflections per hour.■

When the source vessel was placed in front of the ionization chamber, the number of deflections immediately increased. For a distance of 3 cm between the beryllium and the counter the number of deflections was nearly 4 per minute. Since the number of deflections remained sensibly the same when thick metal sheets, even as much as 2 cm of lead,■ were interposed between the source vessel and the counter, it was clear that these deflections were due to a penetrating radiation emitted from the

---

* Chadwick, Constable, and Pollard, *Proc. Roy. Soc.*, A, vol. 130 (1931), p. 463.

† The radium D was obtained from old radon tubes generously presented by Dr. C. F. Burnam and Dr. F. West of the Kelly Hospital, Baltimore.

beryllium. It will be shown later that the deflections were due to atoms of nitrogen set in motion by the impact of the beryllium radiation.

When a sheet of paraffin wax about 2 mm thick was interposed in the path of the radiation just in front of the counter, the number of deflections recorded by the oscillograph increased markedly. This increase was due to particles ejected from the paraffin wax so as to pass into the counter. By placing absorbing screens of aluminium between the wax and the counter an absorption curve was obtained. From this curve it appears that the particles have a maximum range of just over 40 cm of air, assuming that an Al foil of 1.64 mg per square centimeter is equivalent to 1 cm of air. By comparing the sizes of the deflections (proportional to the number of ions produced in the chamber)■ due to these particles with those due to protons of about the same range it was obvious that the particles were protons. From the range-velocity curve for protons we deduce therefore that the maximum velocity imparted to a proton by the beryllium radiation is about 3.3×10^9 cm per second, corresponding to an energy of about 5.7×10^6 electron volts.

Which was a function of the energy of the protons.

The effect of exposing other elements to the beryllium radiation was then investigated. An ionization chamber was used with an opening covered with a gold foil of 0.5 mm air equivalent. The element to be examined was fixed on a clean brass plate and placed very close to the counter opening. In this way lithium, beryllium, boron, carbon, and nitrogen, as paracyanogen, were tested. In each case the number of deflections observed in the counter increased when the element was bombarded by the beryllium radiation. The ranges of the particles ejected from these elements were quite short, of the order of some millimeters in air. The deflections produced by them were of different sizes, but many of them were large compared with the deflection produced even by a slow proton.■ The particles therefore have a large ionizing power and are probably in each case recoil atoms of the elements. Gases were investigated by filling the ionization chamber with the required gas by circulation for several minutes. Hydrogen, helium, nitrogen, oxygen, and argon were examined in this way. Again, in each case deflections were observed which were attributed to the production of recoil atoms in the different gases. For a given position of the beryllium source relative to the counter, the number of recoil atoms was roughly the same for each gas. This point will be referred to later. It appears then that the beryllium radiation can impart energy to the atoms of matter through which it passes and that the chance of an energy transfer does not vary widely from one element to another.

The ionization produced by such charged particles is inversely proportional to the energy and directly proportional to the square of the charge.

It has been shown that protons are ejected from paraffin wax with energies up to a maximum of about 5.7×10^6 electron volts. If the ejection be ascribed to a Compton recoil from a quantum of radiation, then the energy of the quantum must be about 55×10^6 electron volts,■ for the maximum energy which can be given to a mass m by a quantum $h\nu$ is

Determined simply by application of the conservation laws.

$\dfrac{2}{2+mc^2/h\nu} \cdot h\nu$. The energies of the recoil atoms produced by this radiation by the same process in other elements can be readily calculated. For example, the nitrogen recoil atoms should have energies up to a maximum of 450,000 electron volts. Taking the energy necessary to form a pair of ions in air as 35 electron volts,■ the recoil atoms of nitrogen should produce not more than about 13,000 pairs of ions. Many of the deflections observed with nitrogen, however, corresponded to far more ions than this; some of the recoil atoms produced from 30,000 to 40,000 ion pairs. In the case of the other elements a similar discrepancy was noted between the observed energies and ranges of the recoil atoms and the values calculated on the assumption that the atoms were set in motion by recoil from a quantum of 55×10^6 electron volts. The energies of the recoil atoms were estimated from the number of ions produced in the counter, as given by the size of the oscillograph deflections. A sufficiently good measurement of the ranges could be made either by varying the distance between the element and the counter or by interposing thin screens of gold between the element and the counter.

Where c is the velocity of light.

This is an average value, good for most gases.

The nitrogen recoil atoms were also examined, in collaboration with Dr. N. Feather, by means of the expansion chamber. The source vessel was placed immediately above an expansion chamber of the Shimizu type, so that a large proportion of the beryllium radiation traversed the chamber.■ A large number of recoil tracks was observed in the course of a few hours. Their range, estimated by eye, was sometimes as much as 5 or 6 mm in the chamber, or, correcting for the expansion, about 3 mm in standard air. These visual estimates were confirmed by a preliminary series of experiments by Dr. Feather with a large automatic expansion chamber, in which photographs of the recoil tracks in nitrogen were obtained. Now the ranges of recoil atoms of nitrogen of different velocities have been measured by Blackett and Lees. Using their results we find that the nitrogen recoil atoms produced by the beryllium radiation may have a velocity of at least 4×10^8 cm per second, corresponding to an energy of about 1.2×10^6 electron volts. In order that the nitrogen nucleus should acquire such an energy in a collision with a quantum of radiation, it is necessary to assume that the energy of the quantum should be about 90×10^6 electron volts, if energy and momentum are conserved in the collision. It has been shown that a

This form of cloud chamber is operated by reducing the pressure suddenly so as to cool the gas and cause condensation of the vapor onto the ions. Hence, at the time of operation, the pressure in the chamber is below atmospheric and the range of particles is correspondingly greater.

quantum of 55×10^6 electron volts is sufficient to explain the hydrogen collisions. In general, the experimental results show that if the recoil atoms are to be explained by collision with a quantum, we must assume a larger and larger energy for the quantum as the mass of the struck atom increases.

3. The Neutron Hypothesis

It is evident that we must either relinquish the application of the conservation of energy and momentum in these collisions or adopt another hypothesis about the nature of the radiation.■ If we suppose that the radiation is not a quantum radiation, but consists of particles of mass very nearly equal to that of the proton, all the difficulties connected with the collisions disappear, both with regard to their frequency and to the energy transfer to different masses. In order to explain the great penetrating power of the radiation we must further assume that the particle has no net charge. We may suppose it to consist of a proton and an electron in close combination,■ the "neutron" discussed by Rutherford* in his Bakerian Lecture of 1920.

When such neutrons pass through matter they suffer occasionally close collisions with the atomic nuclei and so give rise to the recoil atoms which are observed. Since the mass of the neutron is equal to that of the proton, the recoil atoms produced when the neutrons pass through matter containing hydrogen will have all velocities up to a maximum which is the same as the maximum velocity of the neutrons.■ The experiments showed that the maximum velocity of the protons ejected from paraffin wax was about 3.3×10^9 cm per second. This is therefore the maximum velocity of the neutrons emitted from beryllium bombarded by α particles of polonium. From this we can now calculate the maximum energy which can be given by a colliding neutron to other atoms,■ and we find that the results are in fair agreement with the energies observed in the experiments. For example, a nitrogen atom will acquire in a head-on collision with the neutron of mass 1 and velocity 3.3×10^9 cm per second a velocity of 4.4×10^8 cm per second, corresponding to an energy of 1.4×10^6 electron volts, a range of about 3.3 mm in air, and a production of ions of about 40,000 pairs. Similarly, an argon atom may

This was seriously considered, of course, but the conservation laws had proved to have such universal application that they could be relinquished only with great reluctance.

It is no longer considered as such a combination, but a separate particle entirely.

The maximum results from a head-on collision.

By straightforward *billiard-ball* collisions.

* Rutherford, *Proc. Roy. Soc.*, A, vol. 97 (1920), p. 374. Experiments to detect the formation of neutrons in a hydrogen discharge tube were made by J. L. Glasson, *Phil. Mag.*, vol. 42 (1921), p. 596, and by J. K. Roberts, *Proc. Roy. Soc.*, A, vol. 102 (1922), p. 72. Since 1920 many experiments in search of these neutrons have been made in this laboratory.

acquire an energy of 0.54×10^6 electron volts, and produce about 15,000 ion pairs. Both these values are in good accord with experiment.*

It is possible to prove that the mass of the neutron is roughly equal to that of the proton, by combining the evidence from the hydrogen collisions with that from the nitrogen collisions. In the succeeding paper, Feather records experiments in which about 100 tracks of nitrogen recoil atoms have been photographed in the expansion chamber. The measurement of the tracks shows that the maximum range of the recoil atoms is 3.5 mm in air at 15° C and 760 mm pressure, corresponding to a velocity of 4.7×10^8 cm per second according to Blackett and Lees. If M, V be the mass and velocity of the neutron, then the maximum velocity given to a hydrogen atom is■

By direct application of the conservation laws. The velocities are low enough so that relativistic effects are negligible.

$$v_p = \frac{2M}{M + 1} \cdot V$$

and the maximum velocity given to a nitrogen atom is

$$v_n = \frac{2M}{M + 14} \cdot V$$

whence

$$\frac{M + 14}{M + 1} = \frac{v_p}{v_n} = \frac{3.3 \times 10^9}{4.7 \times 10^8}$$

and

$$M = 1.15.$$

The total error in the estimation of the velocity of the nitrogen recoil atom may easily be about 10 percent, and it is legitimate to conclude that the mass of the neutron is very nearly the same as the mass of the proton.

We have now to consider the production of the neutrons from beryllium by the bombardment of the α particles. We must suppose than an α particle is captured by a Be⁹ nucleus with the formation of a carbon C¹² nucleus and the emission of a neutron. The process is analogous to

Such as found by Rutherford.

the well-known artificial disintegrations,■ but a neutron is emitted instead of a proton. The energy relations of this process cannot be exactly deduced, for the masses of the Be⁹ nucleus and the neutron are not known

They are now known very accurately.

accurately.■ It is, however, easy to show that such a process fits the experimental facts. We have

$$\text{Be}^9 + \text{He}^4 + \text{kinetic energy of } \alpha$$

$$= \text{C}^{12} + n^1 + \text{kinetic energy of C}^{12} + \text{kinetic energy of } n^1.$$

---

* It was noted that a few of the nitrogen recoil atoms produced about 50,000 to 60,000 ion pairs. These probably correspond to the cases of disintegration found by Feather and described in his paper.

If we assume that the beryllium nucleus consists of two α particles and a neutron, then its mass cannot be greater than the sum of the masses of these particles, for the binding energy corresponds to a defect of mass. The energy equation becomes

$$(8.00212 + n^1) + 4.00106 + K.E. \text{ of } \alpha > 12.0003 + n^1$$
$$+ K.E. \text{ of } C^{12} + K.E. \text{ of } n^1$$

or

$$K.E. \text{ of } n^1 < K.E. \text{ of } \alpha + 0.003 - K.E. \text{ of } C^{12}.$$

Since the kinetic energy of the α particle of polonium is 5.25×10^6 electron volts, it follows that the energy of emission of the neutron cannot be greater than about 8×10^6 electron volts. The velocity of the neutron must therefore be less than 3.9×10^9 cm per second. We have seen that the actual maximum velocity of the neutron is about 3.3×10^9 cm per second, so that the proposed disintegration process is compatible with observation.

■ ....

Omitted is a brief discussion of the neutrons emitted in the *backward* direction from the beryllium target.

4. The Nature of the Neutron

It has been shown that the origin of the radiation from beryllium bombarded by α particles and the behavior of the radiation, so far as its interaction with atomic nuclei is concerned, receive a simple explanation on the assumption that the radiation consists of particles of mass nearly equal to that of the proton but which have no charge. The simplest hypothesis one can make about the nature of the particle is to suppose that it consists of a proton and an electron in close combination, giving a net charge 0 and a mass which should be slightly less than the mass of the hydrogen atom.■ This hypothesis is supported by an examination of the evidence which can be obtained about the mass of the neutron.

Because of the greater binding energy between electron and proton in this *assumed* combination.

As we have seen, a rough estimate of the mass of the neutron was obtained from measurements of its collisions with hydrogen and nitrogen atoms, but such measurements cannot be made with sufficient accuracy for the present purpose. We must turn to a consideration of the energy relations in a process in which a neutron is liberated from an atomic nucleus; if the masses of the atomic nuclei concerned in the process are accurately known, a good estimate of the mass of the neutron can be deduced. The mass of the beryllium nucleus has, however, not yet been measured, and, as was shown in part 3, only general conclusions can be drawn from this reaction. Fortunately, there remains the case of boron. It was stated in part 1 that boron bombarded by α particles of polonium also emits a radiation which ejects protons from materials containing hydrogen. Further examination showed that this radiation behaves in all respects like that from beryllium, and it must therefore be assumed to consist of neutrons. It is probable that the neutrons are emitted from the isotope B^{11}, for we know that the isotope B^{10} disintegrates with the emission of a proton.* The process of disintegration will then be

$$B^{11} + He^4 \rightarrow N^{14} + n^1.$$

With the mass spectrograph.

The masses of B^{11} and N^{14} are known from Aston's measurements,■ and the further data required for the deduction of the mass of the neutron can be obtained by experiment.

In the source vessel of Figure 1 the beryllium was replaced by a target of powdered boron, deposited on a graphite plate. The range of the protons ejected by the boron radiation was measured in the same way as with the beryllium radiation. The effects observed were much smaller than with beryllium, and it was difficult to measure the range of the protons accurately. The maximum range was about 16 cm in air, corresponding to a velocity of 2.5×10^9 cm per second. This then is the maximum velocity of the neutron liberated from boron by an α particle of polonium of velocity 1.59×10^9 cm per second. Assuming that momentum is conserved in the collision, the velocity of the recoiling N^{14} nucleus can be calculated, and we then know the kinetic energies of all the particles concerned in the disintegration process. The energy equation of the process is

$$\text{Mass of } B^{11} + \text{mass of } He^4 + K.E. \text{ of } He^4$$
$$= \text{mass of } N^{14} + \text{mass of } n^1 + K.E. \text{ of } N^{14} + K.E. \text{ of } n^1.$$

These masses are somewhat in error, by about 0.05 percent.

The masses are $B^{11} = 11.00825 \pm 0.0016$; $He^4 = 4.00106 \pm 0.0006$; $N^{14} = 14.0042 \pm 0.0028$.■ The kinetic energies in mass units are α particle $= 0.00565$; neutron $= 0.0035$; and nitrogen nucleus $= 0.00061$. We find therefore that the mass of the neutron is 1.0067. The errors quoted for the mass measurements are those given by Aston. They are the maximum errors which can be allowed in his measurements, and the probable error may be taken as about one quarter of these.† Allowing for the errors in the mass measurements it appears that the mass of the neutron cannot be less than 1.003, and that it probably lies between 1.005 and 1.008.■

The presently accepted value is 1.008982 (±3).

---

* Chadwick, Constable, and Pollard, *loc. cit.*

† The mass of B^{11} relative to B^{10} has been checked by optical methods by Jenkins and McKellar (*Phys. Rec.*, vol. 39 (1932), p. 549. Their value agrees with Aston's to 1 part in 10^5. This suggests that great confidence may be put in Aston's measurements.

Such a value for the mass of the neutron is to be expected if the neutron consists of a proton and an electron, and it lends strong support to this view. Since the sum of the masses of the proton and electron is 1.0078, the binding energy, or mass defect, of the neutron is about 1 to 2 million electron volts. This is quite a reasonable value. We may suppose that the proton and electron form a small dipole, or we may take the more attractive picture of a proton embedded in an electron.■ On either view, we may expect the "radius" of the neutron to be a few times 10^{-13} cm.■

■ ⋯⋯⋯⋯⋯⋯⋯⋯⋯⋯

GENERAL REMARKS

It is of interest to examine whether other elements, besides beryllium and boron, emit neutrons when bombarded by α particles. So far as experiments have been made, no case comparable with these two has been found. Some evidence was obtained of the emission of neutrons from fluorine and magnesium, but the effects were very small, rather less than 1 percent of the effect obtained from beryllium under the same conditions. There is also the possibility that some elements may emit neutrons spontaneously, e.g., potassium, which is known to emit a nuclear β radiation accompanied by a more penetrating radiation. Again no evidence was found of the presence of neutrons, and it seems fairly certain that the penetrating type is, as has been assumed, a γ radiation.■

Although there is certain evidence for the emission of neutrons only in two cases of nuclear transformations, we must nevertheless suppose that the neutron is a common constituent of atomic nuclei.■ We may then proceed to build up nuclei out of α particles, neutrons, and protons, and we are able to avoid the presence of uncombined electrons in a nucleus. This has certain advantages for, as is well known, the electrons in a nucleus have lost some of the properties which they have outside, e.g., their spin and magnetic moment.■ If the α particle, the neutron, and the proton are the only units of nuclear structure, we can proceed to calculate the mass defect or binding energy of a nucleus as the difference between the mass of the nucleus and the sum of the masses of the constituent particles. It is, however, by no means certain that the α particle and the neutron are the only complex particles in the nuclear structure, and therefore the mass defects calculated in this way may not be the true binding energies of the nuclei.■ In this connection it may be noted that the examples of disintegration discussed by Dr. Feather are not all of one type, and he suggests that in some cases a particle of mass

There are serious theoretical difficulties involved in such a picture.

This is the correct order of magnitude.

Omitted is a discussion of neutron interactions with nuclei.

This is known definitely to be the case.

Many more (α-n) reactions are now known, i.e., reactions in which neutrons are produced by α-particle bombardment.

That is, if electrons were assumed in the nucleus, their spins and magnetic moments had to be ignored, for they did not then correspond to the observed facts.

The α particle is no longer regarded as a constituent of the nucleus; instead, it loses its identity within the nucleus, becoming simply two protons and two neutrons

2 and charge 1, the hydrogen isotope recently reported by Urey, Brick-wedde, and Murphy, may be emitted. It is indeed possible that this particle also occurs as a unit of nuclear structure.■

Not in the present picture, which regards it simply as a proton and neutron.

It has so far been assumed that the neutron is a complex particle consisting of a proton and an electron. This is the simplest assumption and it is supported by the evidence that the mass of the neutron is about 1.006, just a little less than the sum of the masses of a proton and an electron. Such a neutron would appear to be the first step in the combination of the elementary particles towards the formation of a nucleus. It is obvious that this neutron may help us to visualize the building up of more complex structures, but the discussion of these matters will not be pursued further for such speculations, though not idle, are not at the moment very fruitful. It is, of course, possible to suppose that the neutron may be an elementary particle.■ This view has little to recommend it at present, except the possibility of explaining the statistics of such nuclei as N^{14}.

This actually is the present view, namely, that the neutron is an elementary particle in the same sense as the proton and the electron rather than being a combination of the latter.

There remains to discuss the transformations which take place when an α particle is captured by a beryllium nucleus, Be^9. The evidence given here indicates that the main type of transformation is the formation of a C^{12} nucleus and the emission of a neutron. The experiments of Curie-Joliot and Joliot,* of Auger,† and of Dee show quite definitely that there is some radiation emitted by beryllium which is able to eject fast electrons in passing through matter. I have made experiments using the Geiger point counter to investigate this radiation and the results suggest that the electrons are produced by a γ radiation.■ There are two distinct processes which may give rise to such a radiation. In the first place, we may suppose that the transformation Be^9 to C^{12} takes place sometimes with the formation of an excited C^{12} nucleus which goes to the ground state with the emission of γ radiation. This is similar to the transformations which are supposed to occur in some cases of disintegration with proton emission, e.g., B^{10}, F^{19}, Al^{27}; the majority of transformations occur with the formation of an excited nucleus, only in about one quarter is the final state of the residual nucleus reached in one step. We should then have two groups of neutrons of different energies and a γ radiation of quantum energy equal to the difference in energy of the neutron groups. The quantum energy of this radiation must be less than the maximum energy of the neutrons emitted, about 5.7×10^6 electron volts. In the second place, we may suppose that occasionally the beryllium nucleus changes to a C^{15} nucleus and that all the surplus energy is

Via the Compton effect, or by internal conversion.

* C. R. Acad. Sci. Paris, vol. 194 (1932), p. 708 and p. 876.
† C. R. Acad. Sci. Paris, vol. 194 (1932), p. 877.

emitted as radiation. In this case the quantum energy of the radiation may be about 10×10^6 electron volts.

It is of interest to note that Webster has observed a soft radiation from beryllium bombarded by polonium α particles, of energy about 5×10^5 electron volts. This radiation may well be ascribed to the first of the two processes just discussed, and its intensity is of the right order. On the other hand, some of the electrons observed by Curie-Joliot and Joliot had energies of the order of 2 to 10×10^6 volts, and Auger recorded one example of an electron of energy about 6.5×10^6 volts. These electrons may be due to a hard γ radiation produced by the second type of transformation.*

It may be remarked that no electrons of greater energy than the above appear to be present. This is confirmed by an experiment† made in this laboratory by Dr. Occhialini. Two tube counters were placed in a horizontal plane and the number of coincidences recorded by them was observed by means of the method devised by Rossi.■ The beryllium source was then brought up in the plane of the counters so that the radiation passed through both counters in turn. No increase in the number of coincidences could be detected. It follows that there are few, if any, β rays produced with energies sufficient to pass through the walls of both counters, a total of 4 mm brass; that is, with energies greater than about 6×10^6 volts. This experiment further shows that the neutrons very rarely produce coincidences in tube counters under the usual conditions of experiment.

The Rossi coincidence technique is an electronic means for determining the nearly simultaneous passage of a particle through two detectors.

In conclusion, I may restate briefly the case for supposing that the radiation the effects of which have been examined in this paper consists of neutral particles rather than of radiation quanta. Firstly, there is no evidence from electron collisions of the presence of a radiation of such a quantum energy as is necessary to account for the nuclear collisions. Secondly, the quantum hypothesis■ can be sustained only by relinquishing the conservation of energy and momentum. On the other hand, the neutron hypothesis gives an immediate and simple explanation of the experimental facts; it is consistent in itself and it throws new light on the problem of nuclear structure.

The emission of gamma radiation.

---

* Although the presence of fast electrons can be easily explained in this way, the possibility that some may be due to secondary effects of the neutrons must not be lost sight of.

† *Cf.* also Rasetti, *Naturwiss.*, vol. 20 (1932), p. 252.

SUMMARY

The properties of the penetrating radiation emitted from beryllium (and boron) when bombarded by the α particles of polonium have been examined. It is concluded that the radiation consists, not of quanta as hitherto supposed, but of neutrons, particles of mass 1, and charge 0. Evidence is given to show that the mass of the neutron is probably between 1.005 and 1.008. This suggests that the neutron consists of a proton and an electron in close combination, the binding energy being about 1 to 2 $\times$ 10^6 electron volts. From experiments on the passage of the neutrons through matter the frequency of their collisions with atomic nuclei and with electrons is discussed.

I wish to express my thanks to Mr. H. Nutt for his help in carrying out the experiments.

SUPPLEMENTARY READING

Dampier, W. C., *A History of Science* (New York: Macmillan, 1946), pp. 419ff.

Holton, G., and Roller, D. H. D., *Foundations of Modern Physical Science* (Cambridge, Mass.: Addison-Wesley, 1958), Ch. 38.

Jones, G. O., Rotblat, J., and Whitrow, G. J., *Atoms and the Universe* (New York: Scribner's, 1956), Chs. 1–3.

Rutherford, E., Chadwick, J., and Ellis, C. D., *Radiations from Radioactive Substances* (Cambridge, Eng.: Cambridge University, 1930).

APPENDIX

James Clerk Maxwell
1831-1879

The Electromagnetic Field

OF THE major achievements in physics two may be singled out for the manner in which they served to synthesize great bodies of knowledge. Toward the end of the seventeenth century Newton published his famous *Principia* (see Chapter 4), in which he unified, in terms of a few simple laws, all that was then known of dynamics. Except for the corrections required to accord with the theory of relativity for bodies traveling at very high speed, rarely encountered except in subatomic phenomena, Newton's laws of motion remain the foundation for the science of mechanics. Little was known in Newton's time about the nature of light; electrostatic and magnetostatic phenomena had been observed by Franklin, Gilbert, and others, but the connection between electricity and magnetism, or between these and light, as it later turned out, went unsuspected. Nearly two centuries later Maxwell did for electromagnetic phenomena what Newton had accomplished for mechanics. He summed up everything that was then known concerning light, electricity, and magnetism. But this was not all. He formulated the mathematical structure, now known as *Maxwell's equations,* that pointed up the unity of the "ether" and formed the basis for all of *electromagnetic theory*. He predicted the existence of electric waves propagating through space, discovered later by Hertz (see Chapter 13), and he contributed, no less successfully, to other branches of physics, notably to the kinetic theory of gases.

James Clerk Maxwell was born to a well-to-do family in Edinburgh, Scotland, on June 13, 1831, a time when Faraday was in the midst of his most important electrical discoveries and only a few years before Lenz came upon his principle of electric induction. He grew up in a period noted chiefly, in physics, for progress in electricity, thermodynamics, and kinetic theory, and for the first clear formulation, by Hermann von Helmholtz (1821-1894), of a general principle of energy conservation.

Maxwell was an inquisitive but not precocious youngster who was tutored privately for a time before entering Edinburgh Academy at the age of ten. There, after a slow start, he began to display extraordinary talents, not only in mathematics, in which he excelled, but also in the writing of English verse, a practice by which he delighted his friends all his life. After spending six years at the academy and three years at the University of Edinburgh Maxwell went on to Cambridge, where he was elected a scholar in Trinity College and received his degree with high honors in 1854.

He remained at Trinity for another two years, studying Faraday's works and engaging in his own researches on mathematics, geometrical optics, and a theory of color. In 1856 he was elected to the Professorship in Natural Philosophy at Marischal College, Aberdeen, where he completed the first of his many remarkable contributions in mathematical physics. This was his Adams Prize essay[1] on the stability of the rings of Saturn, a work that placed him among the front ranks of his contemporaries. It was here also that he became interested in the kinetic theory of gases and solved the problem of the distribution of velocities among the molecules of a gas, known generally as the *Maxwellian distribution*. While his formal proof of this important law did not go unchallenged, there is no doubt of the correctness of the final result.

In 1860 Maxwell was appointed Professor of Natural Philosophy in King's College, London, where he remained for the next five years, a period that was his most creative. He completed his work on the theory of color, developed his theory of electricity and magnetism, contributed further to the kinetic theory of gases, and investigated experimentally the viscosity of air at different temperatures and pressures. The last of these formed the subject of a Bakerian Lecture which Maxwell presented to the Royal Society early in 1866. The same year he published a paper on the *Dynamical Theory of Gases,* in which certain errors in his earlier work on kinetic theory, pointed out by Rudolph Clausius (1822-1888), were corrected. It was also during this period that he took active part, together with B. Stewart and F. Jenkin, in experiments to determine the value of the *ohm* in absolute measure.

Maxwell resigned his professorship at the end of the 1865 academic session to devote his time more fully to his scientific studies, and to the study of English literature, which he greatly enjoyed. During the next few years he completed the major part of his classic treatise on electromagnetic

---

[1] A prize established in 1848 to be awarded periodically for the best solution to some problem of great scientific importance proposed by the examiners.

theory, although it was not published until 1873.[2] While in retirement Maxwell partly inspired and lent active support to a movement to establish a chair in experimental physics and a physical laboratory in Cambridge University. In 1871 the university approved such a chair and Maxwell was appointed professor of experimental physics and director of the newly established Cavendish Laboratory, named for one of the most distinguished experimenters ever associated with Cambridge, Henry Cavendish (see Chapter 6).

Maxwell gave much of his time during the next few years to the building and furnishing of the new laboratory, which was officially opened in 1874 and soon became one of the leading physical research laboratories in the world. His own interests during this period were given chiefly to lecturing and to the task of editing the papers of Henry Cavendish,[3] whose unpublished work on theoretical and experimental electricity impressed Maxwell by its originality and by the fact that it anticipated several discoveries later made by others. Maxwell's crowning achievement was, of course, his theory of the *electromagnetic field,* in which, among other things he showed light to be an electromagnetic phenomenon, in the same sense as electricity and magnetism, and pictured the propagation of electric waves through the *ether.* Unfortunately, he did not live to see experimental confirmation of his prediction, for he died in the prime of his career, at the age of forty-eight, in November 1879, eight years before Hertz demonstrated the existence of electric waves. It was then that Maxwell's genius was fully recognized and his lasting fame assured.

Maxwell's contribution to the development of physics goes far beyond the solutions that he found to particular problems. As Einstein pointed out in commemoration of Maxwell's birth:

> We may say that, before Maxwell, Physical Reality, in so far as it was to represent the processes of nature, was thought of as consisting in material particles, whose variations consist only in movements governed by partial differential equations. Since Maxwell's time, Physical Reality has been thought of as represented by continuous fields, governed by partial differential equations, and not capable of any mechanical interpretation. This change in the conception of Reality is the most profound and the most fruitful that physics has experienced since the time of Newton. . . .[4]

---

[2] J. C. Maxwell, *A Treatise on Electricity and Magnetism* (1st ed., 1873; 2d ed. 1881; 3d ed. 1891; 3d ed. reprinted by Dover, New York, 1954).

[3] J. C. Maxwell, *The Electrical Researches of the Hon. Henry Cavendish* (Cambridge, Eng.: Cambridge University, 1879).

[4] A. Einstein, in *James Clerk Maxwell, A Commemoration Volume* (New York: The Macmillan Company, 1931), p. 71.

Maxwell gave mathematical form to Faraday's conceptions of electrical phenomena. He derived a set of equations relating all known electric and magnetic phenomena; that is, the equations give quantitative relations between the electric and magnetic fields, and the charges, currents, and time-varying currents producing these fields. They contain Coulomb's law of force between electric charges, as well as his corresponding law for magnetic poles (Chapter 5), Oersted's discovery of the magnetic effect of an electric current (Chapter 9), Ampère's work in electrodynamics, Ohm's law relating the current in a conductor to the potential difference across it, Faraday's law of electromagnetic induction (Chapter 10), and, of course, Lenz's law (Chapter 11). They contain all this and more, for they include Maxwell's hypothesis that electric waves should proceed from oscillating electric currents and travel through free space with the velocity of light. Given the electric and magnetic forces everywhere in space at some initial time, Maxwell's equations permit one to calculate them for all future time. Faraday found it useful to regard *action at a distance,* typical of electric and magnetic phenomena (and gravitational as well), in terms of *lines of force,* which were thought of somewhat as mechanical linkages in the medium surrounding a magnetic body or an electric charge. The medium thus became the seat of the electric and magnetic field, and since the effects could be transmitted through empty space, it was clear that the medium could not be of ordinary material form. The concept of an elastic *ether* as the medium for transmission of electric and magnetic forces grew by analogy with the physics of fluids, for it was difficult to conceive of a medium without some material form of its own. In fact, there is a similarity in structure and range of application between Maxwell's equations and the fundamental equations of fluid dynamics. Despite the fact that the ether envisoned by Maxwell is no longer considered a useful physical concept because of inconsistencies disclosed by the theory of relativity, the idea of an electromagnetic field associated with the observed phenomena remains an important conceptual feature of electromagnetism.

Beyond the purely electrical consequences of Maxwell's theory may be found several distinctive features. He assumed the continuity of current; that is, that all currents flow in closed circuits. He introduced the idea that energy resides throughout the electromagnetic field, rather than in the conductors alone. He showed the identity of the electromagnetic medium with the *luminiferous ether* (the medium by which light is propagated) and concluded thereby that light is an electromagnetic phenomena. And he introduced the concept of *displacement current,* which in free space implies that when electric charge flows through a medium there results a current in addition to that represented by the motion of the charges. This current he interpreted as being connected with a "displacement" of the electromagnetic medium, which accorded with his view of the ether as an elastic medium subject to stresses and strains; these giving rise to electric and magnetic forces.

The extract that follows is taken from an early paper of Maxwell's, written during his last year at King's College, entitled *A Dynamical Theory of the Electromagnetic Field,* Philosophical Transactions, vol. 155 (1865), page 459.

Maxwell's "Experiment"

PART I. INTRODUCTORY

(1) The most obvious mechanical phenomenon in electrical and magnetical experiments is the mutual action by which bodies in certain states set each other in motion while still at a sensible distance from each other.■ The first step, therefore, in reducing these phenomena into scientific form, is to ascertain the magnitude and direction of the force acting between the bodies, and when it is found that this force depends in a certain way upon the relative position of the bodies and on their electric or magnetic condition, it seems at first sight natural to explain the facts by assuming the existence of something either at rest or in motion in each body, constituting its electric or magnetic state, and capable of acting at a distance according to mathematical laws.

In this way mathematical theories of statical electricity, of magnetism, of the mechanical action between conductors carrying currents, and of the induction of currents have been formed. In these theories the force acting between the two bodies is treated with reference only to the condition of the bodies and their relative position, and without any express consideration of the surrounding medium.■

These theories assume, more or less explicitly, the existence of substances the particles of which have the property of acting on one another at a distance by attraction or repulsion. The most complete development of a theory of this kind is that of W. Weber,[1]■ who has made the same theory include electrostatic and electromagnetic phenomena.

In doing so, however, he has found it necessary to assume that the

The so-called *action-at-a-distance* phenomena.

That is, they were not field theories.

Wilhelm Weber (1804-1891), the first to attempt a comprehensive synthesis of the electrical knowledge then known. In the course of his work he found that the velocity of light was somehow related to electrical units, an idea later developed fully by Maxwell.

[1] *Electrodynamische Maassbestimmungen,* Leipzic trans. (1849), vol. i, and Taylor's *Scientific Memoirs,* vol. v, art. xiv.

Diamagnetic materials have permeability less than 1, and tend to move from stronger to weaker regions of a magnetic field, which is just opposite to the behavior of *paramagnetic* materials. Substances which are strongly paramagnetic, such as iron, are known as *ferromagnetic*.

Maxwell (and Faraday) found it difficult to conceive of action-at-a-distance forces without consideration of the intervening medium.

sensible: appreciable

Heinrich Geissler (1814-1879), a well-known glass blower in Tübingen, constructed tubes containing small traces of gas for electrical discharge experiments.

force between two electric particles depends on their relative velocity, as well as on their distance.

This theory, as developed by W. Weber and C. Neumann,[2] is exceedingly ingenious and wonderfully comprehensive in its application to the phenomena of statical electricity, electromagnetic attractions, induction of currents and diamagnetic phenomena■; and it comes to us with the more authority, as it has served to guide the speculations of one who has made so great an advance in the practical part of electric science, both by introducing a consistent system of units in electrical measurement, and by actually determining electrical quantities with an accuracy hitherto unknown.

(2) The mechanical difficulties, however, which are involved in the assumption of particles acting at a distance with forces which depend on their velocities are such as to prevent me from considering this theory as an ultimate one, though it may have been, and may yet be useful in leading to the coordination of phenomena.

■I have therefore preferred to seek an explanation of the fact in another direction, by supposing them to be produced by actions which go on in the surrounding medium as well as in the excited bodies, and endeavoring to explain the action between distant bodies without assuming the existence of forces capable of acting directly at sensible■ distances.

(3) The theory I propose may therefore be called a theory of the *electromagnetic field,* because it has to do with the space in the neighborhood of the electric or magnetic bodies, and it may be called a *dynamical* theory, because it assumes that in that space there is matter in motion, by which the observed electromagnetic phenomena are produced.

(4) The electromagnetic field is that part of space which contains and surrounds bodies in electric or magnetic conditions.

It may be filled with any kind of matter, or we may endeavor to render it empty of all gross matter, as in the case of Geissler's■ tubes and other so-called vacua.

There is always, however, enough of matter left to receive and transmit the undulations of light and heat, and it is because the transmission of these radiations is not greatly altered when transparent bodies of measurable density are substituted for the so-called vacuum, that we are obliged to admit that the undulations are those of an ethereal substance,

---

[2] "Explicare tentatur quomodo fiat ut lucis planum polarizationis per vires electricas vel magneticas declenetur."—Halis Saxonum, 1858.

and not of the gross matter, the presence of which merely modifies in some way the motion of the ether.■

We have therefore some reason to believe, from the phenomena of light and heat, that there is an ethereal medium filling space and permeating bodies, capable of being set in motion and of transmitting that motion from one part to another, and of communicating that motion to gross matter so as to heat it and affect it in various ways.

(5) Now the energy communicated to the body in heating it must have formerly existed in the moving medium, for the undulations had left the source of heat some time before they reached the body, and during that time the energy must have been half in the form of motion of the medium and half in the form of elastic resilience.■ From these considerations Professor W. Thomson■ has argued,[3] that the medium must have a density capable of comparison with that of gross matter, and has even assigned an inferior limit to that density.

(6) We may therefore receive, as a datum derived from a branch of science independent of that with which we have to deal, the existence of a pervading medium, of small but real density, capable of being set in motion, and of transmitting motion from one part to another with great, but not infinite, velocity.■

Hence the parts of this medium must be so connected that the motion of one part depends in some way on the motion of the rest; and at the same time these connections must be capable of a certain kind of elastic yielding, since the communication of motion is not instantaneous, but occupies time.

The medium is therefore capable of receiving and storing up two kinds of energy, namely, the "actual"■ energy depending on the motions of its parts, and "potential" energy, consisting of the work which the medium will do in recovering from displacement in virtue of its elasticity.

The propagation of undulations consists in the continual transformation of one of these forms of energy into the other alternately, and at any instant the amount of energy in the whole medium is equally divided, so that half is energy of motion and half is elastic resilience.■

(7) A medium having such a constitution may be capable of other kinds of motion and displacement than those which produce the phenomena of light and heat, and some of these may be of such a kind that they may be evidenced to our senses by the phenomena they produce.

(8) Now we know that the luminiferous medium is in certain cases

---

[3] "On the Possible Density of the Luminiferous Medium, and on the Mechanical Value of a Cubic Mile of Sunlight," *Transactions of the Royal Society of Edinburgh* (1854), p. 57.

Despite the fact that the velocity of light was known to decrease in a material such as glass, for example, Maxwell concluded that the change was not great enough (the velocity in glass is about ⅔ that in air) to be accounted for if molecules of the glass itself had to be set in motion.

If the medium is regarded in the nature of an elastic body such as a spring, then the potential energy of the spring when compressed or stretched corresponds to the *elastic resilience.*

Sir William Thomson, Lord Kelvin (1824-1907), known primarily for his discoveries in thermodynamics, also worked extensively in electricity.

The fact that this conclusion was derived from considerations of light and heat propagation, rather than electricity, was regarded as giving it greater validity in view of its wider application.

"actual": kinetic

In any vibrating system there is a

continual inter-
change between
kinetic and po-
tential energy. If
we consider the
case of sound
propagation, for
example, then
the same holds
for every portion
of the medium
through which
the acoustic
energy travels;
at any given time
half the energy
will be kinetic
and half potential.
Maxwell carried
this analogy over
to the electro-
magnetic medium.

The so-called
Faraday effect.

Émile Verdet
(1824-1866).
Verdet's constant
is the rotation
produced when
light travels unit
distance in a unit
magnetic field.

The Faraday
effect, that is.

Actually, the
Faraday effect
depends markedly
on the nature of
the medium, and
is now accounted
for in terms of
atomic structure.
In a magnetic
field the electron
orbits are so
modified as to
have the effect
of rotating the
plane of polariza-
tion. (Polarized
light has its vibra-
tions restricted to
a single plane—
the *plane of
polarization*.)

acted on by magnetism; for Faraday[4] discovered that when a plane polarized ray traverses a transparent diamagnetic medium in the direction of the lines of magnetic force produced by magnets or currents in the neighborhood, the plane of polarization is caused to rotate.■

This rotation is always in the direction in which positive electricity must be carried round the diamagnetic body in order to produce the actual magnetization of the field.

E. Verdet[5]■ has since discovered that if a paramagnetic body, such as solution of perchloride of iron in ether, be substituted for the diamagnetic body, the rotation is in the opposite direction.

Now Professor W. Thomson[6] has pointed out that no distribution of forces acting between the parts of a medium whose only motion is that of the luminous vibrations, is sufficient to account for the phenomena,■ but that we must admit the existence of a motion in the medium depending on the magnetization, in addition to the vibratory motion which constitutes light.

It is true that the rotation by magnetism of the plane of polarization has been observed only in media of considerable density; but the properties of the magnetic field are not so much altered by the substitution of one medium for another, or for a vacuum, as to allow us to suppose that the dense medium does anything more than merely modify the motion of the ether.■ We have therefore warrantable grounds for inquiring whether there may not be a motion of the ethereal medium going on wherever magnetic effects are observed, and we have some reason to suppose that this motion is one of rotation, having the direction of the magnetic force as its axis.

(9) We may now consider another phenomenon observed in the electromagnetic field. When a body is moved across the lines of magnetic force it experiences what is called an electromotive force■; the two extremities of the body tend to become oppositely electrified, and an electric current tends to flow through the body. When the electromotive force is sufficiently powerful, and is made to act on certain compound bodies, it decomposes them, and causes one of their components to pass towards one extremity of the body, and the other in the opposite direction.■

Here we have evidence of a force causing an electric current in spite of resistance; electrifying the extremities of a body in opposite ways, a

---

[4] *Experimental Researches,* Series 19.
[5] *Comptes Rendus* (1856, second half year), p. 529; (1857, first half year), p. 1209.
[6] *Proceedings of the Royal Society,* June 1856 and June 1861.

condition which is sustained only by the action of the electromotive force, and which, as soon as that force is removed, tends, with an equal and opposite force, to produce a counter current through the body and to restore the original electrical state of the body■; and finally, if strong enough, tearing to pieces chemical compounds and carrying their components in opposite directions, while their natural tendency is to combine, and to combine with a force which can generate an electromotive force in the reverse direction.

Induced emf

The electrolytic effects observed by Faraday.

This, then, is a force acting on a body caused by its motion through the electromagnetic field, or by changes occurring in that field itself■; and the effect of the force is either to produce a current and heat the body, or to decompose the body, or, when it can do neither, to put the body in a state of electric polarization—a state of constraint in which opposite extremities are oppositely electrified, and from which the body tends to relieve itself as soon as the disturbing force is removed.

The "counter current" is that predicted by Lenz's law.
The induced emf does not require relative motion of the field and conductor; a changing field is sufficient.

(10) According to the theory which I propose to explain, this "electromotive force" is the force called into play during the communication of motion from one part of the medium to another, and it is by means of this force that the motion of one part causes motion in another part. When electromotive force acts on a conducting circuit, it produces a current, which, as it meets with resistance, occasions a continual transformation of electrical energy into heat, which is incapable of being restored again to the form of electrical energy by any reversal of the process.

(11) But when electromotive force acts on a dielectric it produces a state of polarization of its parts similar in distribution to the polarity of the parts of a mass of iron under the influence of a magnet, and like the magnetic polarization, capable of being described as a state in which every particle has its opposite poles in opposite conditions.[7]

In a dielectric under the action of electromotive force, we may conceive that the electricity in each molecule is so displaced that one side is rendered positively and the other negatively electrical, but that the electricity remains entirely connected with the molecule, and does not pass from one molecule to another.■ The effect of this action on the whole dielectric mass is to produce a general displacement of electricity in a certain direction. This displacement does not amount to a current, because when it has attained to a certain value it remains constant, but it is the commencement of a current, and its variations constitute currents in the positive or the negative direction according as the displace-

Such displacement of charge does not involve a motion of the molecule itself, but only a slight shifting of the electronic orbits. The over-all effect is similar to charging an object by induction.

[7] Faraday, *Exp. Res.,* Series XI.; Mossotti, *Mem. della Soc. Italiana (Modena),* vol. xxiv, part 2, p. 49.

ment is increasing or decreasing. In the interior of the dielectric there is no indication of electrification, because the electrification of the surface of any molecule is neutralized by the opposite electrification of the surface of the molecules in contact with it; but at the bounding surface of the dielectric, where the electrification is not neutralized, we find the phenomena which indicate positive or negative electrification.

The relation between the electromotive force and the amount of electric displacement it produces depends on the nature of the dielectric, the same electromotive force producing generally a greater electric displacement in solid dielectrics, such as glass or sulphur, than in air.■

Equivalent to stating that the dielectric constant of a solid is greater than that of air.

(12) Here, then, we perceive another effect of electromotive force, namely, electric displacement, which according to our theory is a kind of elastic yielding to the action of the force, similar to that which takes place in structures and machines owing to the want of perfect rigidity of the connections.

The *specific inductive capacity* is known as the dielectric constant.

(13) The practical investigation of the inductive capacity■ of dielectrics is rendered difficult on account of two disturbing phenomena. The first is the conductivity of the dielectric, which, though in many cases exceedingly small, is not altogether insensible.■ The second is the phenomenon called electric absorption,[8] in virtue of which, when the dielectric is exposed to electromotive force, the electric displacement gradually increases, and when the electromotive force is removed, the dielectric does not instantly return to its primitive state, but only discharges a portion of its electrification, and when left to itself gradually acquires electrification on its surface, as the interior gradually becomes depolarized.■ Almost all solid dielectrics exhibit this phenomenon, which gives rise to the residual charge in the Leyden jar, and to several phenomena of electric cables described by Mr. F. Jenkin.[9]

If the dielectric contains free, mobile charges, these will move under the action of a field, giving rise to a fictitious dielectric constant.

The molecular forces of friction in some solids prevent the dielectric from depolarizing immediately; in certain waxes the bound charge becomes more or less permanent, much like a permanent magnet. These are known as *electrets*.

(14) We have here two other kinds of yielding besides the yielding of the perfect dielectric, which we have compared to a perfectly elastic body. The yielding due to conductivity may be compared to that of a viscous fluid (that is to say, a fluid having great internal friction),■ or a soft solid on which the smallest force produces a permanent alteration of figure increasing with the time during which the force acts. The yielding due to electric absorption may be compared to that of a cellular elastic body containing a thick fluid in its cavities. Such a body, when subjected to pressure, is compressed by degrees on account of the gradual yielding of the thick fluid; and when the pressure is removed it does not

That is, large viscosity.

---

[8] Faraday, *Exp. Res.,* pp. 1233-1250.
[9] *Reports of British Association,* 1859, p. 248; and *Report of Committee of Board of Trade on Submarine Cables,* pp. 136 and 464.

at once recover its figure, because the elasticity of the substance of the body has gradually to overcome the tenacity of the fluid before it can regain complete equilibrium.■

Several solid bodies in which no such structure as we have supposed can be found, seem to possess a mechanical property of this kind[10]; and it seems probable that the same substances, if dielectrics, may possess the analogous electrical property, and if magnetic, may have corresponding properties relating to the acquisition, retention, and loss of magnetic polarity.■

(15) It appears therefore that certain phenomena in electricity and magnetism lead to the same conclusion as those of optics, namely, that there is an ethereal medium pervading all bodies, and modified only in degree by their presence; that the parts of this medium are capable of being set in motion by electric currents and magnets; that this motion is communicated from one part of the medium to another by forces arising from the connections of those parts■; that under the action of these forces there is a certain yielding depending on the elasticity of these connections; and that therefore energy in two different forms may exist in the medium, the one form being the actual energy of motion of its parts, and the other being the potential energy stored up in the connections, in virtue of their elasticity.

(16) Thus, then, we are led to the conception of a complicated mechanism capable of a vast variety of motion, but at the same time so connected that the motion of one part depends, according to definite relations, on the motion of other parts, these motions being communicated by forces arising from the relative displacement of the connected parts, in virtue of their elasticity.■ Such a mechanism must be subject to the general laws of Dynamics, and we ought to be able to work out all the consequences of its motion, provided we know the form of the relation between the motions of the parts.

(17) We know that when an electric current is established in a conducting circuit, the neighboring part of the field is characterized by certain magnetic properties,■ and that if two circuits are in the field, the magnetic properties of the field due to the two currents are combined. Thus each part of the field is in connection with both currents, and the two currents are put in connection with each other in virtue of their connection with the magnetization of the field. The first result of this connection that I propose to examine, is the induction of one current by another, and by the motion of conductors in the field.

While the mechanical analogues employed by Maxwell no doubt proved useful these phenomena are now accounted for in terms of intermolecular forces and thermal disorienting effects.

While mechanical "memory" effects are observed in some materials there apparently is no simple correlation between these and their electric or magnetic properties.

As in an elastic fluid.

Note that the mysterious action-at-a-distance concept is replaced by a point-to-point mechanical linkage—a more rational approach, of course, but one requiring a medium having certain properties.

As discovered by Oersted.

---

[10] As, for instance, the composition of glue, treacle, etc., of which small plastic figures are made, which after being distorted gradually recover their shape.

Hermann von Helmholtz (1821-1894) was the first to present a convincing argument for the principle of conservation of energy, on which was based his derivation of the "force-equivalent of electromagnetism."

L, M, N are coefficients of induction, which depend upon the geometrical relationship of the conductors.

Lines of force were first introduced by Faraday. By their direction they give the direction of the magnetic field and, by their number passing through unit area, the strength of the field.

The second result, which is deduced from this, is the mechanical action between conductors carrying currents. The phenomenon of the induction of currents has been deduced from their mechanical action by Helmholtz[11] and Thomson[12]. I have followed the reverse order, and deduced the mechanical action from the laws of induction. I have then described experimental methods of determining the quantities $L, M, N,$ on which these phenomena depend.■

(18) I then apply the phenomena of induction and attraction of currents to the exploration of the electromagnetic field, and the laying down of systems of lines of magnetic force which indicate its magnetic properties.■ By exploring the same field with a magnet, I show the distribution of its equipotential magnetic surfaces, cutting the lines of force at right angles.

In order to bring these results within the power of symbolical calculation, I then express them in the form of the General Equations of the Electromagnetic Field. These equations express—

(A) The relation between electric displacement, true conduction, and the total current compounded of both.

(B) The relation between the lines of magnetic force and the inductive coefficients of a circuit, as already deduced from the laws of induction.

(C) The relation between the strength of a current and its magnetic effects, according to the electromagnetic system of measurement.

(D) The value of the electromotive force in a body, as arising from the motion of the body in the field, the alteration of the field itself, and the variation of electric potential from one part of the field to another.

(E) The relation between electric displacement, and the electromotive force which produces it.

(F) The relation between an electric current, and the electromotive force which produces it.

(G) The relation between the amount of free electricity at any point, and the electric displacements in the neighborhood.

(H) The relation between the increase or diminution of free electricity and the electric currents in the neighborhood.

There are twenty of these equations in all, involving twenty variable quantities.■

(19) I then express in terms of these quantities the intrinsic energy

Maxwell's basic equations of the electromagnetic field consist of four interrelated differential equations involving partial derivatives. When written out in component form the number is much greater. But even in abbreviated form they are not simple.

---

[11] "Conservation of Force," *Physical Society of Berlin*, 1847; and Taylor's *Scientific Memoirs*, 1853, p. 114.
[12] *Reports of the British Association*, 1848; *Philosophical Magazine*, Dec. 1851.

of the Electromagnetic Field as depending partly on its magnetic and partly on its electric polarization at every point.

From this I determine the mechanical force acting, 1st, on a moveable conductor carrying an electric current; 2dly, on a magnetic pole; 3dly, on an electrified body.

The last result, namely, the mechanical force acting on an electrified body, gives rise to an independent method of electrical measurement founded on its electrostatic effects. The relation between the units employed in the two methods is shown to depend on what I have called the "electric elasticity" of the medium, and to be a velocity, which has been experimentally determined by Weber and Kohlrausch.■

I then show how to calculate the electrostatic capacity of a condenser and the specific inductive capacity of a dielectric.

The case of a condenser composed of parallel layers of substances of different electric resistances and inductive capacities is next examined, and it is shown that the phenomenon called electric absorption will generally occur, that is, the condenser, when suddenly discharged, will after a short time show signs of *residual* charge.■

(20) The general equations are next applied to the case of a magnetic disturbance propagated through a nonconducting field, and it is shown that the only disturbances which can be so propagated are those which are transverse to the direction of propagation, and that the velocity of propagation is the velocity v, found from experiments such as those of Weber,■ which expresses the number of electrostatic units of electricity which are contained in one electromagnetic unit.

This velocity is so nearly that of light, that it seems we have strong reason to conclude that light itself (including radiant heat, and other radiations if any) is an electromagnetic disturbance in the form of waves propagated through the electromagnetic field according to electromagnetic laws. If so, the agreement between the elasticity of the medium as calculated from the rapid alternations of luminous vibrations, and as found by the slow processes of electrical experiments,■ shows how perfect and regular the elastic properties of the medium must be when not encumbered with any matter denser than air. If the same character of the elasticity is retained in dense transparent bodies, it appears that the square of the index of refraction is equal to the product of the specific dielectric capacity and the specific magnetic capacity.■ Conducting media are shown to absorb such radiations rapidly, and therefore to be generally opaque.

The conception of the propagation of transverse magnetic disturbances to the exclusion of normal ones is distinctly set forth by Professor

Found to be the velocity of light. The fact that this ratio for an electrical quantity equalled the velocity of light was strong evidence to Maxwell of the electromagnetic character of light.

This effect in nonhomogeneous dielectrics was later shown to occur by Wagner, and is now known as the Maxwell-Wagner mechanism.

Wilhelm Weber (1804-1891) devised a method of measuring a resistance in absolute units, i.e., in terms of mass, length, and time, and found it to have the dimensions of velocity.

The frequency of light is much higher than that of "electric waves," etc.; i.e., the wavelength is much shorter.

The magnetic permeability. Maxwell later refers to this quantity as the *coefficient of magnetic induction.*

Faraday[13] in his "Thoughts on Ray Vibrations." The electromagnetic theory of light, as proposed by him, is the same in substance as that which I have begun to develop in this paper, except that in 1846 there were no data to calculate the velocity of propagation.

(21) The general equations are then applied to the calculation of the coefficients of mutual induction of two circular currents and the coefficient of self-induction in a coil. The want of uniformity of the current in the different parts of the section of a wire at the commencement of the current is investigated, I believe for the first time, and the consequent correction of the coefficient of self-induction is found.■

Maxwell added these to demonstrate the use of his equations. The study of transient effects at the start of a current has since grown into a major branch of electricity.

These results are applied to the calculation of the self-induction of the coil used in the experiments of the Committee of the British Association on Standards of Electric Resistance, and the value compared with that from the experiments.

PART II. ON ELECTROMAGNETIC INDUCTION

Electromagnetic Momentum of a Current.■

Called by Faraday the *electrotonic state* of the circuit.

(22) We may begin by considering the state of the field in the neighborhood of an electric current. We know that magnetic forces are excited in the field, their direction and magnitude depending according to known laws upon the form of the conductor carrying the current. When the strength of the current is increased, all the magnetic effects are increased in the same proportion. Now, if the magnetic state of the field depends on motions of the medium, a certain force must be exerted in order to increase or diminish these motions, and when the motions are excited they continue, so that the effect of the connection between the current and the electromagnetic field surrounding it, is to endow the current with a kind of momentum, just as the connection between the driving point of a machine and a flywheel endows the driving point with an additional momentum, which may be called the momentum of the flywheel reduced to the driving point. The unbalanced force acting on the driving point increases this momentum and is measured by the rate of its increase.

This electrical "inertia," which is in keeping with Lenz's law, depends upon the *inductance* of the conductor, which is a function of its geometry.

In the case of electric currents, the resistance to sudden increase or diminution of strength produces effects exactly like those of momentum,■ but the amount of this momentum depends on the shape of the conductor and the relative position of its different parts.

---

[13] *Philosophical Magazine*, May 1846, or *Experimental Researches*, iii, p. 447.

Mutual Action of Two Currents.

(23) If there are two electric currents in the field, the magnetic force at any point is that compounded of the forces due to each current separately,■ and since the two currents are in connection with every point of the field, they will be in connection with each other, so that any increase or diminution of the one will produce a force with or contrary to the other.

■. . . . .

Exploration of the Electromagnetic Field

(47) Let us now suppose a primary circuit *A* to be of invariable form, and let us explore the electromagnetic field about it by means of a secondary circuit *B*, which we shall suppose to be variable in form and position.

We may begin by supposing *B* to consist of a short straight conductor with its extremities sliding on two parallel conducting rails, which are put in connection at some distance from the sliding piece.

Then, if sliding the movable conductor in a given direction increases the value of *M*,■ a negative electromotive force will act in the circuit *B*, tending to produce a negative current in *B* during the motion of the sliding piece.

If a current be kept up in the circuit *B*, then the sliding piece will itself tend to move in that direction, which causes *M* to increase. At every point of the field there will always be a certain direction such that a conductor moved in that direction does not experience any electromotive force in whatever direction its extremities are turned. A conductor carrying a current will experience no mechanical force urging it in that direction or the opposite.

This direction is called the direction of the line of magnetic force through that point.■

Motion of a conductor across such a line produces electromotive force in a direction perpendicular to the line and to the direction of motion, and a conductor carrying a current is urged in a direction perpendicular to the line and to the direction of the current.

(48) We may next suppose *B* to consist of a very small plane circuit capable of being placed in any position and of having its plane turned in any direction. The value of *M* will be greatest when the plane of the circuit is perpendicular to the line of magnetic force.■ Hence if a current is maintained in *B* it will tend to set itself in this position, and will of itself indicate, like a magnet, the direction of the magnetic force.

The magnetic field due to each current is a vector quantity (force per unit pole); hence, the fields add vectorially.

Omitted are a number of illustrations of electromagnetic momentum and other induction phenomena.

M is the coefficient of induction (mutual inductance) between the two conductors.

Or the direction of the magnetic field. The force on a conductor carrying a current in a magnetic field is always at right angles to the field.

For then the *flux* threading the circuit is a maximum.

On Lines of Magnetic Force

(49) Let any surface be drawn, cutting the lines of magnetic force, and on this surface let any system of lines be drawn at small intervals, so as to lie side by side without cutting each other. Next, let any line be drawn on the surface cutting all these lines, and let a second line be drawn near it, its distance from the first being such that the value of M for each of the small spaces enclosed between these two lines and the lines of the first system is equal to unity.

In this way let more lines be drawn so as to form a second system, so that the value of M for every reticulation■ formed by the intersection of the two systems of lines is unity.

Finally, from every point of intersection of these reticulations let a line be drawn through the field, always coinciding in direction with the direction of magnetic force.

(50) In this way the whole field will be filled with lines of magnetic force at regular intervals, and the properties of the electromagnetic field will be completely expressed by them.

For, 1st, if any closed curve be drawn in the field, the value of M for that curve will be expressed by the *number* of lines of force which *pass through* that closed curve.■

2dly. If this curve be a conducting circuit and be moved through the field, an electromotive force will act in it, represented by the rate of decrease of the number of lines passing through the curve.■

3dly. If a current be maintained in the curcuit, the conductor will be acted on by forces tending to move it so as to increase the number of lines passing through it, and the amount of work done by these forces is equal to the current in the circuit multiplied by the number of additional lines.■

4thly. If a small plane circuit be placed in the field, and be free to turn, it will place its plane perpendicular to the lines of force. A small magnet will place itself with its axis in the direction of the lines of force.

5thly. If a long uniformly magnetized bar is placed in the field, each pole will be acted on by a force in the direction of the lines of force. The number of lines of force passing through unit of area is equal to the force acting on a unit pole multiplied by a coefficient depending on the magnetic nature of the medium, and called the coefficient of magnetic induction.■

In fluids and isotropic solids the value of this coefficient μ is the same in whatever direction the lines of force pass through the substance, but in crystallized, strained, and organized solids the value of μ may depend on the direction of the lines of force with respect to the axes of crystalli-zation, strain, or growth.

In all bodies μ is affected by temperature, and in iron it appears to diminish as the intensity of the magnetization increases.■

On Magnetic Equipotential Surfaces

(51) If we explore the field with a uniformly magnetized bar, so long that one of its poles is in a very weak part of the magnetic field, then the magnetic forces will perform work on the other pole as it moves about the field.

If we start from a given point, and move this pole from it to any other point, the work performed will be independent of the path of the pole between the two points; provided that no electric current passes between the different paths pursued by the pole.■

Hence, when there are no electric currents but only magnets in the field, we may draw a series of surfaces such that the work done in passing from one to another shall be constant whatever be the path pursued between them. Such surfaces are called Equipotential Surfaces, and in ordinary cases are perpendicular to the Lines of magnetic force.

If these surfaces are so drawn that, when a unit pole passes from any one to the next in order, unity of work is done, then the work done in any motion of a magnetic pole will be measured by the strength of the pole multiplied by the number of surfaces which it has passed through in the positive direction.

(52) If there are circuits carrying electric currents in the field, then there will still be equipotential surfaces in the parts of the field external to the conductors carrying the currents, but the work done on a unit pole in passing from one to another will depend on the number of times which the path of the pole circulates round any of these currents. Hence the potential in each surface will have a series of values in arithmetical progression, differing by the work done in passing completely round one of the currents in the field.■

The equipotential surfaces will not be continuous closed surfaces, but some of them will be limited sheets, terminating in the electric cir-cuit as their common edge or boundary. The number of these will be equal to the amount of work done on a unit pole in going round the current, and this by the ordinary measurement $= 4\pi\gamma$, where γ is the value of the current.

These surfaces, therefore, are connected with the electric current as soap bubbles are connected with a ring in Plateau's experiments.■ Every

As the iron nears saturation the change in induc-tion for a given change in mag-netic field strength de-creases.

The work is a function only of the difference in magnetic poten-tial between the two points.

The magnetic lines of force about a conductor carrying a current consist of concentric circles. Hence work must be done continuously in moving a magnetic pole about a current. For each revolu-tion of a unit pole the work is $4\pi\gamma$, where γ is the current.

In which the bubbles represent equipotential sur-faces connnected to a wire ring (the current). The soap-bubble analogy is frequently used to demonstrate elec-tric and magnetic potential problems.

The derivation of the general equations of the electromagnetic field, plus various applications, are omitted.

current γ has $4\pi\gamma$ surfaces attached to it. These surfaces have the current for their common edge, and meet it at equal angles. The form of the surfaces in other parts depends on the presence of other currents and magnets, as well as on the shape of the circuit to which they belong.

■ . . . .

SUPPLEMENTARY READING

Cajorie, F., *A History of Physics* (New York: Macmillan, 1899), pp. 251-257.

Dampier, W. C., *A History of Science* (Cambridge, Eng.: Cambridge University, 1949).

Glazebrook, R. T., *James Clerk Maxwell and Modern Physics* (London: Cassell, 1901).

Holton, G., and D. H. D. Roller, *Foundations of Modern Physical Science* (Reading, Mass.: Addison-Wesley, 1958), Ch. 29.

Lenard, P., *Great Men of Science* (New York: Macmillan, 1933), pp. 339 ff.

Magie, W. F., *A Source Book in Physics* (New York: McGraw-Hill, 1935), pp. 528 ff.

James Clerk Maxwell, A Commemoration Volume 1831-1931 (New York: Macmillan, 1931).

Newman, J. R., "James Clerk Maxwell" in *Lives in Science* by the Editors of Scientific American (New York: Simon and Schuster, 1957), pp. 155 ff.

Niven, W. D., ed., *James Clerk Maxwell, Scientific Papers* (New York: Dover, 1952).

Max Planck
1858-1947

The Quantum Hypothesis

BY THE beginning of the twentieth century it was clear that classical concepts, which had proved so successful in dealing with most physical phenomena, were inadequate to account for certain experiences such as the emission and absorption of radiation. At the turn of the century, in December 1900, Max Planck proposed so bold and revolutionary a solution to this problem as to set physics off on a completely new and fruitful course. The electromagnetic theory of light, which had been fully developed by Maxwell and confirmed by Hertz, seemed firmly established; electromagnetic waves were emitted by oscillating electric charges, the frequency of the radiation being the same as that of the oscillating charge producing it. That this applied as well to visible light as to radio waves was demonstrated late in the nineteenth century by Pieter Zeeman (1865-1943), who observed a splitting of the spectral lines when a light source was placed in a magnetic field, and by Hendrick A. Lorentz (1853-1928), who explained the effect in terms of vibrating electric charges. Lorentz accounted for the simpler features of the *Zeeman effect* by assuming that light was emitted by electrons vibrating under the influence of a linear restoring force, much as a mass vibrating at the end of a spring, and then examining the effect of a magnetic field on such vibrations. Here was the first experimental evidence of a connection between the radiation from atoms and the electromagnetic field. But Zeeman's experiments were performed with luminous gases, from which are observed discrete spectral lines. The difficulty arose with efforts to account for the continuous spectrum emitted by a *black body*. In fact, even the Zeeman effect cannot be explained completely in terms of classical concepts; a detailed explanation requires the use of the new quantum ideas.

A black body, which is one that absorbs all the radiation falling on it, was known to emit at any given temperature a characteristic continuous spectrum[1]; the distribution of frequencies in this spectrum depends not on the composition of the body, but only on its temperature. While an ideal black surface is not found in practice it can be approximated very closely by means of a hollow enclosure having a very small opening in its wall. Radiation entering the hole has but a slight chance of escaping; instead, it is gradually absorbed by repeated encounters with the internal surface of the cavity. Similarly, if the enclosure is heated, the radiation issuing from the opening, known as *cavity radiation*, is effectively true black-body radiation.

In 1879 Josef Stefan (1835-1893) pointed out that some measurements of the rate of loss of heat from a hot wire, made by John Tyndall (1820-1893), indicated that the loss by radiation was proportional to the fourth power of the absolute temperature. Several years later Ludwig Boltzmann (1844-1906) was able to derive this relationship, now known as the *Stefan-Boltzmann law*, by applying thermodynamic reasoning to Maxwell's theory. There was another notable success along these lines. In 1893 Wilhelm Wien (1864-1928) concluded on similar theoretical grounds that the wavelength at which maximum energy is radiated from a black body is inversely proportional to the absolute temperature of the body. This is the *Wien displacement law*, which was verified experimentally in the last few years of the nineteenth century, and partly for which Wien was awarded a Nobel prize in 1911.

But here the success of classical methods ended. The principles of Newtonian mechanics, of Maxwell's electromagnetic theory, and of thermodynamics proved insufficient to account for the distribution of energy in black-body radiation; that is, the actual shape of the spectral curve could not be deduced accurately. The classical approach was straightforward: the emission of radiation was believed to be due to vibrating charges within the body. These were assumed to vibrate with different frequencies and it was the variation in relative intensity of these frequencies that determined the characteristic black-body spectrum. The general procedure, therefore, was to consider a very large number of such oscillators having all possible frequencies, impose the condition that they be in thermal equilibrium with one another at the temperature in question, and then determine what distribution of intensities among the different oscillators satisfied these requirements. Wien proposed one variant of this approach which agreed with experiment only at short wavelengths. Lord Rayleigh (J. W. Strutt, 1842-

---

[1] A continuous spectrum contains all frequencies up to a certain maximum.

1919) suggested another based on the modes of vibration of the electro-magnetic field within an enclosure, later developed by James Jeans (1877-1946) in the form known as the *Rayleigh-Jeans formula,* which matched the spectrum at long wavelengths only. Neither one, however, could account for the complete spectrum.

It was in 1900, prior to the Rayleigh-Jeans theory, that Planck made his bold suggestion. He first found an empirical relation that fit the entire spectrum, and then proceeded to search for a physical mechanism that could yield such a result. He retained the classical view that black-body radiation should be accounted for in terms of the laws governing the emission and absorption of electromagnetic energy by electric oscillators. He assumed further that because of the very large number of oscillators present all fre-quencies were represented and hence were to be found in the emitted radia-tion. It was at this point that he broke with tradition, for he concluded that while all frequencies were possible, an oscillator may possess only *dis-crete* amounts of energy. In classical physics, the energy of an oscillator depends only on its frequency and amplitude, with no restriction on either. Yet Planck assumed that the energy of each oscillator must be an integral multiple of the quantity hv, where v is the frequency of the oscillator and h a universal constant *(Planck's constant)* sometimes known as the *quantum of action.* According to Planck the emission and absorption of radiation by an oscillator involved a change in energy "level" of the oscillator from one such "allowed" energy to another.

This seemingly simple suggestion—that energy is emitted or absorbed only in whole quanta h—altered the entire course of physics. To his con-temporaries, and to Planck himself, the idea that natural processes could be discontinuous was difficult to accept wholeheartedly. It was not until the quantum hypothesis proved successful time and again in accounting for a variety of phenomena that it was accepted as part of the basic structure of physics, despite the serious philosophical question it posed. Black-body radiation, the photoelectric effect, atomic spectra, the Compton effect, wave mechanics—these are but a few outgrowths of Planck's hypothesis. It is difficult to imagine what physics might have been like in the first half of the twentieth century without this guiding principle.

Max Karl Ernst Ludwig Planck was born in Kiel, Germany, where his father was professor of constitutional law in the university, on April 23, 1858. He was a young boy when Maxwell developed his electromagnetic theory of light, and a contemporary of Heinrich Hertz (Chapter 13), who demonstrated the existence of Maxwell's "waves." Both of these were to have considerable influence on Planck's own work. Planck lived through several chapters of German history. As Max von Laue pointed out in a memorial address upon Planck's death in 1947: "... The birth and meteoric ascent of the German Empire occurred during his lifetime, and so did its total eclipse and ghastly disaster." [2] Planck's early life, judging from his autobiography,[3] was uneventful. He attended the *Maximilian-Gymnasium* in Munich, where he developed an interest in physics, and then entered, at the age of seventeen, the University of Munich. He remained at Munich for three years, concentrating on the study of physics. He then went to the University of Berlin for a year, where he came under the guidance of the two distinguished physicists, Hermann von Helmholtz (1821-1894) and Gustav Kirchhoff (1824-1887). Planck became interested in thermodynamics while in Berlin, and in 1879 he presented a doctoral dissertation to the University of Munich on the *Second Law of Thermodynamics,* which he managed to interpret in terms of the *entropy* of a system.

He received his doctorate *summa cum laude* and became a *privat dozent* (lecturer or instructor) at Munich until 1885, when he was appointed professor *extraordinarius* (associate professor) in theoretical physics at the University of Kiel. Four years later, upon the death of Kirchhoff, Planck was invited to take his place at the University of Berlin. He accepted the invitation gladly, for Berlin was then a great center for physics, and began thereby an association that lasted for fifty years. In 1892 he became professor *ordinarius* (full professor). All the while he continued his researches in thermodynamics, marked particularly by the publication of his classic treatise on the subject in 1897. By 1900 his studies in thermodynamics and electromagnetic radiation brought Planck to the problem of the distribution of energy in the black-body spectrum, a problem which he could solve only by introducing his revolutionary quantum hypothesis. This was the climax of his work, just as it was a turning point in the history of physics. He continued teaching and writing for many more years and was gratified to see the early opposition to his discovery gradually give way first to cautious acceptance and finally to widespread approval. In 1905 Einstein adopted the quantum hypothesis to account for the photoelectric effect (Chapter 17). Two years later he explained the variation in specific heat of a solid with temperature on the same basis, and not long afterward Bohr used the idea to develop his remarkable theory of atomic spectra. Planck was awarded the Nobel prize in physics for 1918, and in 1926, when he became professor emeritus, he was elected a foreign member of the Royal Society. He held a number of distinguished offices in his professional field. In 1912 he became permanent secretary to the Prussian Academy of Science,

---

[2] M. Planck, *Scientific Autobiography,* trans. by F. Gaynor (New York: Philosophical Library, 1949), p. 7.
[3] *Ibid.*

and in 1930 he was elected president of the Kaiser Wilhelm Institute for the Advancement of Science, a post he retained until 1937.

The last years of his life were unhappy ones for Planck. He suffered several personal losses, including a son who was executed for alleged complicity in the ill-fated plot to overthrow the Nazi regime toward the end of World War II. He witnessed the efforts of this regime, including some of his colleagues, to create a "German physics," in which the discoveries of Einstein and other Jewish scientists were to play a minor role. When he died, on October 4, 1947, he could probably find little consolation in the knowledge that the attempt to establish a German dictatorship in science had failed, for in its madness the Third Reich had all but destroyed the great tradition of physics in Germany.

Professor Niels Bohr, who made such effective use of Planck's hypothesis in his theory of atomic spectra, said of this guiding principle:

> Scarcely any other discovery in the history of science has produced such extraordinary results within the short span of our generation as those which have arisen directly from Max Planck's discovery of the elementary quantum of action. . . . It has shattered the foundations of our ideas not only in the realm of classical science but also in our everyday ways of thinking. It is to this emancipation from inherited traditions of thought that we owe the wonderful progress which has been made in our knowledge of natural phenomena during the past generation. . . .[4]

The following is extracted from Planck's historic paper "On the Law of Distribution of Energy in the Normal Spectrum," which was reported to the German Physical Society on October 19 and December 14, 1900 and later published in *Annalen der Physik,* vol. 4 (1901), page 553.

---

[4] N. Bohr, *Die Naturwissenschaften,* vol. 26 (1938), p. 483.

Planck's "Experiment"

INTRODUCTION

The recent spectral measurements made by O. Lummer and E. Pringsheim[1], and even more notable those by H. Rubens and F. Kurlbaum,[2] which together confirmed an earlier result obtained by H.

---

[1] O. Lummer and E. Pringsheim, *Transactions of the German Physical Society* 2 (1900), p. 163.

[2] H. Rubens and F. Kurlbaum, *Proceedings of the Imperial Academy of Science,* Berlin, October 25, 1900, p. 929.

Beckmann,[3] show that the law of energy distribution in the normal spectrum, first derived by W. Wien from molecular-kinetic considerations and later by me from the theory of electromagnetic radiation, is not valid generally.■

In any case the theory requires a correction, and I shall attempt in the following to accomplish this on the basis of the theory of electromagnetic radiation which I developed. For this purpose it will be necessary first to find in the set of conditions leading to Wien's energy distribution law that term which can be changed; thereafter it will be a matter of removing this term from the set and making an appropriate substitution for it.

In my last article[4] I showed that the physical foundations of the electromagnetic radiation theory, including the hypothesis of "natural radiation,"■ withstand the most severe criticism; and since to my knowledge there are no errors in the calculations, the principle persists that the law of energy distribution in the normal spectrum is completely determined when one succeeds in calculating the entropy S■ of an irradiated, monochromatic, vibrating resonator as a function of its vibrational energy U. Since one then obtains, from the relationship $dS/dU = 1/\theta$, the dependence of the energy U on the temperature θ, and since the energy is also related to the density of radiation at the corresponding frequency by a simple relation,[5] one also obtains the dependence of this density of radiation on temperature.■ The normal energy distribution is then the one in which the radiation densities of all different frequencies have the same temperature.

Consequently, the entire problem is reduced to determining S as a function of U, and it is to this task that the most essential part of the following analysis is devoted. In my first treatment of this subject I had expressed S, by definition, as a simple function of U without further foundation, and I was satisfied to show that this form of entropy meets all the requirements imposed on it by thermodynamics.■ At that time I believed that this was the only possible expression and that consequently Wien's law, which follows from it, necessarily had general validity. In a later, closer analysis,[6] however, it appeared to me that there must be other expressions which yield the same result, and that in any case one needs another condition in order to be able to calculate S uniquely. I believed I had found such a condition in the principle, which at that time

---

[3] H. Beckmann, *Inaugural dissertation,* Tübingen 1898. See also H. Rubens, *Wied. Ann. 69* (1899), p. 582.

[4] M. Planck, *Ann. d. Phys. 1* (1900), p. 719.

[5] Compare with equation (8).

[6] M. Planck, *loc. cit.,* pp. 730 ff.

seemed to me perfectly plausible, that in an infinitely small irreversible change in a system, near thermal equilibrium, of N identical resonators in the same stationary radiation field, the increase in the total entropy $S_N = NS$ with which it is associated depends only on its total energy $U_N = NU$ and the changes in this quantity, but not on the energy U of individual resonators.■ This theorem leads again to Wien's energy distribution law. But since the latter is not confirmed by experience one is forced to conclude that even this principle cannot be generally valid and thus must be eliminated from the theory.[7]

Thus another condition must now be introduced which will allow the calculation of S, and to accomplish this it is necessary to look more deeply into the meaning of the concept of entropy. Consideration of the untenability of the hypothesis made formerly will help to orient our thoughts in the direction indicated by the above discussion. In the following a method will be described which yields a new, simpler expression for entropy and thus provides also a new radiation equation which does not seem to conflict with any facts so far determined.

I. CALCULATION OF THE ENTROPY OF A RESONATOR AS A FUNCTION OF ITS ENERGY

§1. Entropy depends on disorder and this disorder, according to the electromagnetic theory of radiation for the monochromatic vibrations of a resonator when situated in a permanent stationary radiation field,■ depends on the irregularity with which it constantly changes its amplitude and phase, provided one considers time intervals large compared to the time of one vibration but small compared to the duration of a measurement.■ If amplitude and phase both remained absolutely constant, which means completely homogeneous vibrations, no entropy could exist and the vibrational energy would have to be completely free to be converted into work. The constant energy U of a single stationary vibrating resonator accordingly is to be taken as a time average, or what is the same thing, as a simultaneous average of the energies of a large number N of identical resonators,■ situated in the same stationary radiation field, and which are sufficiently separated so as not to influence each other directly. It is in this sense that we shall refer to the average energy U of a single resonator. Then to the total energy

(1) $U_N = NU$

The density of radiation is the amount of energy contained in unit volume.

While meeting the requirements of thermodynamics the original definition led to inconsistencies in regard to blackbody radiation.

That is, Planck first assumed that the change in entropy of a system of resonators depended only on the total change of energy of the resonators, regardless of how this change was distributed among them. It will be seen that the resonators must be treated as individuals rather than as a group.

Planck considered a number of oscillators situated in a space enclosed by reflecting walls. These oscillators were assumed to exchange energy with one another by absorbing or emitting electromagnetic waves until a stationary (equilibrium) radiation field obtained in the cavity.

If the time interval is not large compared with the period, the "degree of irregularity" can-

---

[7] Moreover one should compare the critiques previously made of this theorem by W. Wien (*Report of the Paris Congress 2*, 1900, p. 40) and by O. Lummer (*loc. cit., 2*, 1900, p. 92.).

not be defined; i.e., the result is not statistically significant. If too long, the "disorder" is no longer evident; it is "averaged" out.

The average behavior of a single resonator over a period of time is the same as the instantaneous average of a large number of resonators.

W is the probability that the resonators have total energy E_N. This relationship formed a major step in Planck's development of the quantum hypothesis.

In connection with the kinetic theory of gases, Boltzmann defined the entropy to be proportional to the logarithm of the probability of finding a given distribution of particles (molecules).

The quantum postulate.

Or "arrangement."

of such a system of N resonators there corresponds a certain total entropy

(2) $S_N = NS$

of the same system, where S represents the average entropy of a single resonator and the entropy S_N depends on the disorder with which the total energy U_N is distributed among the individual resonators.

§2.   We now set the entropy S_N of the system proportional to the logarithm of its probability W,■ within an arbitrary additive constant, so that the N resonators together have the energy E_N:

(3) $S_N = k \log W + \text{constant}$

In my opinion this actually serves as a definition of the probability W, since in the basic assumptions of electromagnetic theory there is no definite evidence for such a probability. The suitability of this expression is evident from the outset, in view of its simplicity and close connection with a theorem from kinetic gas theory.[8]■

§3.   It is now a matter of finding the probability W so that the N resonators together possess the vibrational energy U_N. Moreover, it is necessary to interpret U_N not as a continuous, infinitely divisible quantity, but as a discrete quantity composed of an integral number of finite equal parts. Let us call each such part the energy element ϵ; consequently we must set■

(4) $U_N = P\epsilon$

where P represents a large integer generally, while the value of ϵ is yet uncertain.

Now it is evident that any distribution of the P energy elements among the N resonators can result only in a finite, integral, definite number. Every such form of distribution we call, after an expression used by L. Boltzmann for a similar idea, a "complex."■ If one denotes the resonators by the numbers 1, 2, 3, ... N, and writes these side by side, and if one sets under each resonator the number of energy elements assigned to it by some arbitrary distribution, then one obtains for every complex a pattern of the following form:

1	2	3	4	5	6	7	8	9	10
7	38	11	0	9	2	20	4	4	5

Here we assume $N = 10$, $P = 100$. The number R of all possible complexes is obviously equal to the number of arrangements that one can obtain in this fashion for the lower row, for a given N and P. For the sake of clarity we should note that two complexes must be considered different if the

───────────

[8] L. Boltzmann, *Proceedings of the Imperial Academy of Science,* Vienna, (II) 76 (1877), p. 428.

corresponding number patterns contain the same numbers but in a different order.

From combination theory one obtains the number of all possible complexes as:■

$$R = \frac{N(N+1)(N+2)\cdots(N+P-1)}{1\cdot 2\cdot 3\cdots P} = \frac{(N+P-1)\,!}{(N-1)\,!\,P\,!}$$

Now according to Stirling's theorem, we have in the first approximation:■
$$N\,! = N^N$$

consequently, the corresponding approximation is:■

$$R = \frac{(N+P)^{N+P}}{N^N \cdot P^P}$$

§4. The hypothesis which we want to establish as the basis for further calculation proceeds as follows: in order for the N resonators to possess collectively the vibrational energy U_N, the probability W must be proportional to the number R of all possible complexes formed by distribution of the energy U_N among the N resonators; or in other words, any given complex is just as probable as any other.■ Whether this actually occurs in nature one can, in the last analysis, prove only by experience. But should experience finally decide in its favor it will be possible to draw further conclusions from the validity of this hypothesis about the particular nature of resonator vibrations; namely in the interpretation put forth by J. v. Kries[9] regarding the character of the "original amplitudes, comparable in magnitude but independent of each other." As the matter now stands, further development along these lines would appear to be premature.

§5. According to the hypothesis introduced in connection with equation (3), the entropy of the system of resonators under consideration is, after suitable determination of the additive constant:

(5) $S_N = k \log R$
$$= k\{(N+P)\log(N+P) - N\log N - P\log P\}$$

and by considering (4) and (1):

$$S^N = kN\left\{\left(1 + \frac{U}{\epsilon}\right)\log\left(1 + \frac{U}{\epsilon}\right) - \frac{U}{\epsilon}\log\frac{U}{\epsilon}\right\}$$

Thus, according to equation (2) the entropy S of a resonator as a function of its energy U is given by:

(6) $S = k\left\{\left(1 + \frac{U}{\epsilon}\right)\log\left(1 + \frac{U}{\epsilon}\right) - \frac{U}{\epsilon}\log\frac{U}{\epsilon}\right\}$

The number of combinations of P things, divided among N compartments.

Stirling's formula, valid for large N, is: $N! \approx \left(\dfrac{N}{e}\right)^N$ which to a first approximation is simply $N! \approx N^N$ ($N!$ is N factorial $= 1\cdot 2\cdot 3\cdots N$)

Here N and P are assumed large compared to 1.

The assumption that the distribution is random would seem the most reasonable. As is now known, all such phenomena are perfectly random.

---

[9] Joh. v. Kries, *The Principles of Probability Calculation* (Freiburg, 1886), p. 36.

II.   INTRODUCTION OF WIEN'S DISPLACEMENT LAW

§6.   Next to Kirchoff's theorem of the proportionality of emissive and absorptive power, the so-called displacement law, discovered by and named after W. Wien,[10] which includes as a special case the Stefan-Boltzmann law of dependence of total radiation on temperature,∎ provides the most valuable contribution to the firmly established foundation of the theory of heat radiation. In the form given by M. Thiesen[11] it reads as follows:

The total radiation is proportional to the fourth power of the absolute temperature.

$$E \cdot d\lambda = \theta^5 \psi(\lambda\theta) \cdot d\lambda$$

The volume density of radiation is the energy contained in unit volume. $d\lambda$ is an infinitesimal range of wavelengths.

where λ is the wavelength, $E\, d\lambda$ represents the volume density of the "black-body" radiation[12] within the spectral region λ to $\lambda + d\lambda$,∎ θ represents temperature, and $\psi(x)$ represents a certain function of the argument x only.∎

That is, $\psi(\lambda\theta)$ means a function of the product $\lambda\theta$ alone rather than a function of the two variables separately.

§7.   We now want to examine what Wien's displacement law states about the dependence of the entropy S of our resonator on its energy U and its characteristic period, particularly in the general case where the resonator is situated in an arbitrary diathermic medium.∎ For this purpose we next generalize Thiesen's form of the law for the radiation in an arbitrary diathermic medium with the velocity of light c. Since we do not have to consider the total radiation, but only the monochromatic radiation,∎ it becomes necessary in order to compare different diathermic media to introduce the frequency ν instead of the wavelength λ.

A diathermic medium is one that is transparent to thermal radiation.

Monochromatic: single wavelength

Thus, let us denote by $u\, d\nu$ the volume density of the radiation energy belonging to the spectral region ν to $\nu + d\nu$; then we write: $u\, d\nu$ instead of $E\, d\lambda$; c/ν instead of λ, and $cd\nu/\nu^2$ instead of $d\lambda$.∎ From which we obtain:

Since $c = \lambda\nu$, $\lambda = c/\nu$, and by taking the rate of change of λ with respect to ν (derivative) we have $d\lambda = -c\, d\nu/\nu^2$.

$$u = \theta^5 \frac{c}{\nu^2} \cdot \psi\left(\frac{c\theta}{\nu}\right)$$

Now according to the well-known Kirchoff-Clausius law, the energy emitted per unit time at the frequency ν and temperature θ from a black surface in a diathermic medium is inversely proportional to the square of the velocity of propagation c^2; hence the energy density u is inversely proportional to c^3 and we have:

$$u = \frac{\theta^5}{\nu^2 c^3} \cdot f\left(\frac{\theta}{\nu}\right)$$

In optics a continuous spectrum is called a "white-light spectrum."

---

[10] W. Wien, *Proceedings of the Imperial Academy of Science,* Berlin, February 9, 1893, p. 55.

[11] M. Thiesen, *Transactions of the German Physical Society* 2 (1900), p. 66.

[12] Perhaps one should speak more appropriately of a "white" radiation, to generalize what one already understands by total white light.∎

where the constants associated with the function f are independent of c.

In place of this, if f represents a new function of a single argument, we can write:

$$(7) \qquad u = \frac{\nu^3}{c^3} \cdot f\left(\frac{\theta}{\nu}\right)^{\blacksquare}$$

Where now the temperature is contained only in the function f.

and from this we see, among other things, that as is well known, the radiant energy $u \cdot \lambda^3$ at a given temperature and frequency is the same for all diathermic media.$^{\blacksquare}$

Since $\nu/c = 1/\lambda$, we have $u \cdot \lambda^3 = f\left(\frac{\theta}{\nu}\right)$, which is independent of the medium; i.e., independent of c.

§8. In order to go from the energy density u to the energy U of a stationary resonator situated in the radiation field and vibrating with the same frequency ν, we use the relation expressed in equation (34) of my paper on irreversible radiation processes[13]:

$$K = \frac{\nu^2}{c^2} U$$

(K is the intensity of a monochromatic, linearly polarized ray), which together with the well-known equation:$^{\blacksquare}$

$$u = \frac{8\pi K}{c}$$

Relating the energy density to the intensity of the radiation.

yields the relation:

$$(8) \qquad u = \frac{8\pi\nu^2}{c^3} U$$

From this and from equation (7) follows:

$$U = \nu \cdot f\left(\frac{\theta}{\nu}\right)$$

where now c does not appear at all In place of this we may also write:$^{\blacksquare}$

$$\theta = \nu \cdot f\left(\frac{U}{\nu}\right)$$

The symbol f is used simply to designate a functional relationship. The new function is not the same as the original when U and θ are interchanged.

§9. Finally, we introduce the entropy S of the resonator by setting$^{\blacksquare}$

$$(9) \qquad \frac{1}{\theta} = \frac{dS}{dU}$$

By the definition of entropy.

We then obtain:

$$\frac{dS}{dU} = \frac{1}{\nu} \cdot f\left(\frac{U}{\nu}\right)^{\blacksquare}$$

and integrated:

$$(10) \qquad \cdot \quad S = f\left(\frac{U}{\nu}\right)$$

This is a differential equation which is solved by multiplying through by the element of energy dU and summing over the entire range of energy; i.e., by integration.

that is, the entropy of a resonator vibrating in an arbitrary diathermic medium depends only on the variable U/ν, containing besides this only

[13] M. Planck, *Ann. d. Phys. 1* (1900), p. 99.

universal constants. This is the simplest form of Wien's displacement **law** known to me.

§10.   If we apply Wien's displacement law in the latter form to equation (6) for the entropy S, we then find that the energy element ϵ must be proportional to the frequency ν, thus:

$$\epsilon = h\nu \blacksquare$$

and consequently:

$$S = k\left\{\left(1 + \frac{U}{h\nu}\right)\log\left(1 + \frac{U}{h\nu}\right) - \frac{U}{h\nu}\log\frac{U}{h\nu}\right\}$$

here h and k are universal constants.

By substitution into equation (9) one obtains:

$$\frac{1}{\theta} = \frac{k}{h\nu}\log\left(1 + \frac{h\nu}{U}\right)$$

(11) $$U = \frac{h\nu}{e^{h\nu/k\theta} - 1}$$

and from equation (8) there then follows the energy distribution law sought for: $\blacksquare$

(12) $$u = \frac{8\pi h\nu^3}{c^3} \cdot \frac{1}{e^{h\nu/k\theta} - 1}$$

or by introducing the substitutions given in §7, in terms of wavelength λ instead of the frequency:

(13) $$E = \frac{8\pi ch}{\lambda^5} \cdot \frac{1}{e^{ch/k\lambda\theta} - 1}$$

I plan to derive elsewhere the expressions for the intensity and entropy of radiation progressing in a diathermic medium, as well as the theorem for the increase of total entropy in nonstationary radiation processes.

III.   NUMERICAL VALUES

§11.   The values of both universal constants h and k may be calculated rather precisely with the aid of available measurements. $\blacksquare$ F. Kurlbaum,[14] designating the total energy radiating into air from 1 sq cm of a black body at temperature $t°C$ in 1 sec by S_t, found that:

$$S_{100} - S_0 = 0.0731 \text{ watt/cm}^2 = 7.31 \cdot 10^5 \text{ erg/cm}^2 \cdot \text{sec}$$

From this one can obtain the energy density of the total radiation energy in air at the absolute temperature 1:

$$\frac{4 \cdot 7.31 \cdot 10^5}{3 \cdot 10^{10}(373^4 - 273^4)} = 7.061 \cdot 10^{-15} \text{ erg/cm}^3 \cdot \text{deg}^4$$

---

[14] F. Kurlbaum, *Wied. Ann.* 65 (1898), p. 759.

Sidebar notes:

The well-known expression for the energy of a *quantum of light.*

The Planck distribution law. While Planck's hypothesis led to the correct radiation law, his derivation is not entirely satisfactory, for in its use of classical laws of absorption and emission it assumes that an oscillator of frequency ν can have not only the energy $h\nu$ but also integral multiples of this value. But this would mean that a photon could have the energy $n h\nu$, which is contrary to Einstein's quantum hypothesis. The way to a correct derivation was demonstrated later by Bose and Einstein. See, e.g., M. Born, *Atomic Physics* (New York: Stechert, 1936), p. 201.

h is known as *Planck's constant* while k is the Boltzmann constant, although it was apparently not introduced by Boltzmann.

On the other hand, according to equation (12) the energy density of the total radiant energy for $\theta = 1$ is:

$$u^* = \int_0^\infty u d\nu\blacksquare = \frac{8\pi h}{c^3} \int_0^\infty \frac{\nu^3 d\nu}{e^{h\nu/k} - 1}$$

$$= \frac{8\pi h}{c^3} \int_0^\infty \nu^3 (e^{-h\nu/k} + e^{-2h\nu/k} + e^{-3h\nu/k} + \cdots)\, d\nu$$

and by termwise integration:$\blacksquare$

$$u^* = \frac{8\pi h}{c^3} \cdot 6\left(\frac{k}{h}\right)^4 \left(1 + \frac{1}{24} + \frac{1}{34} + \frac{1}{44} + \cdots\right)$$

$$= \frac{48\pi k^4}{c^3 h^3} \cdot 1.0823$$

If we set this equal to $7.061 \cdot 10^{-15}$, then, since $c = 3 \cdot 10^{10}$ cm/sec, we obtain:

(14) $$\frac{k^4}{h^3} = 1.1682 \cdot 10^{15}$$

§12. O. Lummer and E. Pringsheim[15] determined the product $\lambda_m \theta$, where λ_m is the wavelength of maximum energy in air at temperature θ, to be 2940 micron · degree.$\blacksquare$ Thus, in absolute measure:

$$\lambda_m \theta = 0.294 \text{ cm} \cdot \text{deg}$$

On the other hand, it follows from equation (13), when one sets the derivative of E with respect to θ equal to zero,$\blacksquare$ thereby finding $\lambda = \lambda_m$

$$\left(1 - \frac{ch}{5k\lambda_m\theta}\right) \cdot e^{ch/k\lambda_m\theta} = 1$$

and from this transcendental equation:$\blacksquare$

$$\lambda_m\theta = \frac{ch}{4.9651k}$$

consequently:

$$\frac{h}{k} = \frac{4.9651 \cdot 0.294}{3 \cdot 10^{10}} = 4.866 \cdot 10^{-11}$$

From this and from equation (14) the values for the universal constants become:$\blacksquare$

(15) $$h = 6.55 \cdot 10^{-27} \text{ erg} \cdot \text{sec}$$

(16) $$k = 1.346 \cdot 10^{-16} \text{ erg/deg}$$

These are the same numbers that I indicated in my earlier communication.

---

[15] O. Lummer and E. Pringsheim, *Transactions of the German Physical Society* 2 (1900), p. 176.

Margin notes:

Indicates the sum of all the products $u d\nu$ as ν takes on all values from 0 to ∞.

By summing the above expression term by term.

1 micron = 10^{-6} meter

That is, by setting $dE/d\theta = 0$ one finds the wavelength λ_m at which E is a maximum.

An equation other than an algebraic one, such as trigonometric, or as here, exponential.

The present values are: $h = 6.6237.10^{-27}$ erg·sec. $k = 1.3803.10^{-16}$ erg/deg

SUPPLEMENTARY READING

Holton, G., and D. H. D. Roller, *Foundations of Modern Physical Science* (Reading, Mass: Addison-Wesley, 1958), Ch. 31.

Lindsay, R. B., and H. Margenau, *Foundations of Physics* (New York: Dover, 1957), Ch. IX.

Planck, M., *A Survey of Physics,* trans. by R. Jones and D. H. Williams (New York: Dutton, 1925), pp. 159 ff.

———, *Scientific Autobiography,* trans. by F. Gaynor (New York: Philosophical Library, 1949).

———, *Where is Science Going?,* trans. by J. Murphy (London: Allen & Unwin, 1933).

Richtmyer, F. K., R. H. Kennard, and T. Lauritsen, *Introduction to Modern Physics,* 5th ed. (New York: McGraw-Hill, 1955), Ch. 4.

Wilson, W., *A Hundred Years of Physics* (London: Duckworth, 1950), Ch. 13.

Zimmer, E., *The Revolution in Physics* (New York: Harcourt, Brace, 1936), Ch. 3.

Albert Einstein[1]

1879-1955

The Theory of Relativity

THE TRANSITION from classical to twentieth-century physics was sparked chiefly by two independent discoveries: the first, in 1900, was Planck's revolutionary quantum hypothesis concerning the emission and absorption of radiant energy; the second, just five years later, was Einstein's discovery that the usual conceptions of space and time, on which is based all of Newtonian mechanics, are in reality limiting cases that agree with everyday experiences but are not valid generally. It is difficult to classify the theory of relativity according to any of the major branches of physics. It stands apart as a fundamental structure, as basic as the concepts of space and time themselves, yet it embraces and influences all physics to some degree. Unlike the usual sequence of events in physics, relativity did not stem primarily from experiment, or from efforts to reconcile disparate points of view; rather it was the result of a critical examination of well-known and widely accepted physical principles.

The study of motion, historically the first to be developed in the *modern period,* necessarily involves the concepts of space and time. The idea that motion must be *relative,* that is, it involves displacements of objects relative to some reference system or other, goes back to very early times. If one asserts, for example, that a train is at rest he implies that it is motionless relative to an observer alongside the track; as far as a passenger is concerned the train is always at rest relative to him unless he walks along the aisle. But even assuming the train, passenger, and observer have no relative motion with respect to one another, or with the surface of the earth, they nevertheless have the velocity of the earth itself; hence, they would be in motion relative to an observer situated on the moon, for example. To carry this

---

[1] For a biographical sketch of Einstein see Chapter 17.

simple idea to its obvious limit, one would inquire whether there exists some absolute reference system to which *all* motion may be compared. It is largely in this regard that the ideas in the past have varied.

In his *Principia* Newton defined *absolute motion* as ". . . the translation of a body from one absolute place into another,"[2] without specifying what he meant by absolute place, rather leaving it as an intuitive concept. He held a somewhat similar view of time, describing *absolute time* as "absolute, true, and mathematical time, of itself, and from its own nature flows equally without regard to anything external. . . ."[3] Time was regarded as being independent of space, and the meaning of such terms as *past, present,* and *future* seemed perfectly clear: the present was but an infinitely short moment between the past and the future. Here we consider those events as *past* provided one can, in principle, obtain some knowledge about them, and as *future* those events over which, again in principle, we have some measure of influence.[4] But this naive conception of time, on which is based all of classical mechanics, and which seems in accord with our normal experience, is not generally valid, for as Einstein showed, the interval between *past* and *future* has a finite extension in time which depends on the distance in space between an event and its observer.

Despite his efforts to define absolute motion in terms of absolute space and time, Newton never made use of these concepts in his application of mechanical principles; neither one, in fact, can be observed. The latter statement essentially summarizes the *special theory of relativity,* namely, the concept of absolute motion has no meaning. All mechanical experiments involve determining the position of some material body as a function of time. If one wished to verify Galileo's law of falling bodies he would simply drop an object and measure its position at various times afterwards. But the position must be specified with respect to some *coordinate system* or *frame of reference,* and since the position generally is determined against a scale fixed in the laboratory, the earth becomes our frame of reference. This is the same reference frame used by Galileo and Newton; hence one would naturally conclude that the laws of classical mechanics must be valid (i.e., reproducible) in every coordinate system rigidly connected with the earth. Actually this is not strictly correct, for the rotation of the earth, by introducing additional acceleration forces, would prevent one from confirming these laws in minute detail.[5] The effect is so small, however, that

---

[2] See Chapter 4, page 50.

[3] See Chapter 4, page 49.

[4] W. Heisenberg, *The Physicists Conception of Nature* (London: Hutchinson, 1958), p. 47.

[5] It should be apparent that were it not for the fact that small deviations from the mechanical laws are observed on earth, our only evidence for its rotation would be indirect—by inference from astronomical observations.

for practical purposes we may assume the earth to be a suitable frame of reference for the laws of motion.[6] The important question is whether there are other coordinate systems in which the laws of mechanics are the same; that is, the equations which express them have the same form. We know that an observer who is accelerating with respect to the earth, as in an automobile rounding a curve or an elevator starting upward, would not, because of inertial effects, experience the same laws of motion in his "laboratory." On the other hand, if the automobile moves at a constant speed in a straight line, or if the elevator is no longer accelerating, we know from our everyday experiences that measurements in these "laboratories" will yield the same results as those obtained in a laboratory fixed to the surface of the earth. It follows, therefore, that if the laws of mechanics are found to be valid in one coordinate system, they are equally valid in any other system having uniform motion relative to the first. Such reference frames, in which the laws of mechanics are the same, are known as *inertial systems,* after Newton's first law (the law of inertia). The principle that the laws of mechanics hold in all inertial systems, sometimes known as the *Galilean relativity principle,*[7] served as a starting point for Einstein's theory of relativity, for he generalized it to apply to all laws of physics, not only the laws of motion.

Two centuries after Newton speculated on the question of absolute motion the subject arose again in connection with *electrodynamics.* From his electromagnetic theory Maxwell predicted the velocity of light (electromagnetic radiation) to be constant for any given medium, and to have the value $c = 3 \times 10^{10}$ cm/sec in empty space. Since there was no meaning to absolute velocity it was concluded that the reference frame for the propagation of light was the "stationary" *ether.* But a crucial experiment, performed in 1881 by A. A. Michelson (1852-1931) and E. W. Morley (1838-1923),[8] failed to confirm this view and thereby cast serious doubt on the entire ether concept. If there is an ether the earth must be moving through it, and therefore the velocity of light, assumed constant with respect to the ether, should differ depending upon whether it is measured in a direction parallel to the earth's motion, or at right angles to it. The expected difference is small, since the velocity of light is about ten thousand times the velocity of the earth in its orbit (with respect to the sun). Nevertheless, it

---

[6] A better reference frame is one fixed to the distant stars.

[7] In connection with his work on projectile motion Galileo first solved the problem of transforming a given motion from one inertial system to another.

[8] *Philosophical Magazine*, vol. 24, December 1887, p. 449. For an original account of this important experiment see W. F. Magie, *A Source Book in Physics* (New York: McGraw-Hill, 1935), p. 369.

is measurable; yet, the Michelson-Morley experiment gave a completely negative result.

There were many attempts to account for the absence of an "ether drift," [9] but all were unsatisfactory until Einstein, in 1905, proposed his *special* or *restricted theory of relativity*.[10] He suggested that the concept of motion through an ether was meaningless, that the only motion having physical significance was that relative to *material* bodies! His theory was based on two simple postulates:

1. Physical laws and principles have the same form in all inertial systems.
2. The velocity of light in any inertial system is independent of the velocity of that system; i.e., the velocity of light is independent of the motion of its source.

It will be noted that the second postulate essentially states an experimental fact, namely the Michelson-Morley result, while the first generalizes a fairly wide range of experience. From these simple premises Einstein was led ultimately to the remarkable conclusion that the mass of a body varies with its velocity, and, equally surprising, that the energy of a body is proportional to its mass.

The theory of relativity was Einstein's most important discovery, and the one by which he achieved his great popular fame. It stands out as one of the boldest concepts in the history of physics—one that fires the imagination in a way that few others can. The following extract is taken from Einstein's classic paper "On the Electrodynamics of Moving Bodies," *Annalen der Physik 17* (1905), p. 891, translated by Saha and Bose.[11]

---

[9] Notable among these was the explanation put forth independently by H. A. Lorentz (1853-1928) and G. F. Fitzgerald (1851-1901) to the effect that a measuring rod contracts in length along the direction in which it moves through the ether, and this contraction makes it impossible to observe the ether drift. The Lorentz-Fitzgerald "contraction" hypothesis, in modified form, was later incorporated into the special theory of relativity.

[10] So called to distinguish it from his *general theory of relativity*, developed ten years later, which deals with accelerated motion and asserts that absolute acceleration has no meaning.

[11] A. Einstein, *The Principle of Relativity*, trans. by Saha and Bose (Calcutta, India: University Press, 1920), or *The Principle of Relativity*, trans. by W. Perrett and G. B. Jeffrey (New York: Dover, 1958).

Einstein's "Experiment"

INTRODUCTION

The phenomena of electricity in motion.

It is well known that if we attempt to apply Maxwell's electrodynamics,▪ as conceived at the present time, to moving bodies, we are led to assymetry which does not agree with observed phenomena. Let us think of the mutual action between a magnet and a conductor. The observed phenomena in this case depend only on the relative motion of the conductor and the magnet, while according to the usual conception, a distinction must be made between the cases where the one or the other of the bodies is in motion. If, for example, the magnet moves and the conductor is at rest, then an electric field of certain energy value is produced in the neighborhood of the magnet, which excites a current in those parts of the field where a conductor exists.■ But if the magnet be at rest and the conductor be set in motion, no electric field is produced in the neighborhood of the magnet, but an electromotive force which corresponds to no energy in itself is produced in the conductor; this causes an electric current of the same magnitude and in the same direction as the electric force, it being of course assumed that the relative motion in both of these cases is the same.

2.   Examples of a similar kind such as the unsuccessful attempt to substantiate the motion of the earth relative to the "light-medium"■ lead us to the supposition that not only in mechanics, but also in electrodynamics, no properties of observed facts correspond to a concept of absolute rest; but that for all coordinate systems for which the mechanical equations hold, the equivalent electrodynamical and optical equations hold also, as has already been shown for magnitudes of the first order.■ In the following we make these assumptions (which we shall subsequently call the Principle of Relativity) and introduce the further assumption—an assumption which is at the first sight quite irreconcilable with the former one—that light is propagated in vacant space, with a velocity c which is independent of the nature of motion of the emitting body. These two assumptions are quite sufficient to give us a simple and consistent theory of electrodynamics of moving bodies on the basis of the Maxwellian theory for bodies at rest. The introduction of a "Lightäther"■ will be proved to be superfluous, for according to the conception which will be developed, we shall introduce neither a space absolutely at rest, and endowed with special properties, nor shall we associate a velocity vector with a point in which electromagnetic processes take place.■

3.   Like every other theory in electrodynamics, the theory is based on the kinematics of rigid bodies■; in the enunciation of every theory, we have to do with relations between rigid bodies (coordinate system), clocks, and electromagnetic processes. An insufficient consideration of these circumstances is the cause of difficulties with which the electrodynamics of moving bodies have to fight at present.

Note that an observer at rest relative to a magnet detects only a magnetic field, while one in motion with respect to the magnet experiences an electric field as well. The apparent asymmetry, which developed because of the tendency to refer all motion to the earth as a frame of reference, thus disappears if relative motion only is considered.

The "ether." The reference here is to the Michelson-Morley experiment.

Both Fresnel and later Lorentz showed that there should be no *first-order* effects on optical phenomena caused by the velocity of the apparatus through the ether, but that there might be *second-order* effects; i.e., effects proportional to the square of the ratio of the velocity of the apparatus to the velocity of light.

luminiferous ether

That is, absolute motion has no more meaning in electrodynamics than in mechanics.

The science which treats of the motions of rigid bodies.

I.   KINEMATICAL PORTION

§1. *Definition of Synchronism.*

Let us have a coordinate system, in which the Newtonian equations hold.■ For distinguishing this system from another which will be introduced hereafter, we shall always call it "the stationary system."

If a material point be at rest in this system, then its position in this system can be found out by a measuring rod, and can be expressed by the methods of Euclidean geometry, or in Cartesian coordinates.

If we wish to describe the motion of a material point, the values of its coordinates must be expressed as functions of time.■ It is always to be borne in mind that *such a mathematical definition has a physical sense, only when we have a clear notion of what is meant by time. We have to take into consideration the fact that those of our conceptions, in which time plays a part, are always conceptions of synchronism.* For example, we say that a train arrives here at 7 o'clock; this means that the exact pointing of the little hand of my watch to 7, and the arrival of the train are synchronous events.

It may appear that all difficulties connected with the definition of time can be removed when in place of time, we substitute the position of the little hand of my watch. Such a definition is in fact sufficient, when it is required to define time exclusively for the place at which the clock is stationed. But the definition is not sufficient when it is required to connect by time events taking place at different stations, or, what amounts to the same thing, to estimate in terms of time the occurrence of events which take place at stations distant from the clock.

Now with regard to this attempt—the time estimation of events—we can satisfy ourselves in the following manner. Suppose an observer, who is stationed at the origin of coordinates with the clock,■ associates a ray of light which comes to him through space, and gives testimony to the event of which the time is to be estimated—with the corresponding position of the hands of the clock. But such an association has this defect: it depends on the position of the observer provided with the clock, as we know by experience. We can arrive at a more practicable result by the following treatment.

If an observer be stationed at A with a clock, he can estimate the time of events occurring in the immediate neighborhood of A, by looking for the position of the hands of the clock, which are synchronous with the event. If an observer be stationed at B with a clock—we should add that the clock is of the same nature as the one at A—he can estimate the time of events occurring at B. But without further premises, it is not possible to compare, as far as time is concerned, the events at B with the events at A. We have therefore an A time, and a B time, but no time common

An inertial system, to which the earth is a reasonably good approximation.

It will be seen how important was a clear picture of the concept of time to the development of relativity theory. Its role, in fact, is basic to the theory.

The origin of coordinates is specified solely for convenience, since the coordinate system here in question is fixed relative to the observer.

to *A* and *B*. This last time (i.e., common time) can be defined, if we establish by definition that the time which light requires in traveling from *A* to *B* is equivalent to the time which light requires in traveling from *B* to *A*.■ For example, a ray of light proceeds from *A* at *A* time t_A towards *B*, arrives and is reflected from *B* at *B* time t_B and returns to *A* at *A* time t'_A. According to the definition, both clocks are synchronous, if

$$t_B - t_A = t'_A - t_B$$

We assume that this definition of synchronism is possible without involving any inconsistency, for any number of points, therefore the following relations hold:

1. If the clock at *B* be synchronous with the clock at *A*, then the clock at *A* is synchronous with the clock at *B*.

2. If the clock at *A* as well as the clock at *B* are both synchronous with the clock at *C*, then the clocks at *A* and *B* are synchronous.

Thus with the help of certain physical experiences, we have established what we understand when we speak of clocks at rest at different stations, and synchronous with one another; and thereby we have arrived at a definition of synchronism and time.

In accordance with experience we shall assume that the magnitude $\dfrac{2\overline{AB}}{t'_A - t_A} = c$, where *c* is a universal constant.■

We have defined time essentially with a clock at rest in a stationary system. On account of its adaptability to the stationary system, we refer to the time defined in this way as "time of the stationary system."

§2.   *On the Relativity of Length and Time*

The following reflections are based on the Principle of Relativity and on the Principle of Constancy of the velocity of light, both of which we define in the following way:

1. The laws according to which the nature of physical systems alter are independent of the manner in which these changes are referred to two coordinate systems which have a translatory motion relative to each other.■

2. Every ray of light moves in the "stationary coordinate system" with the same velocity *c*, the velocity being independent of the condition whether this ray of light is emitted by a body at rest or in motion. Therefore

$$\text{velocity} = \frac{\text{Path of Light}}{\text{Interval of time}}$$

where, by "interval of time," we mean time as defined in §1.

Let us have a rigid body at rest; this has a length *l*, when measured by a measuring rod at rest; we suppose that the axis of the rod is laid along

This seemingly "obvious" statement is nevertheless important for a full development of the theory. One can imagine, for example, that on the basis of motion through an *ether* a ray proceeding from A to B may be traveling with the direction of motion while a ray from B to A would be opposite to it and hence have different speed.

AB is the mean distance between A and B, and c is the velocity of light.

A rotary motion is ruled out because of the acceleration forces produced. The translatory motion must be at uniform (relative) velocity.

the X axis of the system at rest, and then a uniform velocity v, parallel to the axis of X, is imparted to it. Let us now inquire about the length of the moving rod; this can be obtained by either of these operations.

(*a*)   The observer provided with the measuring rod moves along with the rod to be measured, and measures by direct superposition the length of the rod—just as if the observer, the measuring rod, and the rod to be measured were at rest.■

That is, they have no relative motion.

(*b*)   The observer finds out, by means of clocks placed in a system at rest (the clocks being synchronous as defined in §1), the points of the system where the ends of the rod to be measured occur at a particular time t. The distance between these two points, measured by the previously used measuring rod, this time it being at rest, is a length, which we may call the "length of the rod."

According to the Principle of Relativity, the length found out by the operation (*a*), which we may call "the length of the rod in the moving system" is equal to the length l of the rod in the stationary system.

The length which is found out by the second method, may be called *the length of the moving rod measured from the stationary system.*■ This length is to be estimated on the basis of our principle, and *we shall find it to be different from l.*

The distinction here is very important. The first method, in which the observer is at rest relative to the rod, is the one which we normally experience.

In the generally recognized kinematics, we silently assume that the lengths defined by these two operations are equal, or in other words, that at an epoch■ of time t, a moving rigid body is geometrically replaceable by the same body which can replace it in the condition of rest.

epoch: instant

Relativity of Time

Let us suppose that the two clocks synchronous with the clocks in the system at rest are brought to the ends A, and B of a rod, i.e., the time of the clocks correspond to the time of the stationary system at the points where they happen to arrive; these clocks are therefore synchronous in the stationary system.

We further imagine that there are two observers at the two watches, and moving with them, and that these observers apply the criterion for synchronism to the two clocks. At the time t_A, a ray of light goes out from A, is reflected from B at the time t_B, and arrives back at A at time t'_A. Taking into consideration the principle of constancy of the velocity of light, we have

$$t_B - t_A = \frac{l_{AB}}{c - v}$$

and

$$t'_A - t_B = \frac{l_{AB}}{c + v}$$

where l_{AB} is the length of the moving rod, measured in the stationary system.■ Therefore the observers stationed with the watches will not find the clocks synchronous, though the observer in the stationary system must declare the clocks to be synchronous. We therefore see that we can attach no absolute significance to the concept of synchronism; but two events which are synchronous when viewed from one system, will not be synchronous when viewed from a system moving relative to this system.

§3.   *Theory of Coordinate and Time Transformation from a stationary system to a system which moves relatively to this with uniform velocity.*

Let there be given, in the stationary system, two coordinate systems, i.e., two series of three mutually perpendicular lines issuing from a point. Let the X axes of each coincide with one another, and the Y and Z axes be parallel. Let a rigid measuring rod and a number of clocks be given to each of the systems, and let the rods and clocks in each be exactly alike each other.

Let the initial point of one of the systems (k) have a constant velocity in the direction of the X axis of the other which is a stationary system K, the motion being also communicated to the rods and clocks in the system (k).■ Any time t of the stationary system K corresponds to a definite position of the axes of the moving system, which are always parallel to the axes of the stationary system. By t, we always mean the time in the stationary system.

We suppose that the space is measured by the stationary measuring rod placed in the stationary system, as well as by the moving measuring rod placed in the moving system, and we thus obtain the coordinates (x, y, z) for the stationary system, and (ξ, η, ζ) for the moving system. Let the time t be determined for each point of the stationary system (which are provided with clocks) by means of the clocks which are placed in the stationary system, with the help of light signals as described in §1. Let also the time τ of the moving system be determined for each point of the moving system (in which there are clocks which are at rest relative to the moving system), by means of the method of light signals between these points (in which there are clocks) in the manner described in §1.■

To every value of (x, y, z, t) which fully determines the position and time of an event in the stationary system, there corresponds a system of values (ξ, η, ζ, τ); now the problem is to find the system of equations connecting these magnitudes.

Primarily it is clear that on account of the property of homogeneity which we ascribe to time and space, the equations must be linear.■

If we put $x' = x - vt$, then it is clear that at a point relatively at rest in the system K, we have a system of values (x', y, z) which are independent

v is the velocity of the rod relative to the stationary system.

That is, the entire system (k) is in motion relative to K, keeping in mind that the term "stationary" as applied to K means only that it is an inertial system.

Thus t and τ are each local times of the "conventional" sort, determined with clocks at rest relative respectively to each system.

That is, the properties of space, at least in so far as the propagation of light is concerned, are the same at all points and in every direction.

of time. Now let us find out τ as a function of (x, y, z, t). For this purpose we have to express in equations the fact that τ is none other than the time given by the clocks which are at rest in the system k which must be made synchronous in the manner described in §1.

Let a ray of light be sent at time τ_0 from the origin of the system k along the X axis towards x' and let it be reflected from that place at time τ_1 towards the origin of moving coordinates and let it arrive there at time τ_2; then we must have

$$\tfrac{1}{2}(\tau_0 + \tau_2) = \tau_1$$

If we now introduce the condition that τ is a function of coordinates, and apply the principle of constancy of the velocity of light in the stationary system, we have■

$$\tfrac{1}{2}\left[\tau(0, 0, 0, t) + \tau\left(0, 0, 0, \left\{t + \frac{x'}{c - v} + \frac{x'}{c + v}\right\}\right)\right]$$
$$= \tau\left(x', 0, 0, t + \frac{x'}{c - v}\right)$$

It is to be noticed that instead of the origin of coordinates, we could select some other point as the exit point for rays of light and therefore the above equation holds for all values of $(x', y, z, t,)$.

A similar conception, being applied to the y and z axis gives us, when we take into consideration the fact that light when viewed from the stationary system, is always propagated along those axes with the velocity $\sqrt{c^2 - v^2}$■

We have the equations:

$$\frac{\delta\tau}{\delta y} = 0 \qquad \frac{\delta\tau}{\delta z} = 0\,■$$

From these equations it follows that τ is a linear function of x and t. From the equation given earlier we obtain

$$\tau = a\left(t - \frac{vx'}{c^2 - v^2}\right)$$

where a is an unknown function of v.

With the help of these results it is easy to obtain the magnitudes (ξ, η, ζ), if we express by means of equations the fact that light, when measured in the moving system is always propagated with the constant velocity c (as the principle of constancy of light velocity in conjunction with the principle of relativity requires). For time $\tau = 0$, if the ray is sent in the direction of increasing ξ, we have

$$\xi = c\tau, \text{ i.e., } \xi = ac\left(t - \frac{vx'}{c^2 - v^2}\right)$$

Now the ray of light moves relative to the origin of k with a velocity $c - v$, measured in the stationary system; therefore we have

$$\frac{x'}{c - v} = t$$

Substituting these values of t in the equation for ξ, we obtain

$$\xi = a \frac{c^2}{c^2 - v^2} x'$$

In an analogous manner, we obtain by considering the ray of light which moves along the y axis,

$$\eta = c\tau = ac\left(t - \frac{vx'}{c^2 - v^2}\right)$$

where $\dfrac{v}{\sqrt{c^2 - v^2}} = t$, $x' = 0$■

Since $\sqrt{c^2 - v^2}$ is the velocity in the y direction.

Therefore $\eta = a \dfrac{c}{\sqrt{c^2 - v^2}} y$, $\zeta = a \dfrac{c}{\sqrt{c^2 - v^2}} z$.

If for x', we substitute its value $x - tv$, we obtain■

The *transformation equations*.

$$\tau = \phi(v) \cdot \beta\left(t - \frac{vx}{c^2}\right),$$

$$\xi = \phi(v) \cdot \beta(x - vt),$$

$$\eta = \phi(v)y$$

$$\zeta = \phi(v)z$$

where $\beta = \dfrac{1}{\sqrt{1 - \dfrac{v^2}{c^2}}}$, and $\phi(v) = \dfrac{ac}{\sqrt{c^2 - v^2}} = \dfrac{a}{\beta}$ is a function of v.

If we make no assumption about the initial position of the moving system and about the null point■ of t, then an additive constant is to be added to the right-hand side.

The zero point.

We have now to show that every ray of light moves in the moving system with a velocity c (when measured in the moving system), in case, as we have actually assumed, c is also the velocity in the stationary system; for we have not as yet adduced any proof in support of the assumption that the principle of relativity is reconcilable with the principle of constant light velocity.

At a time $\tau = t = 0$ let a spherical wave be sent out from the common origin of the two systems of coordinates,■ and let it spread with a velocity c in the system K. If (x, y, z) be a point reached by the wave, we have

As from a point source of light.

Which is the equation of a sphere having radius ct.

$$x^2 + y^2 + z^2 = c^2t^2■$$

With the aid of our transformation-equations, let us transform this equation, and we obtain by a simple calculation,

$$\xi^2 + \eta^2 + \zeta^2 = c^2\tau^2$$

Therefore the wave is propagated in the moving system with the same velocity c, and as a spherical wave. Therefore we show that the two principles are mutually reconcilable.

In the transformations we have an undetermined function $\phi(v)$ which we now proceed to find.

Let us introduce for this purpose a third coordinate system k', which is set in motion relative to the system k, the motion being parallel to the ξ axis.■ Let the velocity of the origin be $(-v)$. At the time $t = 0$, all the initial coordinate points coincide, and for $t = x = y = z = 0$, the time t' of the system $k' = 0$. We shall say that $(x'y'z't')$ are the coordinates measured in the system k', then by a twofold application of the transformation equations, we obtain

Which, in turn, is parallel to the x axis in the stationary system.

$$t' = \phi(-v)\beta(-v)\left\{\tau + \frac{v}{c^2}\xi\right\} = \phi(v)\phi(-v)t,$$

$$x' = \phi(v)\beta(v)(\xi + v\tau) = \phi(v)\phi(-v)x, \text{ etc.}$$

Since the relations between (x', y', z', t') and (x, y, z, t) do not contain time explicitly, therefore K and k' are relatively at rest.■

Which agrees, of course, with our usual conceptions, since k' was given a velocity (−v) relative to k, which in turn has a velocity (+v) with respect to K.

It appears that the systems K and k' are identical.

$$\therefore \; \phi(v)\phi(-v) = 1$$

Let us now turn our attention to the part of the y axis between ($\xi = 0$, $\eta = 0$, $\zeta = 0$), and ($\xi = 0$, $\eta = l$, $\zeta = 0$). Let this piece of the y axis be covered with a rod moving with the velocity v relative to the system K and perpendicular to its axis,■ the ends of the rod having therefore the coordinates

The rod is parallel to the y axis but moves perpendicular to it.

$$\begin{aligned} x_1 &= vt, & y &= \frac{l}{\phi(v)}, & z_1 &= 0 \\ x_2 &= vt, & y_2 &= 0, & z_2 &= 0 \end{aligned}\Bigg\}$$

Therefore the length of the rod measured in the system $K = \dfrac{l}{\phi(v)}$.

For the system moving with velocity $(-v)$, we have on grounds of symmetry,■

Since it should make no difference if the system moves right or left with respect to K.

$$\frac{l}{\phi(v)} = \frac{l}{\phi(-v)}$$

$$\therefore \; \phi(v) = \phi(-v)$$

$$\therefore \; \phi(v) = 1■$$

Since, from above, $\phi(v)\,\phi(-v) = 1$.

§4.   *The Physical Significance of the Equations obtained concerning moving rigid bodies and moving clocks.*

Let us consider a rigid sphere (i.e., one having a spherical figure when tested in the stationary system) of radius R which is at rest relative to the system (K), and whose center coincides with the origin of K, then the equation of the surface of this sphere, which is moving with a velocity v relative to K, is

$$\xi^2 + \eta^2 + \zeta^2 = R^2$$

At time $t = 0$, the equation is expressed in terms of $(x, y, z, t,)$ as■

$$\frac{x^2}{\left(\sqrt{1 - \dfrac{v^2}{c^2}}\right)^2} + y^2 + z^2 = R^2$$

From the transformation equations, since $\phi(v) = 1$,
$\xi = \beta x$ (at $t = 0$)
$\eta = y$
$\zeta = z$

A rigid body which has the figure of a sphere when measured in the moving system, has therefore in the moving condition, when considered from the stationary system, the figure of a rotational ellipsoid■ with semi-axes

An ellipse rotated about its minor axis.

$$R\sqrt{1 - \frac{v^2}{c^2}},\ R,\ R$$

Therefore the y and z dimensions of the sphere (therefore of any figure also) do not appear to be modified by the motion, but the x dimension is shortened in the ratio $1 : \sqrt{1 - \dfrac{v^2}{c^2}}$; the shortening is the larger, the larger is v. For $v = c$, all moving bodies, when considered from a stationary system shrink into planes. For a velocity larger than the velocity of light, our propositions become meaningless; in our theory c plays the part of infinite velocity.

It is clear that similar results hold about stationary bodies in a stationary system when considered from a uniformly moving system.■

It is only the relative motion that must be taken into account.

Let us now consider that a clock which is lying at rest in the stationary system gives the time t, and lying at rest relative to the moving system is capable of giving the time τ; suppose it to be placed at the origin of the moving system k, and to be so arranged that it gives the time τ. How much does the clock gain when viewed from the stationary system K? We have,

$$\tau = \frac{1}{\sqrt{1 - \dfrac{v^2}{c^2}}}\left(t - \frac{v}{c^2}x\right), \text{ and } x = vt,$$

$$\therefore\ \ \tau - t = \left[1 - \sqrt{1 - \frac{v^2}{c^2}}\right]t$$

When expanded,
$\left(\dfrac{1-v^2}{c^2}\right)^{\frac{1}{2}} \approx 1 - \dfrac{1}{2}\dfrac{v^2}{c^2},$

Therefore the clock loses by an amount $\dfrac{1}{2}\dfrac{v^2}{c^2}$ per second of motion, to the second order of approximation.■

which for small $\dfrac{v}{c}$ is a good approximation.

Since the clock loses by the amount $\frac{1}{2} \frac{v^2}{c^2}$ each second.

From this, the following peculiar consequence follows. Suppose at two points A and B of the stationary system two clocks are given, which are synchronous in the sense explained in §3 when viewed from the stationary system. Suppose the clock at A to be set in motion in the line joining it with B, then after the arrival of the clock at B, they will no longer be found synchronous, but the clock which was set in motion from A will lag behind the clock which had been all along at B by an amount $\frac{1}{2} t \frac{v^2}{c^2}$■, where t is the time required for the journey.

If the motion is not in a straight line it may be resolved into components and treated accordingly. When the segments of the polygonal line are short, it approximates a continuous curve.

We see forthwith that the result holds also when the clock moves from A to B by a polygonal line, and also when A and B coincide.■

If we assume that the result obtained for a polygonal line holds also for a curved line, we obtain the following law. If at A, there be two synchronous clocks, and if we set in motion one of them with a constant velocity along a closed curve till it comes back to A, the journey being completed in t seconds, then after arrival, the last mentioned clock will be behind the stationary one by $\frac{1}{2} t \frac{v^2}{c^2}$ seconds. From this, we conclude that a clock placed at the equator must be slower by a very small amount than a similarly constructed clock which is placed at the pole, all other conditions being identical.■

The relative velocity here is small, about 1000 mph, but with the precision clocks now available, the effect should be detectable.

Omitted are sections dealing with relativistic addition of velocities and applications to electrodynamics, including the derivation for relativistic mass of a moving body.

■   .   .   .   .

SUPPLEMENTARY READING

Barnett, L., *The Universe and Dr. Einstein* (New York: William Sloane, 1948).

Born, M., *Einstein's Theory of Relativity,* trans. by H. L. Brose (London: Methuen, 1924).

Einstein, A., *The Meaning of Relativity,* 6th ed. (London: Methuen, 1956).

———, and L. Infeld, *The Evolution of Physics* (New York: Simon and Schuster, 1938), pp. 129ff.

Frank, P., *Relativity and Its Astronomical Implications* (Cambridge, Mass.: Sky Publishing, 1943).

Sherwin, C. W., *Basic Concepts of Physics* (New York: Dryden, 1957), Ch. 11.

Wilson, W., *A Hundred Years of Physics,* (London: Duckworth, 1950).

Niels Bohr

1885 - 1962

The Hydrogen Atom

PLANCK'S quantum hypothesis, by which he accounted for black-body radiation, opened the door to a new kind of physics: twentieth-century *quantum physics*. Introduced in 1900, it soon formed the basis for several significant developments. Einstein, in 1905, used the new concept to explain the photoelectric effect, and two years later he applied it to account for the variation of specific heat of a solid with temperature. Each of these added great strength to Planck's theory, but perhaps the most remarkable application was discovered in 1913, when Niels Bohr proposed this theory of atomic structure and the origin of optical spectra.

Experimental spectroscopy became a highly developed branch of physics during the latter part of the nineteenth century, when accurate wavelength measurements were made possible through the invention of the prism spectroscope, in 1859, by R. W. Bunsen (1811-1899) and G. R. Kirchhoff (1824-1887). It had long been known, of course, that different substances, when suitably "excited" in a flame or by an electric discharge, emitted light that was characteristic of the substance. This was first observed by the Scottish physicist Thomas Melvill (1726-1753), who mixed various salts with "burning spirits" and examined the light through a prism. He found the spectrum of a hot gas (the vaporized salt) to be markedly different from that of an incandescent solid or liquid. The latter gave a continuous spectrum while the former consisted of discrete patches of color having the shape of the aperture used to define the beam of light. In Melvill's experiments these were circular spots, corresponding to a circular aperture placed before the prism, but as the use of narrow rectangular slits came into practice the characteristic patterns were called *line spectra*.

Following the work of Bunsen and Kirchhoff, and the impetus given by it to experimental spectroscopy, attention was directed primarily toward the

329

search for regularities in line spectra. Since the emission of light was thought to result from vibrations of some sort, harmonic relations were sought for among the lines in a given spectrum, reasoning by analogy with the overtones found in vibrating mechanical systems. Simple, harmonic series were not observed but other useful relationships were, the most significant being the discovery in 1885 by Johann Balmer (1825-1898) that the wavelengths of all the lines then known in the spectrum of hydrogen could be expressed by the simple formula:

$$\lambda_n = b \left(\frac{n^2}{n^2 - 2^2} \right)$$

where b is a constant (determined empirically by Balmer) and n takes the values 3, 4, 5, . . . for the various lines. Of the nine lines known to Balmer, all of which were in excellent agreement with his formula (within 1 part in 1,000), four in the visible region had been measured accurately by A. J. Ångström (1814-1874), the remaining five, in the ultraviolet region, were measured by Sir William Huggins (1824-1910). Regularities among other spectra were discovered shortly afterward, notably by J. R. Rydberg (1854-1919), who found a general formula valid for a great many spectral series, and W. Ritz (1878-1909) who suggested that the frequency of any spectral line could be expressed as the difference between two terms characteristic of the emitting atom. This suggestion, which was essentially a generalization of the Rydberg formula, is now known as the Ritz *Combination Principle*. When expressed in terms of frequencies, or as is generally done, in terms of *wave numbers* ($\bar{\nu} = 1/\lambda$), the Balmer formula becomes a special case of the more general Rydberg equation:

$$\bar{\nu} = \frac{1}{\lambda} = R_H \left(\frac{1}{2^2} - \frac{1}{n^2} \right) \qquad\qquad n = 3, 4, 5, . . .$$

where R_H is a universal constant known as the *Rydberg constant for hydrogen*. If the first term in brackets is replaced by $1/n_f$, and n is allowed to take on values $(n_f + 1)$, $(n_f + 2)$, $(n_f + 3)$, . . ., we have an expression valid for each of the known series in the hydrogen spectrum; thus, $n_f = 1$ yields the Lyman series, $n_f = 2$ the Balmer series, $n_f = 3$ the Paschen series, $n_f = 4$ the Brackett series, and $n_f = 5$ the Pfund series. It can be seen that for hydrogen the terms in the Ritz principle have the simple form R/n^2, where n is an integer.

However useful these empirical formulae were in other respects, particularly in predicting new spectral lines, they provided no direct clue to the basic mechanism of atomic spectra. The guiding idea came, instead, from another direction, when Rutherford (see Chapter 19) proposed, in

1911, a nuclear model of the atom as a result of his experiments on the scattering of alpha particles. Viewed on classical grounds Rutherford's picture of a positively charged nucleus with electrons somehow associated with it presented grave difficulties. If the charges were assumed at rest relative to one another no stable configuration could be found; if in motion the electrons might be thought to revolve about the nucleus. But since an orbital electron would be subject to centripetal acceleration it should continue to radiate energy and eventually spiral into the nucleus. Furthermore, the radiated energy should be of constantly increasing frequency as the electron orbit becomes smaller, which did not agree with the discrete spectral lines observed.

It was at this point that Bohr combined Rutherford's idea with Planck's quantum hypothesis to develop a theory of atomic structure that accounted for the observed spectrum of hydrogen and hydrogenlike atoms (that is, singly ionized helium, doubly ionized lithium, and so on). He avoided the classical difficulties of instability simply by denying that classical electrodynamics applies in detail to atomic systems. Instead, he assumed that the electron could revolve about the nucleus in certain fixed orbits (either circular or elliptical), called *stationary states,* in which it is perfectly stable. Each such orbit represents a definite energy and when the atom makes a transition from a state of higher energy W_1 to one of lower energy W_2, Bohr postulated that the difference in energy is radiated as a photon of frequency ν given by:

$$h\nu = W_1 - W_2 \qquad (h = \text{Planck's constant})$$

in accordance with Planck's hypothesis. Bohr went still further by providing a means of calculating the "allowed" orbits; he postulated that the stationary states are those in which the angular momentum of the electron about the nucleus is an integral multiple of $h/2\pi$. He then proceeded to calculate the orbits by setting the electrostatic force of attraction between electron and nucleus equal to the centripetal reaction of the electron resulting from its circular motion. Thus, while abandoning traditional concepts on the one hand Bohr nevertheless resorted to classical methods to calculate the *energy levels* of the atom, the energy in each state being the sum of the kinetic energy of rotation of the electron and the potential energy due to its electrostatic "binding" by the nucleus. From these he derived the general Rydberg formula for hydrogen.

Despite the seemingly arbitrary nature of the Bohr postulates, and the uneasy feeling one derived from the mixture of classical and quantum physics, the theory accounted successfully for a wide range of experience. Moreover it provided a "picture" of atomic processes which the new *quantum mechanics* that ultimately replaced the Bohr theory does not give. The new theory is a completely abstract formalism which does not permit ready interpretation in terms of everyday experiences or of meaningful mental images. Hence the usefulness of such concepts as electron orbits and energy levels, and transitions pictured as instantaneous "jumps" from one orbit to another, lies primarily in the sense of satisfaction they give to our classically oriented minds. Largely because of this the "Bohr atom" no doubt will remain one of the most widely used models in the field of quantum physics.

Niels Henrik David Bohr was born on October 7, 1885, in Copenhagen, Denmark, where his father was professor of physiology at the University of Copenhagen. Bohr attended public schools in Copenhagen and then the university, where he distinguished himself by winning, at the age of 22, the gold medal of the Royal Danish Academy of Science. After receiving his doctorate from Copenhagen in 1911 he went to the Cavendish Laboratory at Cambridge to study with J. J. Thomson. The following year found Bohr in Manchester, where Rutherford was then proposing his nuclear model of the atom. When he returned to Copenhagen in 1913 Bohr applied Planck's quantum hypothesis to the Rutherford atom, thereby developing his ingenious interpretation of atomic spectra. He was then 28 years old. In 1916 Bohr was appointed professor of theoretical physics at the University of Copenhagen, and four years later was made director of the newly established Institute for Theoretical Physics, which he was instrumental in founding and which shortly became one of the leading theoretical centers in Europe. For his remarkable contribution to man's understanding of atomic structure Bohr was awarded a Nobel prize in 1922.

He was visiting in Princeton in 1939 when word was received from Germany of the fission of uranium. Following a suggestion of Lise Meitner and Otto Frish that in the fission of uranium one might draw a crude analogy between the heavy unstable nucleus and a water drop about to break up, Bohr and John Wheeler set the foundation for the "liquid drop" model of nuclear phenomena.

Bohr was in Denmark in 1940 when the Nazis invaded that country. He remained until 1943, when he fled to Sweden and then to England. Later that year he reached the United States, where he spent the remaining years of the war at Los Alamos, lending his great talents to the task of developing atomic weapons.

With the end of the war Bohr returned to Denmark and his beloved Institute. He was knighted by King Frederik in 1947, and among the many honors that came to him was the first Atoms for Peace award, in 1957. He died in Copenhagen on November 18, 1962.

The following is from Bohr's paper "On the Constitution of Atoms and Molecules," *Philosophical Magazine,* Series 6, vol. 26, July 1913, pages 1-25.

Bohr's "Experiment"

INTRODUCTION

In order to explain the results of experiments on scattering of α rays by matter Prof. Rutherford[1] has given a theory of the structure of atoms.■ According to this theory, the atoms consist of a positively charged nucleus surrounded by a system of electrons kept together by attractive forces from the nucleus; the total negative charge of the electrons is equal to the positive charge of the nucleus. Further, the nucleus is assumed to be the seat of the essential part of the mass of the atom, and to have linear dimensions exceedingly small compared with the linear dimensions of the whole atom. The number of electrons in an atom is deduced to be approximately equal to half the atomic weight.■ Great interest is to be attributed to this atom-model; for, as Rutherford has shown, the assumption of the existence of nuclei, as those in question, seems to be necessary in order to account for the results of the experiments on large angle scattering of the α rays.[2]■

In an attempt to explain some of the properties of matter on the basis of this atom-model we meet, however, with difficulties of a serious nature arising from the apparent instability of the system of electrons:■ difficulties purposely avoided in atom-models previously considered, for instance, in the one proposed by Sir J. J. Thomson.[3] According to the theory of the latter the atom consists of a sphere of uniform positive electrification, inside which the electrons move in circular orbits.■

The principal difference between the atom-models proposed by Thomson and Rutherford consists in the circumstance that the forces acting on the electrons in the atom-model of Thomson allow of certain configurations and motions of the electrons for which the system is in a stable equilibrium; such configurations, however, apparently do not exist for the second atom-model. The nature of the difference in question will perhaps be most clearly seen by noticing that among the quantities characterizing the first atom a quantity appears—the radius of the positive sphere—of dimensions of a length and of the same order of magnitude as the linear extension of the atom, while such a length does not appear among the quantities characterizing the second atom, viz. the

---

[1] E. Rutherford, *Phil. Mag.*, xxi, p. 669 (1911).
[2] See also Geiger and Marsden, *Phil. Mag.*, April 1913.
[3] J. J. Thomson, *Phil. Mag.*, vii, p. 237 (1904).

The nuclear model of the atom.

Deduced from x-ray scattering. Excluding hydrogen, the ratio Z/A, where Z is the atomic number and A the mass number, ranges from ½ to about ⅖ for all isotopes.

That is, the large Coulomb force required for such scattering could be understood if the positive charge of the atom were concentrated in a very small region (the nucleus).

The chief difficulty was that on classical grounds accelerating electrons must radiate energy. Therefore, except when disturbed, the electrons had to be assumed at rest. But even here there is a difficulty, for it can be shown that there is no stable arrangement of positive and negative charges at rest (Earnshaw's theorem).

Normally, in the Thomson model, the electrons were assumed to occupy certain equilibrium positions.

The Rutherford model, except for assigning numerical values to the nuclear charge, provided no basis for relating this to the distribution or mass of the electrons.

Planck's theory.

That is, a value for the diameter of an atom.

Such as the Rydberg formula.

charges and masses of the electrons and the positive nucleus;■ nor can it be determined solely by help of the latter quantities.

The way of considering a problem of this kind has, however, undergone essential alterations in recent years owing to the development of the theory of the energy radiation,■ and the direct affirmation of the new assumptions introduced in this theory, found by experiments on very different phenomena such as specific heats, photoelectric effect, Röntgen rays, etc. The result of the discussion of these questions seems to be a general acknowledgement of the inadequacy of the classical electrodynamics in describing the behavior of systems of atomic size.[4] Whatever the alteration in the laws of motion of the electrons may be, it seems necessary to introduce in the laws in question a quantity foreign to the classical electrodynamics, i.e., Planck's constant, or as it often is called, the elementary quantum of action. By the introduction of this quantity the question of the stable configuration of the electrons in the atoms is essentially changed, as this constant is of such dimensions and magnitude that it, together with the mass and charge of the particles, can determine a length of the order of magnitude required.■

This paper is an attempt to show that the application of the above ideas to Rutherford's atom-model affords a basis for a theory of the constitution of atoms. It will further be shown that from this theory we are led to a theory of the constitution of molecules.

In the present first part of the paper the mechanism of the binding of electrons by a positive nucleus is discussed in relation to Planck's theory. It will be shown that it is possible from the point of view taken to account in a simple way for the law of the line spectrum of hydrogen.■ Further, reasons are given for a principal hypothesis on which the considerations confined in the following parts are based.

I wish here to express my thanks to Prof. Rutherford for his kind and encouraging interest in this work.

§1.   GENERAL CONSIDERATIONS

The inadequacy of the classical electrodynamics in accounting for the properties of atoms from an atom-model as Rutherford's, will appear very clearly if we consider a simple system consisting of a positively charged nucleus of very small dimensions and an electron describing closed orbits around it. For simplicity, let us assume that the mass of the

---

[4] See f. inst., "Theorie du ravonnement et les quanta." *Rapports de la reunion a Bruxelles*, November 1911, Paris, 1912.

electron is negligibly small in comparison with that of the nucleus, and further, that the velocity of the electron is small compared with that of light.∎

Let us at first assume that there is no energy radiation. In this case the electron will describe stationary elliptical orbits.∎ The frequency of revolution ω and the major-axis of the orbit $2a$ will depend on the amount of energy W which must be transferred to the system in order to remove the electron to an infinitely great distance apart from the nucleus. Denoting the charge of the electron and of the nucleus by $-e$ and E respectively and the mass of the electron by m, we thus get∎

$$\omega = \frac{\sqrt{2}}{\pi} \frac{W^{3/2}}{eE\sqrt{m}}, \; 2a = \frac{eE}{W} \quad \cdot \quad \cdot \quad \cdot \quad \cdot \quad (1)$$

Further, it can easily be shown that the mean value of the kinetic energy of the electron taken for a whole revolution is equal to W.∎ We see that if the value of W is not given, there will be no values of ω and a characteristic for the system in question.

Let us now, however, take the effect of the energy radiation into account, calculated in the ordinary way from the acceleration of the electron. In this case the electron will no longer describe stationary orbits.∎ W will continuously increase, and the electron will approach the nucleus describing orbits of smaller and smaller dimensions, and with greater and greater frequency; the electron on the average gaining in kinetic energy at the same time as the whole system loses energy. This process will go on until the dimensions of the orbit are of the same order of magnitude as the dimensions of the electron or those of the nucleus. A simple calculation shows that the energy radiated out during the process considered will be enormously great compared with that radiated out by ordinary molecular processes.∎

It is obvious that the behavior of such a system will be very different from that of an atomic system occurring in nature. In the first place, the actual atoms in their permanent state seem to have absolutely fixed dimensions and frequencies.∎ Further, if we consider any molecular process, the result seems always to be that after a certain amount of energy characteristic for the system in question is radiated out, the system will again settle down in a stable state of equilibrium, in which the distances apart of the particles are of the same order of magnitude as before the process.

Now the essential point in Planck's theory of radiation is that the energy radiation from an atomic system does not take place in the continuous way assumed in the ordinary electrodynamics, but that it, on the contrary, takes place in distinctly separated emissions, the amount of

The first assumption places the center of mass of the system at the nucleus; the second avoids the need for relativistic corrections to the electron mass.

The most general type of orbit in *central motion.*

W is the *binding energy* of the electron. The expressions follow directly from Kepler's laws.

In circular orbits the potential and kinetic energies are equal; so too in elliptic orbits, provided the average kinetic energy is considered.

Orbits in which the electron does not radiate.

Molecular processes involve energies of the order of fractions of an electron volt to several volts.

From spectral observations.

energy radiated out from an atomic vibrator of frequency ν in a single emission being equal to $\tau h\nu$, where τ is an entire number, and h is a universal constant.[5] ■

Returning to the simple case of an electron and a positive nucleus considered above, let us assume that the electron at the beginning of the interaction with the nucleus was at a great distance apart from the nucleus, and had no sensible velocity relative to the latter. Let us further assume that the electron after the interaction has taken place has settled down in a stationary orbit around the nucleus. We shall, for reasons referred to later, assume that the orbit in question is circular; this assumption will, however, make no alteration in the calculations for systems containing only a single electron.

Let us now assume that, during the binding of the electron, a homogeneous radiation is emitted of frequency ν, equal to half the frequency of revolution of the electron in its final orbit; then, from Planck's theory, we might expect that the amount of energy emitted by the process considered is equal to $\tau h\nu$, where h is Planck's constant and τ an entire number. If we assume that the radiation emitted is homogeneous, the second assumption concerning the frequency of the radiation suggests itself, since the frequency of revolution of the electron at the beginning of the emission is 0.■ The question, however, of the rigorous validity of both assumptions, and also of the application made of Planck's theory, will be more closely discussed in §3.

Putting

$$W = \tau h \frac{\omega}{2}, \quad \cdots \quad \cdots \quad (2)■$$

we get by help of the formula (1)

$$W = \frac{2\pi^2 m e^2 E^2}{\tau^2 h^2}, \quad \omega = \frac{4\pi^2 m e^2 E^2}{\tau^3 h^3}, \quad 2a = \frac{\tau^2 h^2}{2\pi^2 m e E} \quad \cdots \quad (3)$$

If in these expressions we give τ different values, we get a series of values for W, ω, and a corresponding to a series of configurations of the system. According to the above considerations, we are led to assume that these configurations will correspond to states of the system in which there is no radiation of energy; states which consequently will be stationary as long as the system is not disturbed from outside. We see that the value of W is greatest if τ has its smallest value 1. This case will therefore correspond to the most stable state of the system,■ i.e., will correspond to the binding of the electron for the breaking up of which the greatest amount of energy is required.

Margin notes:

τ is an integer; h is Planck's constant.

That is, the average frequency of revolution is taken to be the frequency of the emitted radiation.

Here ω is the angular frequency.

The state that is most tightly bound.

---

[5] See f. inst., M. Planck, *Ann. d. Phys.*, xxxi, p. 758 (1910); xxxvii, p. 642 (1912); *Verh. deutsch. Phys. Ges.*, 1911, p. 138.

Putting in the above expressions $\tau = 1$ and $E = e$, and introducing the experimental values

$$e = 4.7 \cdot 10^{-10}, \frac{e}{m} = 5.31 \cdot 10^{17}, h = 6.5 \cdot 10^{-27}$$

we get:

$$2\alpha = 1.1 \cdot 10^{-8}\text{cm}, \omega = 6.2 \cdot 10^{15} \frac{1}{\text{sec}}, \frac{W}{e} = 13 \text{ volts}$$

We see that these values are of the same order of magnitude as the linear dimensions of the atoms, the optical frequencies, and the ionization-potentials.■

The general importance of Planck's theory for the discussion of the behavior of atomic systems was originally pointed out by Einstein.[6]■ The considerations of Einstein have been developed and applied on a number of different phenomena, especially by Stark, Nernst, and Sommerfeld.■ The agreement as to the order of magnitude between values observed for the frequencies and dimensions of the atoms, and values for these quantities calculated by considerations similar to those given above, has been the subject of much discussion. It was first pointed out by Haas,[7] in an attempt to explain the meaning and value of Planck's constant on the basis of J. J. Thomson's atom-model, by help of the linear dimensions and frequency of an hydrogen atom.

Systems of the kind considered in this paper, in which the forces between the particles vary inversely as the square of the distance, are discussed in relation to Planck's theory by J. W. Nicholson.[8]■ In a series of papers this author has shown that it seems to be possible to account for lines of hitherto unknown origin in the spectra of the stellar nebulae and that of the solar corona, by assuming the presence in these bodies of certain hypothetical elements of exactly indicated constitution. The atoms of these elements are supposed to consist simply of a ring of a few electrons surrounding a positive nucleus of negligibly small dimensions. The ratios between the frequencies corresponding to the lines in question are compared with the ratios between the frequencies corresponding to different modes of vibration of the ring of electrons. Nicholson has obtained a relation to Planck's theory showing that the ratios between the wave-

---

[6] A. Einstein, *Ann. d. Phys.*, xvii, p. 132 (1905); xx, p. 199 (1906); xxii, p. 180 (1907).

[7] A. E. Haas, *Jahrb. d. Rad. u. El.*, vii, p. 261 (1910). See further, A. E. Schidlof, *Ann. d. Phys.*, xxxv, p. 90 (1911); E. Wertheimer, *Phys. Zeitschr.*, xii, p. 409 (1911), *Verh. deutsch. Phys. Ges.*, 1912, p. 431; F. A. Lindemann, *Verh. deutsch. Phys. Ges.*, 1911, pp. 482, 1107; F. Haber, *Verh. deutsch. Phys. Ges.*, 1911, p. 1117.

[8] J. W. Nicholson, *Month. Not. Roy. Astr. Soc.* lxxii, pp. 49, 139, 677, 693, 729 (1912).

Atomic diameters are of the order of 10^{-8} cm and the ionization potential of hydrogen (the energy required to remove the electron to infinity) is 13.5 electron volts.

By his theories of the photoelectric effect and specific heats.

Johannes Stark (1874-1957) discovered the splitting of spectral lines when the emitting atoms are in a strong electric field. W. H. Nernst (1864-1941) confirmed experimentally the decrease of specific heat of a solid with temperature. A. Sommerfeld (1868-1951), who later generalized the Bohr theory for elliptic orbits, was one of the pioneers in the field of quantum physics.

Then professor of mathematics at King's College, London. Like Bohr, he too was influenced by Rutherford's work. Using classical methods he was able to calculate the ratios of the possible frequencies of vibration. He anticipated Bohr by his discovery that the angular momen-

length of different sets of lines of the coronal spectrum can be accounted for with great accuracy by assuming that the ratio between the energy of the system and the frequency of rotation of the ring is equal to an entire multiple of Planck's constant. The quantity Nicholson refers to as the energy is equal to twice the quantity which we have denoted above by W.■ In the latest paper cited Nicholson has found it necessary to give the theory a more complicated form, still, however, representing the ratio of energy to frequency by a simple function of whole numbers.

The excellent agreement between the calculated and observed values of the ratios between the wavelengths in question seems a strong argument in favor of the validity of the foundation of Nicholson's calculations. Serious objection, however, may be raised against the theory. These objections are intimately connected with the problem of the homogeneity of the radiation emitted. In Nicholson's calculations the frequency of lines in a line-spectrum is identified with the frequency of vibration of a mechanical system in a distinctly indicated state of equilibrium.■ As a relation from Planck's theory is used, we might expect that the radiation is sent out in quanta; but systems like those considered, in which the frequency is a function of the energy, cannot emit a finite amount of a homogeneous radiation; for, as soon as the emission of radiation is started, the energy and also the frequency of the system are altered. Further, according to the calculation of Nicholson, the systems are unstable for some modes of vibration. Apart from such objections—which may be only formal—it must be remarked that the theory in the form given does not seem to be able to account for the well-known laws of Balmer and Rydberg connecting the frequencies of the lines in the line-spectra of the ordinary elements.■

It will now be attempted to show that the difficulties in question disappear if we consider the problems from the point of view taken in this paper. Before proceeding it may be useful to restate briefly the ideas characterizing the preceding calculations. The principal assumptions used are:

(1) That the dynamical equilibrium of the systems in the stationary states can be discussed by help of the ordinary mechanics, while the passing of the systems between different stationary states cannot be treated on that basis.■

(2) That the latter process is followed by the emission of a *homogeneous* radiation, for which the relation between the frequency and the amount of energy emitted is the one given by Planck's theory.

The first assumption seems to present itself; for it is known that the ordinary mechanics cannot have an absolute validity, but will only hold

---

Side notes (left margin):

tum of an atom is an integral multiple of

$$\frac{h}{2\pi}.$$

That is, Nicholson used the *total* energy of the system.

Nicholson employed mathematical techniques very similar to those used earlier by Maxwell in his theoretical study of the rings of Saturn.

The theory did not yield the expression found empirically by Balmer and Rydberg, although it did provide accurate frequency ratios.

There is no picture indicating *how* a system passes between stationary states.

in calculations of certain mean values of the motion of the electrons. On the other hand, in the calculations of the dynamical equilibrium in a stationary state in which there is no relative displacement of the particles, we need not distinguish between the actual motions and their mean values. The second assumption is in obvious contrast to the ordinary ideas of electrodynamics, but appears to be necessary in order to account for experimental facts.■

In the aforementioned calculations we have made further use of the more special assumptions, viz. that the different stationary states correspond to the emission of a different number of Planck's energy quanta, and that the frequency of the radiation emitted during the passing of the system from a state in which no energy is yet radiated out to one of the stationary states, is equal to half the frequency of revolution of the electron in the latter state.■ We can, however (see 3), also arrive at the expressions (3) for the stationary states by using assumptions of somewhat different form. We shall, therefore, postpone the discussion of the special assumptions, and first show how by the help of the above principal assumptions, and of the expressions (3) for the stationary states, we can account for the line-spectrum of hydrogen.

> According to ordinary electro-dynamics the frequency emitted by a vibrating system is the same as the vibration frequency.
>
> See equation (2).

§2. EMISSION OF LINE-SPECTRA

Spectrum of hydrogen. General evidence indicates that an atom of hydrogen consists simply of a single electron rotating round a positive nucleus of charge e.[9] The reformation■ of a hydrogen atom, when the electron has been removed to great distances from the nucleus—e.g., by the effect of electrical discharge in a vacuum tube—will accordingly correspond to the binding of an electron by a positive nucleus considered earlier. If in (3) we put $E = e$,■ we get for the total amount of energy radiated out by the formation of one of the stationary states,

> Recombination
>
> Since the nuclear charge equals the charge on an electron.

$$W_\tau = \frac{2\pi^2 me^4}{h^2\tau^2}$$

The amount of energy emitted by the passing of the system from a state corresponding to $\tau = \tau_1$ to one corresponding to $\tau = \tau_2$ is consequently

$$W_{\tau_2} - W_{\tau_1} = \frac{2\pi^2 me^4}{h^2}\left(\frac{1}{\tau_2^2} - \frac{1}{\tau_1^2}\right)$$

---

[9] See f. inst. N. Bohr, *Phil. Mag.*, xxv, p. 24 (1913). The conclusion drawn in the paper cited is strongly supported by the fact that hydrogen, in the experiments on positive rays of Sir J. J. Thomson, is the only element which never occurs with a positive charge corresponding to the loss of more than one electron■ [comp. *Phil. Mag.*, xxiv, p. 672 (1912)].

> That is, *doubly ionized* hydrogen is never observed, whereas doubly ionized helium, for example, is.

According to the
Planck hypothesis.

If now we suppose that the radiation in question is homogeneous, and that the amount of energy emitted is equal to $h\nu$,■ where ν is the frequency of the radiation, we get

$$W_{\tau_2} - W_{\tau_1} = h\nu$$

and from this

$$\nu = \frac{2\pi^2 m e^4}{h^3}\left(\frac{1}{\tau_2^2} - \frac{1}{\tau_1^2}\right) \quad \cdots \quad (4)$$

τ_1 then takes on
the values 3, 4,
5, . . .

We see that this expression accounts for the law connecting the lines in the spectrum of hydrogen. If we put $\tau_2 = 2$ and let τ_1 vary,■ we get the ordinary Balmer series. If we put $\tau_2 = 3$, we get the series in the ultrared observed by Paschen[10] and previously suspected by Ritz. If we put $\tau_2 = 1$ and $\tau_2 = 4, 5, \ldots$, we get series respectively in the extreme ultraviolet and the extreme ultrared, which are not observed, but the existence of which may be expected.■

Two of these
series were dis-
covered later;
that for $\tau_2 = 1$ is
known as the
Lyman series, and
for $\tau_2 = 4$, the
Brackett series.
The first is in the
ultraviolet region,
the second in the
infrared.

The agreement in question is quantitative as well as qualitative. Putting:

$$e = 4.7 \cdot 10^{-10},\ \frac{e}{m} = 5.31 \cdot 10^{17},\ \text{and}\ h = 6.5 \cdot 10^{-27}$$

we get:

$$\frac{2\pi^2 m e^4}{h^3} = 3.1 \cdot 10^{15}$$

The observed value for the factor outside the bracket in the formula (4) is

$$3.290 \cdot 10^{15}$$

Using more
accurate values
for m, e, and h,
the theoretical
value becomes
3.26×10^{15},
which is in closer
agreement with
the experimental
value.

The agreement between the theoretical and observed values is inside the uncertainty due to experimental errors in the constants entering in the expression for the theoretical value.■ We shall in §3 return to consider the possible importance of the agreement in question.

Thus, if an atom
does absorb an
amount of energy
corresponding to
an electron orbit
greater than the
mean separation
between it and its
neighbors, it will
transfer the
energy to them
rather than re-
radiate.

It may be remarked that the fact, that it has not been possible to observe more than 12 lines of the Balmer series in experiments with vacuum tubes, while 33 lines are observed in the spectra of some celestial bodies, is just what we should expect from the above theory. According to the equation (3) the diameter of the orbit of the electron in the different stationary states is proportional to τ^2. For $\tau = 12$ the diameter is equal to $1.6 \cdot 10^{-6}$ cm, or equal to the mean distance between the molecules in a gas at a pressure of about 7 mm mercury; for $\tau = 33$ the diameter is equal to $1.2 \cdot 10^{-5}$ cm, corresponding to the mean distance of the molecules at a pressure of about 0.02 mm mercury.■ According to the theory the necessary condition for the appearance of a great number of lines is therefore a very small density of the gas; for simultaneously to

---

[10] F. Paschen, *Ann. d. Phys.*, xxvii, p. 565 (1908).

obtain an intensity sufficient for observation the space filled with the gas must be very great.■ If the theory is right, we may therefore never expect to be able in experiments with vacuum tubes to observe the lines corresponding to high numbers of the Balmer series of the emission spectrum of hydrogen; it might, however, be possible to observe the lines by investigation of the absorption spectrum of this gas (see §4).

It will be observed that we in the above way do not obtain other series of lines, generally ascribed to hydrogen; for instance, the series first observed by Pickering[11] in the spectrum of the star ζ Puppis, and the set of series recently found by Fowler[12]■ by experiments with vacuum tubes containing a mixture of hydrogen and helium. We shall, however, see that, by help of the above theory, we can account naturally for these series of lines if we ascribe them to helium.

A neutral atom of the latter element consists, according to Rutherford's theory, of a positive nucleus of charge $2e$ and two electrons. Now considering the binding of a single electron by a helium nucleus, we get, putting $E = 2e$ in the expressions (3), and proceeding in exactly the same way as above,■

$$\nu = \frac{8\pi^2 me^4}{h^3}\left(\frac{1}{\tau_2^2} - \frac{1}{\tau_1^2}\right) = \frac{2\pi^2 me^4}{h^3}\left(\frac{1}{\left(\frac{\tau_2}{2}\right)^2} - \frac{1}{\left(\frac{\tau_1}{2}\right)^2}\right)$$

If in this formula we put $\tau_2 = 1$ or $\tau_2 = 2$, we get series of lines in the extreme ultraviolet. If we put $\tau_2 = 3$, and let τ_1 vary, we get a series which includes 2 of the series observed by Fowler, and denoted by him as the first and second principal series of the hydrogen spectrum. If we put $\tau_2 = 4$, we get the series observed by Pickering in the spectrum of ζ Puppis. Every second of the lines in this series is identical with a line in the Balmer series of the hydrogen spectrum■; the presence of hydrogen in the star in question may therefore account for the fact that these lines are of a greater intensity than the rest of the lines in the series. The series is also observed in the experiments of Fowler, and denoted in his paper as the Sharp series of the hydrogen spectrum. If we finally in the above formula put $\tau_2 = 5, 6, \ldots$, we get series, the strong lines of which are to be expected in the ultrared.

The reason why the spectrum considered is not observed in ordinary helium tubes may be that in such tubes the ionization of helium is not so complete as in the star considered or in the experiments of Fowler,■ where a strong discharge was sent through a mixture of hydrogen and helium. The condition for the appearance of the spectrum is, according

Otherwise, the number of radiating atoms is too small to be detected; this condition, while difficult to satisfy in the laboratory, is met by large celestial bodies.

A. Fowler (1868-1940) later pointed out that the value of the Rydberg constant was slightly different for different elements, which added greatly to the confirmation of Bohr's theory. The difference resulted from differences in the reduced mass of the atom.

That is, for ionized helium.

Except for a slight shift owing to the change in reduced mass; but this amounts only to 1 part in 1000 and its significance went unrecognized at the time.

That is, unless a heavy discharge is passed through the gas, or sufficient thermal energy is supplied to it, relatively few of the atoms will exist in the ionized state.

---

[11] E. C. Pickering, *Astrophys. J.*, iv, p. 369 (1896); v, p. 92 (1897).
[12] A. Fowler, *Month. Not. Roy. Astr. Soc.*, lxxiii, Dec. 1912.

to the above theory, that helium atoms are present in a state in which they have lost both their electrons. Now we must assume that the amount of energy to be used in removing the second electron from a helium atom is much greater than that to be used in removing the first.■ Further, it is known from experiments on positive rays, that hydrogen atoms can acquire a negative charge; therefore the presence of hydrogen in the experiments of Fowler may have the effect that more electrons are removed from some of the helium atoms than would be the case if only helium were present.■

Spectra of other substances. In case of systems containing more electrons we must—in conformity with the result of experiments—expect more complicated laws for the line-spectra than those considered. I shall try to show that the point of views taken above allows, at any rate, a certain understanding of the laws observed.

According to Rydberg's theory—with the generalization given by Ritz[13]■—the frequency corresponding to the lines of the spectrum of an element can be expressed by

$$\nu = F_r(\tau_1) - F_s(\tau_2)$$

where τ_1 and τ_2 are entire numbers,■ and F_1, F_2, F_3, . . . are functions of τ which approximately are equal to $\dfrac{K}{(\tau + \alpha_1)^2}$, $\dfrac{K}{(\tau + \alpha_2)^2}$, . . . K is a universal constant, equal to the factor outside the bracket in the formula (4) for the spectrum of hydrogen.■ The different series appear if we put τ_1 or τ_2 equal to a fixed number and let the other vary.

The circumstance that the frequency can be written as a difference between two function of entire numbers suggests an origin of the lines in the spectra in question similar to the one we have assumed for hydrogen; i.e., that the lines correspond to a radiation emitted during the passing of the system between two different stationary states. For systems containing more than one electron the detailed discussion may be very complicated, as there will be many different configurations of the electrons which can be taken into consideration as stationary states. This may account for the different sets of series in the line-spectra emitted from the substances in question. Here I shall only try to show how, by help of the theory, it can be simply explained that the constant K entering in Rydberg's formula is the same for all substances.

Let us assume that the spectrum in question corresponds to the radiation emitted during the binding of an electron; and let us further assume that the system including the electron considered is neutral.■ The force on the electron, when at a great distance apart from the nucleus

The first ionization potential is 24.46 ev; the second is 54.14 ev.

Because of the great difference in ionization potentials, a hydrogen ion *cannot* remove an electron from a helium atom.

The Ritz Combination Principle.

Integers.

$K = cR$, where R is the Rydberg constant and c the velocity of light. For hydrogen α_1 and α_2 are zero.

That is, the system alone has a net charge of +1.

---

[13] W. Ritz, *Phys. Zeitschr.*, ix, p. 521 (1908).

and the electrons previously bound, will be very nearly the same as in the above case of the binding of an electron by a hydrogen nucleus. The energy corresponding to one of the stationary states will therefore, for τ great, be very nearly equal to that given by the expression (3), if we put $E = e$. For τ great we consequently get

$$\lim (\tau^2 \cdot F_1(\tau)) = \lim (\tau^2 \cdot F_2(\tau)) = \ldots \frac{2\pi^2 m e^4 \blacksquare}{h^3}$$

in conformity with Rydberg's theory.■

§3. GENERAL CONSIDERATIONS CONTINUED

We shall now return to the discussion of the special assumptions used in deducing the expressions (3) for the stationary states of a system consisting of an electron rotating round a nucleus.

For one, we have assumed that the different stationary states correspond to the emission of a different number of energy quanta. Considering systems in which the frequency is a function of the energy, this assumption, however, may be regarded as improbable; for as soon as one quantum is sent out the frequency is altered. We shall now see that we can leave the assumption used and still retain the equation (2), and thereby the formal analogy with Planck's theory.

Firstly, it will be observed that it has not been necessary, in order to account for the law of the spectra by help of the expressions (3) for the stationary states, to assume that in any case a radiation is sent out corresponding to more than a single energy quantum, $h\nu$. Further information on the frequency of the radiation may be obtained by comparing calculations of the energy radiation in the region of slow vibrations based on the above assumptions with calculations based on the ordinary mechanics. As is known, calculations on the latter basis are in agreement with exper ments on the energy radiation in the named region.■

Let us assume that the ratio between the total amount of energy emitted and the frequency of revolution of the electron for the different stationary states is given by the equation $W = f(\tau) \cdot h\omega$,■ instead of by the equation (2). Proceeding in the same way as above, we get in this case instead of (3)

$$W = \frac{\pi^2 m e^2 E^2}{2h^2 f^2(\tau)}, \quad \omega = \frac{\pi^2 m e^2 E^2}{2h^3 f^3(\tau)}$$

Assuming as above that the amount of energy emitted during the passing of the system from a state corresponding to $\tau = \tau_1$ to one for which $\tau = \tau_2$ is equal to $h\nu$, we get instead of (4)

$$\nu = \frac{\pi^2 m e^2 E^2}{2h^3} \left(\frac{1}{f^2(\tau_2)} - \frac{1}{f^2(\tau_1)} \right)$$

---

lim: the limit of

Since $F_1 \sim$ $\dfrac{K}{(\tau + a_1)^2}$, etc. which approaches $\dfrac{K}{\tau^2}$ for large τ, and $K = \dfrac{2\pi^2 m e^4}{h^3}$.

This is known as the *Bohr Correspondence Principle;* essentially it asserts that in the region of large quantum number (large orbit) classical and quantum physics lead to the same predictions. In this case the frequency of the radiated light should be the same as the rotational frequency of the electron in moving between adjacent (large) orbits.

It can be seen that this is a more general functional dependence.

Where c is a constant, not to be confused with the velocity of light.

Let $\tau_2 = N$ and $\tau_1 = N-1$.

We see that in order to get an expression of the same form as the Balmer series we must put $f(\tau) = c\tau$. ■

In order to determine c let us now consider the passing of the system between two successive stationary states corresponding to $\tau = N$ and $\tau = N - 1$; introducing $f(\tau) = c\tau$, we get for the frequency of the radiation emitted ■

$$\nu = \frac{\pi^2 m e^2 E^2}{2c^2 h^3} \cdot \frac{2N - 1}{N^2(N - 1)^2}$$

For the frequency of revolution of the electron before and after emission we have

$$\omega_N = \frac{\pi^2 m e^2 E^2}{2c^3 h^3 N^3} \text{ and } \omega_{N-1} = \frac{\pi^2 m e^2 E^2}{2c^3 h^3 (N - 1)^3}$$

For N>>1, (N−1)→N.

If N is great the ratio between the frequency before and after the emission will be very near equal to 1; ■ and according to the ordinary electrodynamics we should therefore expect that the ratio between the frequency of radiation and the frequency of revolution also is very nearly

Since $\nu \approx$ $\dfrac{\pi^2 m e^2 E^2}{c^2 h^3 N^3}$ and $\omega \approx$ $\dfrac{\pi^2 m e^2 E^2}{2c^3 h^3 N^3}$

equal to 1. This condition will only be satisfied if $c = \frac{1}{2}$. ■ Putting $f(\tau) = \dfrac{\tau}{2}$, we, however, again arrive at the equation (2) and consequently at the expression (3) for the stationary states.

If we consider the passing of the system between two states corresponding to $\tau = N$ and $\tau = N - n$, where n is small compared with N, we get with the same approximation as above, putting $f(\tau) = \dfrac{\tau}{2}$,

$$\nu = n\omega$$

The possibility of an emission of a radiation of such a frequency may also be interpreted from analogy with the ordinary electrodynamics, as an electron rotating around a nucleus in an elliptical orbit will emit a radiation which according to Fourier's theorem can be resolved into homogeneous components, the frequencies of which are $n\omega$, if ω is the frequency of revolution of the electron. ■

According to Fourier's theorem any harmonic function can be expressed as the sum of a series of terms which are functions of integral multiples of the fundamental frequency, that is, $n\omega$, where $n = 1, 2, 3, \ldots$.

We are thus led to assume that the interpretation of the equation (2) is not that the different stationary states correspond to an emission of different numbers of energy quanta, but that the frequency of the energy emitted during the passing of the system from a state in which no energy is yet radiated out to one of the different stationary states, is equal to different multiples of $\dfrac{\omega}{2}$ where ω is the frequency of revolution of the electron in the state considered. From this assumption we get exactly the same expressions as before for the stationary states, and from these by help of the principal assumptions given earlier the same expression for the law of the hydrogen spectrum. Consequently we may regard our preliminary considerations only as a simple form of representing the results of the theory.■

Before we leave the discussion of this question, we shall for a moment return to the question of the significance of the agreement between the observed and calculated values of the constant entering in the expressions (4) for the Balmer series of the hydrogen spectrum. From the above consideration it will follow that, taking the starting point in the form of the law of the hydrogen spectrum and assuming that the different lines correspond to a homogeneous radiation emitted during the passing between different stationary states, we shall arrive at exactly the same expression for the constant in question as that given by (4), if we only assume (1) that the radiation is sent out in quanta $h\nu$, and (2) that the frequency of the radiation emitted during the passing of the system between successive stationary states will coincide with the frequency of revolution of the electron in the region of slow vibrations.■

As all the assumptions used in this latter way of representing the theory are of what we may call a qualitative character, we are justified in expecting—if the whole way of considering it is a sound one—an absolute agreement between the values calculated and observed for the constant in question, and not only an approximate agreement. The formula (4) may therefore be of value in the discussion of the results of experimental determinations of the constants e, m, and h.

While there obviously can be no question of a mechanical foundation of the calculations given in this paper,■ it is, however, possible to give a very simple interpretation of the result of the calculation leading to equation (3) by help of symbols taken from the ordinary mechanics. Denoting the angular momentum of the electron round the nucleus by M, we have immediately for a circular orbit $\pi M = \dfrac{T}{\omega}$, where ω is the frequency of revolution and T the kinetic energy of the electron■; for a circular orbit we further have $T = W$ and from (2), we consequently get

$$M = \tau M_o,$$

where

$$M_o = \frac{h}{2\pi} = 1.04 \times 10^{-27} ■$$

If we therefore assume that the orbit of the electron in the stationary states is circular, the result of the calculation can be expressed by the simple condition: that the angular momentum of the electron round the nucleus in a stationary state of the system is equal to an entire multiple of a universal value, independent of the charge on the nucleus. The

It should be noted that Bohr approached the problem essentially through the correspondence idea.

That is, for large orbits.

Bohr meant here that he could give no classical picture of the process.

Note that ω is *not* the angular velocity, as in common usage, but the angular frequency.

More accurately, $\dfrac{h}{2\pi} = 1.055 \times 10^{-27}$

possible importance of the angular momentum in the discussion of atomic systems in relation to Planck's theory is emphasized by Nicholson.[14]■

The great number of different stationary states we do not observe except by investigation of the emission and absorption of radiation. In most of the other physical phenomena, however, we only observe the atoms of the matter in a single distinct state, i.e., the state of the atoms at low temperature. From the preceding considerations we are immediately led to the assumption that the "permanent"■ state is the one among the stationary states during the formation of which the greatest amount of energy is emitted. According to the equation (3), this state is the one which corresponds to $\tau = 1$.

It was pointed out earlier that in this postulate Bohr actually was anticipated by Nicholson, although the latter evidently did not recognize its full significance.

Or "ground" state.

§4. ABSORPTION OF RADIATION

In order to account for Kirchhoff's law■ it is necessary to introduce assumptions on the mechanism of absorption of radiation which correspond to those we have used considering the emission. Thus we must assume that a system consisting of a nucleus and an electron rotating round it under certain circumstances can absorb a radiation of a frequency equal to the frequency of the homogeneous radiation emitted during the passing of the system between different stationary states. Let us consider the radiation emitted during the passing of the system between two stationary states A_1 and A_2 corresponding to values for τ equal to τ_1 and τ_2, $\tau_1 > \tau_2$. As the necessary condition for an emission of the radiation in question was the presence of the system in the state A_1, we must assume that the necessary condition for an absorption of the radiation is the presence of the system in the state A_2.■

These considerations seem to be in conformity with experiments on absorption in gases. In hydrogen gas at ordinary conditions for instance there is no absorption of a radiation of a frequency corresponding to the line-spectrum of this gas; such an absorption is only observed in hydrogen gas in a luminous state.■ This is what we should expect according to the above. We have previously assumed that the radiation in question was emitted during the passing of the systems between stationary states corresponding to $\tau \geqq 2$.■ The state of the atoms in hydrogen gas at ordinary conditions should, however, correspond to $\tau = 1$; furthermore, hydrogen atoms at ordinary conditions combine into molecules, i.e., into systems in which the electrons have frequencies different from those in the atoms.■ From the circumstance that certain substances in a non-luminous state, as, for instance, sodium vapor, absorb radiation corre-

Of black-body radiation.

The absorption of *mechanical* energy by atomic systems was beautifully demonstrated in 1914 by J. Franck and G. Hertz, who determined the *excitation potentials* (energy levels) of gases by observing the energies at which electrons interacted with the atoms of gas.

Since under normal conditions most atoms would be in the ground state.

For the Balmer series, see equation (4).

There are, in addition to the atomic lines, spectral lines due to molecular vibrations and rotations.

[14] J. W. Nicholson, *loc. cit.*, p. 679.

sponding to lines in the line-spectra of the substances, we may, on the other hand, conclude that the lines in question are emitted during the passing of the system between two states, one of which is the permanent state.■

■. . . . .

In sodium, for example, the prominent spectral lines, of which there are two, do result from transitions to the ground state.

Omitted are some applications to absorption phenomena and a consideration of more complex systems than hydrogen.

SUPPLEMENTARY READING

Born, M., *The Restless Universe* (New York: Dover, 1957), Ch. IV.

Cajori, F., *A History of Physics* (New York: The Macmillan Co., 1899). See pp. 153-177 for an account of early spectroscopic studies.

Holton, G., and Roller, D. H. D., *Foundations of Modern Physical Science* (Reading, Mass.: Addison-Wesley, 1958), Chs. 33, 34, 35.

Pauli, W. (ed.), *Niels Bohr and the Development of Physics* (New York: McGraw-Hill, 1955).

Peierls, R. E., *The Laws of Nature* (New York: Scribner, 1956), Chs. 7, 8.

Richtmyer, F. K., Kennard, E. H., and Lauritsen, T., *Introduction to Modern Physics* (New York: McGraw-Hill, 1955), Ch. 5.

Wilson, W. A., *A Hundred Years of Physics* (London: Duckworth, 1950), Ch. 16.

Zimmer, E., *The Revolution in Physics* (New York: Harcourt, Brace, 1936), Ch. IV.

Arthur Compton

1892 - 1962

The Compton Effect

TWO experiments provide convincing evidence of the quantum nature of light: the photoelectric effect, which was accounted for by Einstein on the basis of Planck's quantum hypothesis (see Chapter 17), and the scattering of photons by electrons, known as the *Compton effect,* which was explained in much the same way. The latter is particularly significant, however, because in its explanation the photons are assigned a momentum, which is a classical property of matter implying mass, thereby adding new depth to the *wave-particle dilemma.*

On classical grounds one would expect light, or electromagnetic radiation generally, to be scattered by electrons because of the electrical nature of each. If the electrons are those associated with atoms, rather than free electrons, an electromagnetic "wave" exerts a force on them and sets them into oscillation, the restoring force being the "elastic bonds" between the electrons and their respective atomic nuclei. The oscillating electrons, because of their acceleration, then radiate in all directions, thus giving the effect of scattering. Unless the frequency of the incident wave is high the amount of energy absorbed from it by the electrons is small, as is the scattered radiation. J. J. Thomson derived an expression for the scattering of x-rays from classical electromagnetic theory, which gave results in qualitative agreement with experiment. Its chief defect was the prediction, as one would expect, that the scattered radiation should have the same frequency as the incident wave. This follows from the fact that the electrons, on this view, are set into oscillation with the frequency of the incident wave and hence radiate the same frequency.

Following their discovery in 1895 (see Chapter 14) the investigation of x-ray phenomena became one of the most active fields of physics during the early part of the twentieth century. It was soon found that the scattered radiation was "softer" than the incident beam, that is, it was absorbed more readily. This was accounted for initially by assuming that the scattered radiation contained some *fluorescent radiation;* that is, characteristic x-rays emitted by atoms when excited by radiation of higher frequency.[1] But on closer study this explanation was found inadequate, and it remained for Compton, in 1923, to propose his bold solution based on the quantum hypothesis.

Compton assumed that the scattering could be regarded as a billiard-ball-like collision between the photon and electron, and that the ordinary conservation laws of mechanics, conservation of energy and momentum, should apply. By assigning to the photon an energy $h\nu$, according to the Planck hypothesis, and a momentum $h\nu/c$ obtained by analogy with the Einstein mass-energy relation, $E = mc^2$, Compton found the energy transferred by the photon to the electron and hence the wavelength of the scattered radiation. These predictions were confirmed by his own measurements of the wavelengths of scattered x-rays, using a crystal spectrometer, and shortly afterward, the predicted recoil electrons were observed in cloud chambers by C. T. R. Wilson and W. Bothe.

The Compton effect seemed to establish the *corpuscular* nature of radiation. The picture of the photon that emerged was that of a "particle" having no charge but nevertheless able to exert a force on an electron (it is somehow accompanied by an electric field), and having mass properties. Several years later, however, L. deBroglie suggested that particles (electrons, atoms, and so forth) sometimes exhibit the properties of waves. When this was confirmed experimentally the wave-particle duality became complete, with the implication that our concepts of "wave" and "particle" may be no more than complementary ways of viewing one and the same process.

Arthur Holly Compton was born in Wooster, Ohio, where his father was professor of philosophy at the College of Wooster, on September 10, 1892. He was born into a family of distinguished educators; he and two brothers, Karl T. and Wilson M., ultimately became college presidents. After graduating from the College of Wooster in 1913, Compton entered the Graduate School of Princeton University, where he received his doctorate in 1916. Following a year as instructor in physics at the University of Minnesota he joined the Westinghouse Light Company as a research physicist. In 1919 Compton went to Cambridge University, where he engaged in research for one year before going to Washington University in St. Louis as

---

[1] The fluorescent radiation is produced when electrons are ejected from the innermost shells of atoms. Hence it is always of lower energy, that is, lower frequency or longer wavelength, than the incident radiation.

professor and head of the physics department. He left Washington University for the University of Chicago in 1923, where he was instrumental in establishing one of the strongest physics departments in the United States. Compton retained his association with the University of Chicago until 1946, when he returned to Washington University as chancellor. Shortly before going to Chicago Compton published his famous papers on the scattering of x-rays, for which he was awarded the Nobel prize jointly with C. T. R. Wilson in 1927. He died in Berkeley, California on March 15, 1962.

The following extracts are taken from two papers, published a few months apart. The first, entitled "A Quantum Theory of the Scattering of X-Rays by Light Elements," appeared in *Physical Review*, vol. 21, May 1923, page 483. The second, "The Spectrum of Scattered X-Rays," *Physical Review*, vol. 22, November 1923, page 409, contains Compton's experimental results on the scattering from graphite.

Phys. Rev., 21, May 1923, p. 483

C. G. Barkla (1877-1944), professor of physics at King's College, London, and later at Edinburgh, discovered the *characteristic* (or fluorescent) *radiation.* Barkla used Thomson's formula as a basis for estimating the number of electrons in atoms of the light elements, concluding that this number was about half the atomic weight. The result, however, was fortuitous, for had he used other wavelengths in his experiments the Thomson formula would not have given the same value.

According to Thomson's theory the scattering

Compton's Experiment

■J. J. Thomson's classical theory of the scattering of x-rays, though supported by the early experiments of Barkla and others,■ has been found incapable of explaining many of the more recent experiments. This theory, based upon the usual electrodynamics, leads to the result that the energy scattered by an electron traversed by an x-ray beam of unit intensity is the same whatever may be the wavelength of the incident rays. Moreover, when the x-rays traverse a thin layer of matter, the intensity of the scattered radiation on the two sides of the layer should be the same. Experiments on the scattering of x-rays by light elements have shown that these predictions are correct when x-rays of moderate hardness are employed; but when very hard x-rays or γ-rays are employed, the scattered energy is found to be decidedly less than Thomson's theoretical value, and to be strongly concentrated on the emergent side of the scattering plate.■

Several years ago the writer suggested that this reduced scattering of the very short wavelength x-rays might be the result of interference between the rays scattered by different parts of the electron, if the electron's diameter is comparable with the wavelength of the radiation. By assuming the proper radius for the electron, this hypothesis supplied a quantitative explanation of the scattering for any particular wavelength.

But recent experiments have shown that the size of the electron which must thus be assumed increases with the wavelength of the x-rays employed,[1][■] and the conception of an electron whose size varies with the wavelength of the incident rays is difficult to defend.

Recently an even more serious difficulty with the classical theory of x-ray scattering has appeared. It has long been known that secondary γ-rays are softer than the primary rays which excite them, and recent experiments have shown that this is also true of x-rays. By a spectroscopic examination of the secondary x-rays from graphite,[■] I have, indeed, been able to show that only a small part, if any, of the secondary x-radiation is of the same wavelength as the primary.[*] While the energy of the secondary x-radiation is so nearly equal to that calculated from Thomson's classical theory that it is difficult to attribute it to anything other than true scattering,[2] these results show that if there is any scattering comparable in magnitude with that predicted by Thomson, it is of a greater wavelength than the primary x-rays.

Such a change in wavelength is directly counter to Thomson's theory of scattering, for this demands that the scattering electrons, radiating as they do because of their forced vibrations when traversed by a primary x-ray, shall give rise to radiation of exactly the same frequency as that of the radiation falling upon them. Nor does any modification of the theory such as the hypothesis of a large electron suggest a way out of the difficulty. This failure makes it appear improbable that a satisfactory explanation of the scattering of x-rays can be reached on the basis of the classical electrodynamics.

THE QUANTUM HYPOTHESIS OF SCATTERING

According to the classical theory, each x-ray affects every electron in the matter traversed, and the scattering observed is that due to the combined effects of all the electrons. From the point of view of the quantum theory, we may suppose that any particular quantum of x-rays is not scattered by all the electrons in the radiator, but spends all of its energy upon some particular electron.[■] This electron will in turn scatter

Marginal notes:

would be *isotropic*; that is, distributed uniformly about the direction of the incident beam.

Since the scattering was found to increase with longer wavelengths.

Using a crystal or Bragg spectrometer.

Multiple scattering can occur, but if the scattering medium is sufficiently thin the chance of double scattering will be negligible.

---

[1] A. H. Compton, *Bull. Nat. Research Council*, No. 20, p. 10 (Oct. 1922).

[*] In previous papers (*Phil. Mag.*, 41, 749, 1921; *Phys. Rev.*, 18, 96, 1921) I have defended the view that the softening of the secondary x-radiation was due to a considerable admixture of a form of fluorescent radiation. Gray (*Phil. Mag.*, 26, 611, 1913; *Frank, Inst. Journ.*, Nov. 1920, p. 643) and Florance (*Phil. Mag.*, 27, 225, 1914) have considered that the evidence favored true scattering, and that the softening is in some way an accompaniment of the scattering process. The considerations brought forward in the present paper indicate that the latter view is the correct one.

[2] A. H. Compton, *loc. cit.*, p. 16.

the ray in some definite direction, at an angle with the incident beam. This bending of the path of the quantum of radiation results in a change in its momentum. As a consequence, the scattering electron will recoil with a momentum equal to the change in momentum of the x-ray. The energy in the scattered ray will be equal to that in the incident ray minus the kinetic energy of the recoil of the scattering electron; and since the scattered ray must be a complete quantum, the frequency will be reduced in the same ratio as is the energy. Thus on the quantum theory we should expect the wavelength of the scattered x-rays to be greater than that of the incident rays.

The effect of the momentum of the x-ray quantum is to set the scattering electron in motion at an angle of less than 90° with the primary beam.■ But it is well known that the energy radiated by a moving body is greater in the direction of its motion. We should therefore expect, as is experimentally observed, that the intensity of the scattered radiation should be greater in the general direction of the primary x-rays than in the reverse direction.

Since the system must preserve its forward momentum, the electron must acquire a forward component of momentum.

The change in wavelength due to scattering.

Imagine, as in Fig. 1A, that an x-ray quantum of frequency ν_o is scattered by an electron of mass m. The momentum of the incident ray

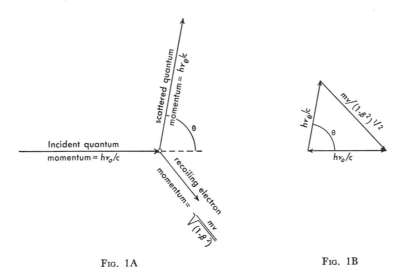

Fig. 1A   Fig. 1B

will be $h\nu_o/c$,■ where c is the velocity of light and h is Planck's constant, and that of the scattered ray is $h\nu_\theta/c$ at an angle θ with the initial momentum. The principle of the conservation of momentum accordingly demands that the momentum of recoil of the scattering electron shall

Since $E = mc^2$ the momentum $mc = \dfrac{E}{c}$, or, for a photon, $\dfrac{h\nu}{c}$.

equal the vector difference between the momenta of these two rays, as in Fig. 1B. The momentum of the electron, $m\beta c/\sqrt{1 - \beta^2}$,■ is thus given by the relation

The relativistic momentum is $mv/\sqrt{1-\beta^2}$, where m is the rest mass. But $v = \beta c$.

$$\left(\frac{m\beta c}{\sqrt{1 - \beta^2}}\right)^2 = \left(\frac{h\nu_o}{c}\right)^2 + \left(\frac{h\nu_\theta}{c}\right)^2 + \frac{2h\nu_o}{c} \cdot \frac{h\nu_\theta}{c} \cos \theta \blacksquare \tag{1}$$

Conservation of momentum.

where β is the ratio of the velocity of recoil of the electron to the velocity of light. But the energy $h\nu_\theta$ in the scattered quantum is equal to that of the incident quantum $h\nu_o$ less the kinetic energy of recoil of the scattering electron, i.e.,

$$h\nu_\theta = h\nu_o - mc^2 \left(\frac{1}{\sqrt{1 - \beta^2}} - 1\right) \blacksquare \tag{2}$$

Conservation of energy.

We thus have two independent equations containing the two unknown quantities β and ν_θ. On solving the equations we find

$$\nu_\theta = \nu_o/(1 + 2a \sin^2 \tfrac{1}{2}\theta) \tag{3}$$

where

$$a = h\nu_o/mc^2 = h/mc\lambda_o \tag{4}$$

Or in terms of wavelength instead of frequency,

$$\lambda_\theta = \lambda_o + (2h/mc) \sin^2 \tfrac{1}{2}\theta \tag{5}$$

It follows from Eq. (2) that $1/(1 - \beta^2) = \{1 + a[1 - (\nu_\theta/\nu_o)]\}^2$, or solving explicitly for β

$$\beta = 2a \sin \tfrac{1}{2}\theta \frac{\sqrt{1 + (2a + a^2) \sin^2 \tfrac{1}{2}\theta}}{1 + 2(a + a^2) \sin^2 \tfrac{1}{2}\theta} \tag{6}$$

Eq. (5) indicates an increase in wavelength due to the scattering process which varies from a few percent in the case of ordinary x-rays to more than 200 percent in the case of γ-rays scattered backward.■ At the same time the velocity of the recoil of the scattering electron, as calculated from Eq. (6), varies from zero when the ray is scattered directly forward to about 80 percent of the speed of light when a γ-ray is scattered at a large angle.

Note that $\lambda_\theta - \lambda_o$ is constant for a given angle θ. Thus, the smaller is λ_o the greater is the fractional change in wavelength.

It is of interest to notice that according to the classical theory, if an x-ray were scattered by an electron moving in the direction of propagation at a velocity $\beta'c$, the frequency of the ray scattered at an angle θ is given by the Doppler principle■ as

$$\nu_\theta = \nu_o \left/ \left(1 + \frac{2\beta'}{1 - \beta'} \sin^2 \tfrac{1}{2}\theta\right)\right. \tag{7}$$

The apparent change in frequency of a source due to relative motion between source and observer.

It will be seen that this is of exactly the same form as Eq. (3), derived on the hypothesis of the recoil of the scattering electron. Indeed, if $a = \beta'/(1 - \beta')$ or $\beta' = a/(1 + a)$, the two expressions become identical.

Omitted are some calculations of the spatial distribution of the scattered radiation and certain initial experimental results. More detailed results were reported in the second paper, which follows below.

Phys. Rev. 22, Nov. 1923, p. 409.

Since the energy is $h\nu$.

It is clear, therefore, that so far as the effect on the wavelength is concerned, we may replace the recoiling electron by a scattering electron moving in the direction of the incident beam at a velocity such that

$$\bar{\beta} = a/(1 + a) \tag{8}$$

We shall call $\bar{\beta}c$ the "effective velocity" of the scattering electrons.

▪ . . . .

▪The writer has recently proposed a theory of the scattering of x-rays, based upon the postulate that each quantum of x-rays is scattered by an individual electron.[1,2] The recoil of this scattering electron, due to the change in momentum of the x-ray quantum when its direction is altered, reduces the energy and hence also the frequency of the quantum of radiation.▪ The corresponding increase in the wave-length of the x-rays due to scattering was shown to be

$$\lambda - \lambda_o = \delta(1 - \cos \theta) \tag{1}$$

where λ is the wavelength of the ray scattered at an angle θ with the primary ray whose wavelength is λ_o, and

$$\delta = h/mc = 0.0242 \text{ Å} ▪$$

Angstrom units:
1 Å = 10^{-8} cm

where h is Planck's constant, m is the mass of the electron, and c the velocity of light. It is the purpose of this paper to present more precise experimental data than has previously been given regarding this change in wavelength when x-rays are scattered.

Apparatus and method. For the quantitative measurement of the change in wavelength it was clearly desirable to employ a spectroscopic method.▪

Because of the precision afforded by such methods.

In view of the comparatively low intensity of scattered x-rays, the apparatus had to be designed in such a manner as to secure the maximum intensity in the beam whose wavelength was measured. The arrangement of the apparatus is shown diagrammatically in Fig. 1. Rays proceeded from the molybdenum target T of an x-ray tube to the graphite scattering block R, which was placed in line with the slits 1 and 2. Lead diaphragms, suitably disposed, prevented stray radiation from leaving the lead box that surrounded the x-ray tube. Since the slit 1 and the diaphragms were mounted upon an insulating support, it was possible to place the x-ray tube close to the slit without danger of puncture.▪ The x-rays, after passing through the slits, were measured by a Bragg spectrometer in the usual manner.

That is, without danger of puncturing the x-ray tube, as there would be if the slit were grounded, thereby providing a large potential difference between the slit and one of the tube elements.

The x-ray tube was of special design. A water-cooled target was mounted in the narrow glass tube, as shown in Fig. 2, so as to shorten as

---

[1] A. H. Compton, *Bull. Nat. Res. Coun.*, No. 20, p. 18 (October 1922); *Phys. Rev.*, 21, 207 (abstract) (Feb. 1923); *Phys. Rev.*, 21, 483 (May 1923).
[2] Cf. also P. Debye, *Phys. Zeitschr.*, 24, 161 (April 15, 1923).

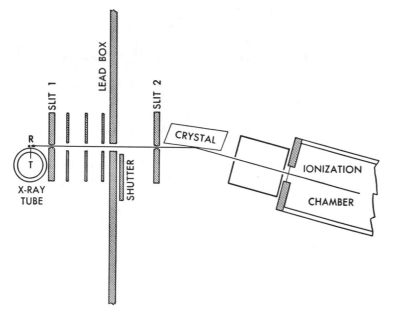

FIG. 1. Measuring the wavelength of scattered x-rays.

much as possible the distance between the target T and the radiator R.■
This distance in the experiments was about 2 cm. When 1.5 kw was
dissipated in the x-ray tube, the intensity of the rays reaching the radiator
was thus 125 times as great as it would have been if a standard Coolidge
tube with a molybdenum target had been employed. The electrodes for
this tube were very kindly supplied by the General Electric Company.

In the final experiments the distance between the slits was about
18 cm, their length about 2 cm, and their width about 0.01 cm. Using a
crystal of calcite, this made possible a rather high resolving power even
in the first-order spectrum.■

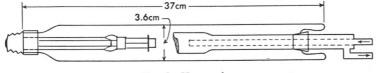

FIG. 2. X-ray tube.

Spectra of scattered molybdenum rays. Results of the measurements, using
slits of two different widths, are shown in Figs. 3 and 4. Curves A represent
the spectrum of the $K\alpha$ line,■ and curves B, C, and D are the spectra of
this line after being scattered at angles of 45°, 90°, and 135° respectively
with the primary beam. While in Fig. 4 the experimental points are a

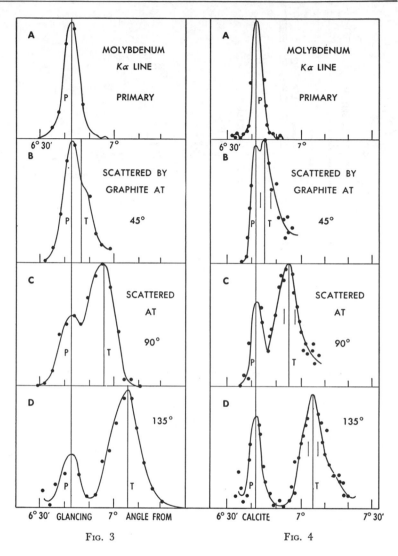

FIG. 3

FIG. 4

Because of a
narrowed slit.

little erratic, it may be noted that in this case the intensity of the x-rays is only about 1/25,000 as great as if the spectrum of the primary beam were under examination,■ so that small variations produce a relatively large effect.

It is clear from these curves that when a homogeneous x-ray is scattered by graphite it is separated into two distinct parts, one of the same wavelength as the primary beam, and the other of increased wavelength. Let us call these the *modified* and the *unmodified* rays respectively. In each curve the line *P* is drawn through the peak of the curve representing the primary line, and the line T is drawn at the angle at which the scattered line should appear according to Eq. (1). In Fig. 4, in which the settings were made with the greater care, within an experimental error of less than 1 minute of arc, or about 0.001 Å, the peak of the unmodified ray falls upon the line P and the peak of the modified ray falls upon the line T. The wavelength of the modified ray thus increases with the scattering angle as predicted by the quantum theory, while the wavelength of the unmodified ray is in accord with the classical theory.

There is a distinct difference between the widths of the unmodified and the modified lines. A part of the width of the modified line is due to the fact that the graphite radiator R subtends a rather large angle as viewed from the target T, so that the angles at which the rays are scattered to the spectrometer crystal vary over an appreciable range. As nearly as I can estimate, the width at the middle of the modified line due to this cause is that indicated in Fig. 4 by the two short lines above the letter T. It does not appear, however, that this geometrical consideration is a sufficient explanation for the whole increased width of the modified line, at least for the rays scattered at 135°. It seems more probable that the modified line is heterogeneous, even in a ray scattered at a definite angle.■

The broadening is due to the motion of atomic electrons; that is, if the electron has a component of velocity in the direction of the incident photon, the change in wavelength would be altered. The actual amount of broadening may be calculated by wave-mechanical methods.

The unmodified ray is usually more prominent in a beam scattered at a small angle with the primary beam, and the modified ray more prominent when scattered at a large angle. A part of the unmodified ray is doubtless due to regular reflection from the minute crystals of which the graphite is composed. If this were the only source of the unmodified ray, however, we should expect its intensity to diminish more rapidly at large angles than is actually observed.■ The conditions which determine the distribution of energy between these two rays are those which determine whether an x-ray shall be scattered according to the simple quantum law or in some other manner. I have studied this distribution experimentally by another method, and shall discuss it in another paper; but the reasons underlying this distribution are puzzling.

The unmodified line results from scattering from bound electrons; that is, the incident photon interacts with an atomic electron in such a manner that it is not ejected, but rather the effect is one in which the photon appears to scatter from the entire atom, the mass being so great as to produce virtually no shift in frequency. Since the Compton equations assume free electrons, they do not take this effect into account.

Experiments with shorter wavelengths. These experiments have been performed using a single wavelength, $\lambda = 0.711$ Å. In this case we find for the modified ray a change in wavelength which increases with the angle of scattering exactly in the manner described by Eq. (1). While these experiments seem conclusive, the evidence would of course be more complete if similar experiments had been performed for other wavelengths. Preliminary experiments similar to those here described have been performed using the K radiation from tungsten, of wavelength about 0.2 Å. This work has shown a change in wavelength of the same order of magnitude as that observed using the molybdenum $K\alpha$ line. Furthermore,

The cross section for Compton scattering was first formulated in 1929 by O. Klein and Y. Nishina. Known as the *Klein-Nishina* *formula* it gives the dependence on energy and angle of the scattered quanta.

as described in earlier papers,[3] absorption measurements have confirmed these results as to order of magnitude over a very wide range of wavelengths. This satisfactory agreement between the experiments and the theory gives confidence in the quantum formula (1) for the change in wavelength due to scattering.■ There is, indeed, no indication of any discrepancy whatever, for the range of wavelengths investigated, when this formula is applied to the wavelength of the modified ray.

---

[3] Cf. e.g., A. H. Compton, *Phys. Rev.,* 21, pp. 494-6 (1923).

SUPPLEMENTARY READING

Compton, A. H., *Atomic Quest* (New York: Oxford University Press, 1956).

Holton, G., and Roller, D. H. D., *Foundations of Modern Physical Science* (Reading, Mass.: Addison-Wesley, 1958), Ch. 32.

Peierls, R. E., *The Laws of Nature* (New York: Scribner, 1956), Ch. 7.

Richtmyer, F. K., Kennard, E. H., and Lauritsen, T., *Introduction to Modern Physics* (New York: McGraw-Hill, 1955), Ch. 8.

Wilson, W., *A Hundred Years of Physics* (London: Duckworth, 1950), Ch. 15.

Zimmer, E., *The Revolution in Physics* (New York: Harcourt, Brace, 1936), Ch. III.

Index

absolute motion, Newton, 46, 216
absolute time, Newton's definition of, 49, 316
absolute truths, 2, 4, 29
Academie des Sciences, 10, 60, 109, 110
academies, scientific, role of in modern physics, 9-10
Academy of Sciences, Prussia, 186, 304
Academy of Sciences, Royal Danish, 332
Accademia dei Lincei, 9, 10
Accademia del Cimento, 9, 10, 59-60
accelerated motion, Galileo's work on, 14, 17-18, 20-35; uniform, defined, 21, 27
acceleration, due to gravity, Galileo, 32, 34
"action-at-a-distance" forces, 21, 130; and Faraday, 286; and Maxwell, 287-288, 293; in Newton's definition of gravity, 48, 58
air, Lenard's measurement of wavelength to ionize, 237; Boyle's work on properties of, 37-38. See also Boyle, Robert
air, resistance of, 48, 55, 83, 231, 243
air pressure. See atmospheric pressure
air pump, invented, 11
alchemy, 7, 8; and Newton, 42, 44
alpha particle, of polonium, kinetic energy of, 275; Rutherford's experiment on collision of with light atoms, 259-265; Rutherford's identification of, 250, 252. See also specific experiments
alpha radiation, 250
alpha rays, 211, 214; scattering of, 333. See also Bohr, Niels
American Physical Society, 239
Ampère, André Marie, 122, 129, 130-131, 138, 161, 163, 164
Anaxagoras, 5, 37
Anderson, Carl D., 266
Ångström, A. J., 330
Angstrom unit, 354
animal electricity, 121
anion, in Faraday's experiment, 148, 153-155
anode, in Faraday's experiment, 147, 153-155
anomalous effect in nitrogen, Rutherford's experiment on, 259-263
Antinori, 141

apparent time, Newton, 49
applied science, defined, 1
Arago, François, 110, 112, 131, 141
arithmetical progression, in optical theory, 97, 99
Archimedes, 3, 6, 13
Aristotle, and alchemy, 7; criticisms of, 4-5, 8; and deductive logic, 2, 3-4; views on matter, 5, 6; views on violent motion, 25
Aristotelian philosophy, 7-8; attacked by Galileo, 14, 15, 16; and Gilbert, 59
Aston, Francis William, 276
astronomy, Aristotle's influence on, 4; Galileo's contributions to, 15, 17; and Kepler, 15
atmospheric pressure, Boyle's measurement of, 38-41; early hypothesis of, 36. See also Boyle, Robert
atom, quantity of electricity in, Faraday, 155; electrons in, 333; energy levels of, 331; structure of, 210. See also Bohr, Niels
atom, nuclear model of, 333
atomic diameter, 334, 337
atomic physics, beginnings of, 250
atomic spectra, 330-331, 332
atomic stopping power, 253
atomic structure, Bohr's theory of, 331. See also Bohr, Niels
atomic systems, mechanical energy and, 346
atomic theory of matter, 216, 218, 239
atomic weapons, 233
atomic weights, and Faraday, 155
atomism, and Aristotle, 4
atomistic view of nature, 5, 6, 8
Atoms for Peace award, first, 332
Auger, 278, 279
average speed, defined, 30
average velocity, Galileo, 19
Avogadro number, 235

Bacon, Sir Francis, 8
Bakerian Lectures, 95, 96, 273, 284
Balmer, Johann, 330, 338
Balmer series in hydrogen spectrum, 330, 340, 341, 345, 346

barium platinocyanide, 201; fluorescence of, 203
Barkla, C. G., 350
barometer, invented, 11
Barrow, Isaac, 43
Becker, 267
Beckmann, H., 305, 306
Becquerel, Henri, 198, 216, 250; experiment on natural radioactivity, 212-215; life and work of, 210-211
Beddoes' [Dr.] West Country, Contributions, 173
Berlin, University of, 185, 240, 304
Bern, University of, 234
beryllium nucleus, structure of, 275
beryllium radiation, properties of, 268-273, 276, 278
Berzelius, Jöns Jakob, 122, 130
beta radiation, 250, 277
beta rays, 211, 214, 258, 259
billiard-ball collisions, Chadwick, 273
binomial theorem, Newton's discovery of, 43
binding energy of neutron, 264, 275, 277, 280, 335
Biot, Jean, 109n, 110
Birkeland, 223
Black, Joseph, 166-167
black-body radiation, energy density of, 312; and Planck, 301, 303, 304, 307, 329. See also Kirchhoff's law; Zeeman effect
Blackett, 272, 274
Bohr, Niels, 251, 252, 253, 304, 305; "experiment" on hydrogen atom, 333-347; life and work of, 329-332
Bohr Correspondence Principle, 343
Boltzmann, L., 302, 306, 308n, 310
Boltzmann constant, 235; values of, 312-313
Bonn, University of, 186
boron radiation, properties of, 268-269, 276
Boscovitch, R. G., 102
Bothe, W., 267, 347
Boyle, Robert, experiment on pressure-volume relationships, 38-41; life and work of, 36-38; and theology, 38
Boyle Lectures, 38
Boyle's law, 37
Boys, C. V., 91
Brackett series in hydrogen spectrum, 330, 340
Bragg, William Henry, 253
Bragg crystal spectrometer, 351, 354, 355
Braun electrometer, 248
Brickwedde, 278
Broglie, L. de, 241, 349
Brownian movements, 242n

Bunsen, R. W., 329

calculating instrument, made by Galileo, 15
calculus, Newton's contributions to, 43, 44
California Institute of Technology, 240
calorimetry, 166-167. See also heat
Cambridge Philosophical Society, 219n
Cambridge University, 76, 77, 94, 95, 251, 252, 267, 284, 285, 349
Capra, Baldassar, 15
Cartesian coordinates, 320
Cassini, 80
cathode, in Faraday's experiment, 147, 153-155
cathode luminescence, in Einstein's "experiment," 236
cathode rays, absorption of, 231; deflection of by magnet, 205; deflection of by electrostatic field, 221-231; Lenard's work on, 199; named by Goldstein, 217; nature of and charge carried by, 219-221, 223; origin of, 235; and radiation, 216-217; Roentgen's work on, 198, 199, 201-209, 210
cations, in Faraday's experiment, 148, 153-155
caustic lime, Faraday's analysis of, 129
Cavendish, Henry, 283; and composition of water, 76; and electrical phenomena, 77; experiment on gravitation, 77-93; life and work of, 75-76; and measurement of density of earth, 66, 77-92
Cavendish Laboratory, 77, 218, 267, 285, 332
cavity radiation, defined, 302
Celsius, Anders, 166
center of gravity, Newton, 51
centrifugal force, 171
centripetal force, Newton, 48
Cesi, Duke Federigo, 9
Chadwick, James, experiment on the neutron, 267-280; life and work of, 266-267
Charlottenburg Institute, 267
chemical affinity, 152, 156, 173, 174
chemical balance, resting point of, 82
chemical effects of electricity, 135
chemical equivalent, 153, 155
chemistry, Boyle's contribution to modern, 37
Chicago, University of, 240, 242, 350
chlorine, Faraday's liquefaction of, 130
chromatic aberrations, 43
chronograph, 248
Clausius, Rudolph, 284, 306, 310, 346
clocks. See time
Cockcroft, John, 266
coefficient of magnetic induction, 295
coefficient of viscosity, Stokes' law, 245
coherent light, 102

cohesion, attraction by, 91
color, fringes of, 96; in optical theory, 96, 99, 100, 101, 103, 106, 107
colors of thin plates, 99
Columbia University, 239, 240
combination theory, 309
combustion, phlogiston theory of, 76
common electricity, 149
common time, 49
compass variation, 80
Compton, Arthur, experiment on Compton effect, 350-358; life and work of, 348-350
Compton, Karl T., 349
Compton, Wilson M., 349
Compton effect, 269, 272, 278, 348-349, 350-358
conductor, in magnetic field, 124-125
conservation of energy, principle of, 159, 160, 168n, 272, 273, 274, 279, 284, 294, 353
conservation of momentum, principle of, 273, 274, 279, 352-353
contact emf's, 121
continuists, 5
continuous spectrum, 310
Coolidge tube, 355
coordinate systems. See time, relativity of
Copenhagen, Polytechnic Institution of, 123
Copenhagen, University of, 122, 123, 337
Copernican theory, Galileo as a supporter of, 15
Copernicus, 9
corona discharge, 196
coronae, 104, 105, 106
corpuscular nature of radiation, 349
corpuscular theory of light, 93-94, 97, 100, 102, 108, 349; and diffraction, 110-112. See also light; wave-particle theory of light
cosmic rays, 240, 270
Coulomb, Charles, 78n; experiment on electric and magnetic force, 61-74; life and work of, 60-61
Coulomb force, 333
Coulomb law for like charges, 63
covalence, 155
Crooke, Sir William, 216, 217, 230
Crooke's tube, 199, 201, 212
Crown Memoir, 110
cubit, defined, 22
Curie, Marie, 210, 211
Curie, Pierre, 210, 211
cyclotron, 266

Dalton, Sir Francis, 67
Dalton, John, 168
Dampier, Sir William, 5n

dark adaption, 256
Darwin, Charles, 253
Davy, Sir Humphry, 95, 129, 173
De Magnete, 59
Debye, P., 354n
declination, 124, 125
deductive logic, defined, 2; compared to induction, 4
Dee, John, 278
Democritus, 5
density, of the earth, 78. See also Cavendish, Henry
density, Newton's definition of, 47
deuterium, discovery of, 266
dialogue, Galileo's use of, 16
Dialogue concerning the two chief Systems of the World, 16-17
Dialogue concerning Two New Sciences, 17, 18-35
diamagnetic materials, 290; permeability of, 288
diamagnetism, 130
diathermic media, 310, 312; radiant energy of, 311
dichroism, 191
dielectric constant, 144, 291, 292, 293, 295
differential calculus, Newton's discovery of, 43
diffraction, of light, 93, 99, 100, 101, 104; corpuscular theory of, 110-112; Fresnel's wavelength theory of, 112-120
diffraction effects, 204
diffraction grating, 105
dipping needle, 138n
displacement, Planck on Wien's law of, 310-313
displacement current, Maxwell's theory of, 286-287
Doppler principle, 353
double-mirror experiment, Fresnel, 111-112
double-slit experiment, Young, 102-107; Fresnel's reference to, 110-111
Dulong, Pierre, 173
dynamics, 4, 11, 52

Earnshaw's theorem, 333
earth, density of, 61, 75, 77-92, compared to water, 91, 192; and laws of motion, 316-317
École des Ponts et Chausées, 109
École Polytechnique, 211
eddy currents, 141
Edinburgh, University of, 284
Ehrenhaft, 241
Einstein, Albert, 337; and electrodynamics of moving bodies, 318-328; and ether theory of light, 101; "experiment" on photoelectric effect, 235-237; life and work of, 232-234;

on Maxwell, 283; and Planck's quantum hypothesis, 304, 329, 348; and principle of relativity, 317, 318-328

Einstein's photoelectric equation, 236; confirmed by Millikan, 240

elastic fluid, vibrations of, 112, 114

elastic limit, 86

elasticity, Boyle's definition of, 38; Young's work on, 95

electret, 292

electric balance, 65, 67

electric charge, inertia of, 218

electric conflict, 124, 125, 126, 127

electric fluid, and magnetic fluid, 66

electric motor, Faraday, 129

electric waves, Hertz' experiment on, 184-185, 186; Maxwell's prediction of, 283, 285

electrical charge, Millikan's experiment on, 240-249

electrical impulses, 114

electrical nature of matter, 216

electricity, 63; early experiments in, 59; evolution of, from magnetism, 136-148; induction of static, 130; measurement of force of, 62-74; in motion, 316; one-fluid theory of, 60, 65, 148; quantity of, in atoms, 152, 155-157. *See also* Coulomb, Charles; Faraday, Michael; Franklin, Benjamin

electricity of tension, 131, 136

electrified-particle theory of radiation, 219

electrochemical decomposition, 146-148

electrochemical equivalent, 154

electrode, and Faraday, 146, 148, 149, 150, 151; and ionization potential of gas, 237

electrodynamic induction, Hertz, 161, 162, 164, 165

electrolysis, Faraday's experiment on, 130, 148-157; and Grove, 173

electrolytes, Faraday, 147, 153-157

electromagnetic field, 184; Maxwell's work on, 283, 285, 287-300; Oersted's work on, 123-127

electromagnetic induction, Faraday's experiment on, 130, 131-148; Maxwell's work on, 296-300

electromagnetic radiation, 192

electromagnetic waves, Hertz' experiment on, 187-197

electromagnetism, theory of, 184, 283

electrometer, Faraday's gold-leaf, 142

electromotive force, 173

electron, determining its charge and mass, 238-239; emission of, 233; energy of, 233; identified as the unit of electric charge, 217;

Thomson's discovery of, 219-231, 238. *See also* Millikan, Robert

electron mass, compared to mass of hydrogen ion, 231

electron radius, 264

electroscope, gold-leaf, 220

electrostatic effects, Gilbert's experiments on, 59

electrostatic field, and deflection of cathode rays, 221-231

electrostatic voltmeter, 248

electrovalence, 153

element, in chemistry, Boyle's definition of, 37

elementary particle, 266

elements, doctrine of four basic, 5, 6, 8

elixir of life, 45

Ellis, C. D., 267

e/m experiment on chathode rays, Thomson, 228-231, 233

emission theory of light. *See* corpuscular theory of light

Emmanuel College, 94

energy, of a body, 318; conservation principle, Lenz's law, 159, 160; conservation principle, and von Helmholtz, 284; of fluids, 171; Galileo's notion of, 22; Young's work on, 95. *See also* conservation of energy

energy density, 308

energy distribution law, Planck, 310-313; Wien, 306, 307, 310-313

energy levels of gases, 344

energy radiation, density of in air, 312-313

Enlightenment, Age of, and physics, 42

entropy, Boltzmann's definition of, 308; Planck's concept of, 304, 306, 307-309, 311, 312; thermodynamic definition of, 306-307; and Wien's displacement law, 310-313

equilibrium, motion of a body in, 53

ether, light vibrations of, 112; Maxwell's work in unity of, 283, 285, 286, 287; physical properties assigned to, 113

"ether drift" and velocity of light, 317-318

ether theory of light, 100-101, 102, 108, 113, 115, 220, 221

Euclid, 3, 320

experimentation, planned, 2

eye, accommodation of, Young's view on, 94

Factitious Airs, 76

Fahrenheit, Gabriel, 166

Fahrenheit scale, 88

falling bodies and Galileo, 18, 20-35; law of,

314, 316. *See* accelerated motion; Galileo, Galilei

Faraday, Michael, 127, 159, 160, 184, 286, 290, 291, 294, 295; concept of field, 130; experiment on electromagnetic induction, 130, 131-148, 163; experiment on electrolysis, 130, 148-157, 173; experiment on rotation, 162; life and work of, 128-131

Faraday effect, 130, 235

fathom, defined, 22

Feather, D., 272, 274

Fermi, Enrico, 266

field, Faraday's concept of, 130. *See also* electromagnetic field

Fitzgerald, G. F., 318n

Florance, 351

fluoresence of glass, and x-rays, 201-203, 205, 210, 216

fluorescent or characteristic lines in x-ray spectrum, 355

fluorescent radiation, 349, 350

flux, 137, 297, 298

fluxions, Newton's discovery of, 43

Forbes, James D., 172

force, lines of, 184, 286, 294, 298-300

force, Newton's concept of, 52, 53

Fourier, Jean Baptiste Joseph, 113, 344

Fowler, A., 340, 341

Frank, J., 346

Franklin, Benjamin, 60, 65, 121, 148

free, or natural, motion, 18, 23

Fresnel, Augustin, 95, 319; experiment on diffraction of light, 110-120; life and work of, 108-110

Fresnel lenses, 110

Fresnel zones, 109n

friction, molecular forces of, 292; of water, Joule's experiment on, 178-182

frictional effects, 130

frictional electricity, 60

frictional machines, in early electrical experiments, 60

fringes of light, crested, 97, 100

Frisch, Otto, 332

Fundamental Law of Electricity, Cavendish, 63ff

funiculus, 40

Galileo, Galilei, 9, 11, 54, 166, 316; "anticlerical" views of, 15-16; experiments on accelerated motion, 18-35; life and work of, 13-18

Galvani, Luigi, 121

galvanic apparatus, in Oersted's experiment, 123, 124, 126

galvanometer, Faraday, 132, 133, 135

gamma radiation, 267

gamma rays, 211, 214, 259, 350

gas discharges, 216-217

gases, conduction of, 217; electrical conduction in, 251; Maxwell's work on, 285

Geiger, Hans, 267, 333

Geiger counter, 252, 267, 268, 278

Geissler, Heinrich, 288

geometrical proof, Galileo, 30

German Physical Society, 305, 313

"German physics," 305

Gilbert, William, 59

Goldstein, Eugen, 216, 217

Göttingen, University of, 240

grain, 64

gravitation, Galileo, 24; universal, and Newton, 55, 75

gravitational force, measurement of, 75

gravity, force of, 91; Newton's laws of, 42, 43-44, 48-49, 55-58. *See also* Galileo, Galilei

Gray, Stephen, 351n

Greece, and origins of modern science, 1-6

green fluorescence, 216

Grimaldi, Francesco, 93, 97, 100, 105, 110

Grove, Sir William, 173

Guericke, Otto von, 37, 60

Haas, A. E., 337

Halley, Edmund, 44

Hallwachs, Wilhelm, 233

Hahn, Otto, 266

Hankel electroscope, 208

harmonic analysis, 113

heat, capacity for, 166, 167-168, 170, 180, 181; conduction of, 178; Joule's mechanical theory of, 168-183; radiation, 178; Rumford's work in, 167-168; specific, 170, 180. *See also* calorimetry

heating effect, of cathode rays, 226; of x-rays, 203

Heidelberg, University of, 290

Heisenberg, W., 316n

helium nucleus, identified as alpha particle, 252

Hellenistic science. *See* Greece

Helmholtz, Hermann von, 159, 185, 284, 294, 304

Helmholtz coils, 225, 229

heliocentric theory, Copernicus, 9

Henry, Charles, 212

Henry, Joseph, 130, 159

Hero of Alexandria, 6
Hertz, G., 346
Hertz, Heinrich, 205, 222, 232-233, 283, 285; experiment on electromagnetic waves, 187-197; life and work of, 184-186
Hittorf, J. W., 216-217
Hittorf-Crookes tubes, 199, 201
Hooke, Robert, 44, 93
horror vacui, 36
Huggins, Sir William, 328
Huygens, Christian, 93, 107, 108, 112-114, 117, 119
hydrogen, as component of water, 76; and Rutherford's experiment on alpha particles, 253-259; spectrum of, 252, 330, 334-348. *See also* Rydberg constant
hydrogen atom, Bohr's work on, 251, 333-348; Rutherford's experiment on, 253-259
hydrogen ion, mass of, 231
hydrogen isotope, 278
hydrostatics, Galileo's treatise on, 15
hyle, 5
hypotheses non fingo, 46n

ice-pail experiment, Faraday, 142-146
Imperial Academy of Sciences, 160
inclination, or dip, 125
inclined plane, Galileo's definition of height of, 27
induced currents, 134, 188
induction, defined, 4; mutual, 159; Newton's first generalization of, 43, 57; principle of, *see* Lenz's law
induction, electromagnetic, Faraday's experiment on, 132-136; Maxwell's work on, 294, 296-300
induction, static electric, Faraday's letter on, 142-146
inductive charge, 145
Industrial Revolution, and science, 121, 168
inertia, early principle of, 14; and Galileo, 24; moment of, 83; and Newton, 47-48, 52, 56
inertial systems, 317, 320, 323; defined, 50
Institute for Advanced Studies, Princeton, 234
L'Institut de France, 10, 211
Institute for Theoretical Physics, 332
instruments, development of, and modern physics, 9, 10-11
integral calculus, Newton's discovery of, 43
interference phenomena of stationary light waves, 189
interference principle of light, Fresnel, 110-112, 115, 116, 119

inverse-square law, Coulomb, 61, 65, 121; Cavendish, 77n, 84
ionization, 223, 241
ionization chamber, 268, 270, 271
ionization potential, of helium atom, 342; of hydrogen, 337, 342; of gas, 237
ionizing power of radiation, 215
isotopes, discovery of by Thomson, 218; and neutron, Chadwick, 276, 278

Jeans, James, 303
Jenkins, F., 284, 292
Joliot-Curie, Frédéric, 266, 267-268, 278, 279
Joliot-Curie, Irène, 266, 267-268, 278, 279
Johns Hopkins University, 239
Joule, James, 159; experiment on mechanical equivalent of heat, 169-183; life and work of, 166-169

Kaiser Wilhelm Institute, 234, 305
Kay, W., 256, 265
Kelvin, Lord. *See* Thomson, Sir William
Kepler, Johannes, 15, 44; and binding energy of electron, 335
Kiel, University of, 186, 304
kinetic energy of an electron, 235, 236; of the alpha particle of polonium, 275
kinetic theory and Einstein, 234
kinetic theory of gases, 171, 216, 284, 308
kinetic theory of matter, and Einstein, 232; and Millikan, 238, 242
King's College, 284
Kirchhoff, Gustav R., 185, 304, 310, 329
Kirchhoff-Clausius law, 310, 346
Klein, O., 358
Klein-Nischina formula, 269, 358
Kolaček, F., 192
Kries, Johan v., 309
Kundt, August, 199
Kurlbaum, F., 305, 306, 312

Laplace, Pierre, 109n, 110
latent charge, 145
latent heat, 166, 167-168
Laue, Max von, 303-304
Lavoisier, Antoine, 40
Lawrence, Ernest, 266
lead, atomic number of, 203
Lee, J. Y., 242n
Lees, Charles Herbert, 272, 274
Legendre, Adrien Marie, 109
Leibnitz, Gottfried Wilhelm von, 44
Lenard, Philipp, 74, 122, 186, 200, 205, 207, 214, 231, 233, 235, 236, 237

Lenard tube, 201

Lenz, Heinrich, experiment on induction, 160-165; life and work of, 159-160

Lenz's law, 159-160, 161-162, 296

Leonard, W. E., 6n

Leucippos, 5

Leyden jar, 133, 135, 142, 292

light, Becquerel's work on, 211; coherent, 102; double-slit experiment on, Young, 102-107, 110-111; interference of, Young, 93, 94, 95, 96-107, 110; and sound, 100; theories of, 93-94; wave nature of, Hertz, 185, 186, 187-197; wave-particle theory of, 93-94, 97, 100, 101-102, 108, 110-112; wavelength theory of, Fresnel, 117-120. *See also* Compton effect; line spectra; Maxwell, James Clerk; Newton, Isaac; optics; spectroscopy

light, diffraction of, Fresnel's experiment on, 110-120

light, velocity of, 317-318, 321. *See also* electromagnetic radiation

Light and Colors, On the Theory of, 95

Lincean Academy, 9

line, value of, in Coulomb's experiment, 62

line spectra, 329-330; emission of, 339-343; of hydrogen, Bohr, 332-348

lines of force, 126, 130, 140, 143, 184, 286, 294, 299

Linus, Franciscus, 40

"liquid drop" model of nuclear phenomena, 332

logic, 2; deductive, 2, 3-4; inductive, 4

London Institution, 173

Lorentz, Hendrick A., 108, 301, 318n, 319

Lorentz-Fitzgerald "contraction" hypothesis, 318n

Louis XIV, and scientific academies, 10

Lucasian chair of mathematics, 43, 44

Lucretius, 6

luminiferous ether, 100, 286

Lummer, O., 305, 306, 307n, 311

Lyman series in hydrogen spectrum, 330, 340

magnetic balance, 67

magnetic bodies, attraction of, 66

magnetic effect, of a current, 129; of earth, 125

magnetic equator, 147

magnetic field, 122, 123, 161, 164; Fresnel's measurement of, 123-127; Newton's work on, 60. *See also* electromagnetism

magnetic fluid, 67, 69, 70, 71; and electric fluid, 66

magnetic flux, 297-298

magnetic force, 59, 66-73

magnetic meridian, 68, 69, 70, 125, 126

magnetic moment, 277

magnetic permeability, 295

magnetic spectrum, 223

magnetism, evolution from electricity, Faraday's experiment, 136-146; induced, 86; and Newton, 48; terrestrial, 59

magnetometer, 68

Manchester, University of, 252, 267

Mariotte, Edmé, 38, 55

Marsden, 253-254, 255, 259, 333n

Maskelyne, Nevil, 92

mass, of an ion, compared to oil drop, 244; Newton's definition of, 46-47, 53

mass defect, 269, 277

mass-energy relation, $E = mc^2$, 347. *See also* Einstein, Albert

mathematics, and deduction, Galileo, 17, 34

matter, Aristotle's view of, 5; atomic theory of, 216; Coulomb's view of, 66-67; electrical theory of, 216; modern view of, 56, 280; Newton's view of, 50; quantity of, Newton, 46

Maxwell, James Clerk, 77, 108, 127, 184-185, 186, 192, 318, 319, 338; "experiment" on electromagnetic field, 287-296; experiment on electromagnetic induction, 296-300; life and work of, 283-287; theory of velocity of light, 317-318. *See also* Hertz, Heinrich

Maxwell-Wagner mechanism, 295

Maxwellian distribution, 284

Maxwell's equations, 283, 294

Mayer, Julius R., 168, 173, 174

McGill University, 252

mean free path, of cathode-rays, 223, 231, 242

mechanical equivalent of heat, Joule's experiment on, 169-183

mechanics, development of, 128; of fluids, 36; Galileo's work in, 15; Newton's interest in, 44

Medici, Grand Duke Ferdinand II, 9

Medici, Grand Duke Leopold, 9

Medici family, and scientific academies, 9-10

Meitner, Lise, 332

Melvill, Thomas, 329

mercury pump, 230

meson, discovery of, 266

Meyer, O. E., 238

Michell, Rev. John, 61, 77, 78, 80

Michelson, A. A., 317-318, 319

micrometer, 63, 64, 66

micron, size of, 313

microphysics, age of, 250

microscope, invented, 11
Middle Ages, and science, 6-8
Millikan, Robert A., 236; experiment on electrical charge, 240-249; life and work of, 238-240
Milton, John, on Galileo, 17
Minnesota, University of, 349
modern physics, origins of, 1-8; beginnings of, 9
modified rays, 356
Molar gas constant, 235
molecular theory. *See* matter
momentum, of an electron, Compton's experiment, 351; defined, 25; Galileo's notion of, 22, 25, 27; rate of change of, Newton, 47, 52. *See also* motion *entries*
monochromatic radiation, 308
Morley, E. W., 317-318, 319
motion, accelerated. *See* accelerated motion
motion, of earth, and Einstein, 319
motion, instantaneous, 26; physics of, 9; planetary, 15
motion, laws of and Galileo, 14, 17, 18, 223. *See also* Galileo's experiment
motion, laws of and Newton, 52-55; absolute, 50-52, 316; definitions of, 46-58; relative, 47, 50-52
motion, of liquids, 36
motion, simple harmonic, formula for, 71
motion, and theory of relativity, 313-318
movimenti locali, Galileo, 17-18
moving bodies, electrodynamics of. *See* Einstein, Albert
Müller, W., 267
Munich, Physical Institute at, 200
Munich, University of, 304
Murphy, G., 278
Musée d'Histoire Naturelle, 211
mysticism, and science, 6

natural motion, 18, 19, 24
natural philosophy, 7, 10
natural places, doctrine of, 5
natural radioactivity, Becquerel's experiment on, 198, 212-215
nature, classical views of, 5; scientific explanation of, 20
Nerst, W. H., 335
neutron, Chadwick's experiment on, 266, 267-280; compared to proton, 274; energy of, 275; mass, 274, 277, 278, 280; measured, 275-278, 280; hypothesis, 273-275; nature of, 275-280; penetration of, 270
Newton, Isaac, 15, 283; "experiment" on motion, 46-58; and gravity, definition of, 42, 43-44, 48-49, 55-58; and Einstein, compared, 232, 311; and induction, defense of, 57-58; and inertia, description of, 47-48, 52, 317; life and work of, 42-46; and magnetic field, 60; and mass, definition of, 46-47, 53; and motion, laws of, 52-55; and optics, theory of, 93, 97-99, 100, 102, 104, 106; philosophic reasoning of, 55-58; and theory of relativity, 315; and time, space, and motion, 50-52, 316
Newtonian equations, and Einstein's definition of synchronism, 321
Newtonian physics, 108
Newton's rings, 106
Niagara Falls, heat of, 171
Nicholson, J. W., 337, 338, 346
Nicol, William, 192
Nicol prisms, 192
Niewenglowski, 212
Nishina, Y., 358
nitric acid, Cavendish's discovery of, 76
nitrogen, anomalous effect in, Rutherford, 259-263
nitrogen atoms, hydrogen atoms from, 263
nitrogen nucleus, composition of, 264
Nobel prize, 186, 217, 218, 232, 240, 250, 252, 267, 304, 350
Nobili, Leopoldo, 141, 161, 163
Nobili galvanometer, 163
Norman Bridge Laboratory of Physics, 240
nuclear atom, Rutherford's model, 211, 250-251, 252
nuclear equation, 264
nuclear physics, development of, 211, 266; Thomson's contribution to, 218
Nutt, H., 280

Occhialini, 279
Oersted, Hans Christian, 129, 164, 184, 293; experiment on electromagnetism, 123-127; life and work of, 121-123
Ohm, George Simon, 184n
ohm, absolute, 284
oil-drop experiment, Millikan, 238-239, 240-249
one-fluid theory of electricity, Franklin, 60, 65, 148
optic nerve, and sight, Fresnel, 113
optics, Newton's contributions to, 43, 95, 97-99, 100, 102, 104
orbits, "allowed," 331
oscillations, 63
Oxford University, 76
oxide, scale of, and light, 106

oxygen, as component of water, 76

paddle-wheel experiment, Joule's, 169, 170
Padua, University of, 15
paramagnetic materials, 288, 290
Pascal, Blaise, 36
Paschen, F., 340
pendulum clocks, 11, 21, 55, 82, 113
permeability, 298
Perrin, Jean Baptiste, 216, 217, 219-220
Peterhouse College, 76
Pfund series, in hydrogen spectrum, 328
philosopher's stone, 7
Philosophical Transactions, 10
phlogiston theory, 76
phosphorescence, 210n, 214, 221, 223; of
   uranium salts, 212, 213; of zinc sulphide, 212
photoelectric current, 233
photoelectric effect, 187, 232-233, 304; Einstein's work on, 235-237, 329; Lenard's
   observations of, 235-236
photons, concept of, 240; energy and momentum of, 347. *See also* Compton effect
physical optics, 95
physical reality, 285, 287
physical theories, defined, 4
physics of motion, 9
Pickering, E. C., 341
Pictet, Marc, 174
pile. *See* voltaic cell
pinhole camera, 206
Pisa, University of, 16
Planck, Max, 237, 329, 331, 335-339, 343,
   346, 348; "experiment" on quantum hypothesis, 305-313; life and work of, 301-305;
   and Wien's displacement law, 310-313
Planck's constant, 235, 236, 303, 331, 335, 336,
   337, 349, 352, 354; determined by Millikan,
   240; applied by Bohr to Rutherford atom,
   252; values of, 312-313
Planck's distribution law, 312
planetary motion, 113
Plateau's experiments, 299
platinum, absorption peak of, 203; atomic
   number of, 203
platinum cathode for x-rays, 209
Plato, 23; and scholasticism, 7
Poisson, Simeon, 109n, 110
polarity, Cavendish, 86-87
polarity of light, 114n, 127, 191-192, 291
Polytechnic Institution, Copenhagen, 123
porous-plug experiment, Thomson, 169
positron, discovery of, 266

"practical" science. *See* applied science
Prague, University of, 234
Priestley, Joseph, 76
primary wave, in Fresnel's light theory, 115,
   116
Princeton University, 232, 349
Principia, 44, 45-48, 283; definition of motion
   in, 316
Pringsheim, E., 305, 306, 313
prism, and radioactive rays, 213; and refraction
   of light, 194-195, 329
projectiles, Galileo's theory of, 18, 317n

quantum of action, 303, 305. *See also* Planck's
   constant
quantum hypothesis, 301, 303, 304, 305, 308,
   315, 348; of scattering, Compton, 349-356.
   See also Compton effect; Planck, Max;
   Planck's constant
quantum mechanics, 331
quantum physics, and Planck, 279, 329, 331,
   346
quantum theory of light, 314; and Einstein,
   232, 233, 234, 236, 237

radiant energy. *See* quantum hypothesis;
   Planck, Max
radiation, Maxwell's principle of, 185; density
   of, 306, 307; nature of, 216-217; rectilinear
   propagation of, 216
radioactivity, and atomic structure, 250;
   Becquerel's work on, 212-215; discovery of,
   210-211; Rutherford's explanation of, 252
radium, discovery of, 210
Radium C, decay of, 254-255, 258, 259
radium-beryllium mixtures, as neutron sources,
   270
radon, 253, 254
rainbows, and interference of light, 99, 107
random phenomena, 309
Rayleigh, Lord. *See* Strutt, J. W.
Rayleigh-Jeans formula, 303
Réaumur, Ferchault de, 166
recoil atoms, Chadwick, 271-280
rectilinear propagation of light, 190-191
reference system, for theory of motion, 316
reflection of light waves, 192-193
refraction, in light theory, 100, 106, 107,
   194-195
refractive index of x-rays, 195-204
relative motion, 315-318; Newton's definition
   of, 47
relative time, Newton, 49

relativity, 47; Einstein's theories of, 232, 233, 234, 315, 319, 321, 332; general theory of, 318n; special theory of, 318-328

relativity principle, Galilean, 317

Renault, Henri Victor, 175, 180

Renaissance, and science in Italy, 8-9, 13

repulsion, law of, 64

repulsive force, 63

resistance, measurement of, 295

retrograde motion of particle, 114

right-hand rule, 126, 139, 164

Ritz, W., 330, 340

Ritz Combination Principle, 330, 342

Roemer, Olaus, 93

Roentgen, Wilhelm K., 210; and Becquerel's radiation, 214, 215, 216, 217; and cathode rays, 199; experiment on x-rays, 201-209; life and work of, 198-200; and magnetic effect, 199

Roentgen rays. See x-rays

Rome, and development of science, 6

Rossi coincidence technique, 279

Royal Institution, 94, 95, 126, 130, 218, 219n

Royal Society, 10, 37, 38, 44, 45, 76, 94, 95, 96, 122, 129, 130, 131, 169, 171, 174, 175, 186, 211, 218

Royds, T., 252

Rubens, H., 305, 306

Ruhmkorff, Heinrich, 188

Rules of Reasoning, Newton, 46

Rumford, Count. See Thompson, Benjamin

Rumford Medal, 110, 186

Rutherford, Ernest, 211, 266, 273, 274, 332, 333, 334; and atomic spectrum, 330-331; development of Geiger counter, 267; experiment on induced transmission, 253-265; life and work of, 250-252

Rydberg, J. R., 330

Rydberg constant for hydrogen, 330, 331, 334, 341, 342, 343

Ryerson Laboratory, 240, 242n

St. Petersburg, University of, 160

Saturn, Maxwell's study of rings of, 284, 338

Sceptical Chymist, The, 37

Scholasticism, 7-8

science, and common sense, 6; defined by da Vinci, 4-5; and Greek culture, 1; history of, 1-11; in the Middle Ages, 6-8; origin of modern, 1-6; in the Renaissance, 8-9; revolution in, 8-11. See also chemistry; modern physics

science, and religion, 6, 7, 8; Boyle's views on, 38; Galileo's views on, 16-17

science education, lack of in England, 76

scientific explanation, 3, 8, 26; economy of, 55

scientific method, Galileo's contributions to, 17, 29

scientific notation, 78, 172

scientific revolution, 6

secondary light waves, 117

sector, made by Galileo, 15

Séguin, Marc, 174

shadow, and light-wave theory, 96, 97, 99, 104, 105, 109, 115, 118, 119; formation of, 204, 205, 206

Sharp series. See Fowler, A.

Shimizu-type expansion chamber, 272

shock waves, 114

soap-bubble analogy, 299

Socrates, 2, 3; and Scholasticism, 7

Soddy, F., 250, 252

Sommerfeld, A., 337

sound waves, nature of, and light, 100, 112, 114, 189

space, absolute and relative, Newton, 50

span, 22

Spanish wax, 63

spark gap, 233

specific heat, 166, 170; of ice and water, 173

spectrograph, mass, 218, 278

spectroscopy, experimental, 329-332; in Compton's experiment, 351, 354

spectrum of hydrogen, emission of, 339-343. See also Bohr, Niels

Sprengel pump, 230

Stark, Johannes, 237, 337

standing waves, 189

stars, motion of, 51; and light, 100, 101

static electric induction, Faraday's letter on, 142-146

stationary states, 331, 341, 345, 346

Stefan, Josef, 302

Stefan-Boltzmann law, 302, 310

Stevin, Simon, 14

Steward, B., 284

Stirling's formula, 309

Stokes, George, 108

Stokes' law, 241, 242, 245, 249

Stoney, G. J., 238

stopping power, 259, 260, 270

Strasbourg, University of, 199

Strassmann, Fritz, 266

Strutt, J. W., 302-303

sun's spectrum, 236

swings, method of, 82

syllogism, defined, 2

synchronism, Einstein's definition of, 320-321. *See also* time, relativity of

Tartaglia, 13
Technische Hochschule, Karlsruhe, 186
teleological argument, 4
telescope, invented, 11, 15; reflecting, invented by Newton, 43
telescopes, Newton's work on, 43
temperature effect, 88; and Cavendish, 90
theology, and Robert Boyle, 38
thermal capacity of thermometer, 181
thermal radiation, and diathermic medium, 310
thermodynamics, kinetic theory of, 234
Thermodynamics, Planck's Second Law of, 304, 306-307
thermometer, Cavendish's use of, 89, 90; invented, 11; in Joule's experiment, 181
thermometry, 166
Thiesen, M., 310
Thompson, Benjamin, 167-168, 172, 174
Thomson, Sir J. J., 186, 217, 238, 241, 251, 332, 333, 337, 339n, 348, 350-351; experiment on electron, 219-231; experiment on e/m, 228-231; life and work of, 216-219
Thomson, Sir William, 169, 289, 290, 294
thorium, radiation properties of, 210
Townsend, J. S. E., 238
tides, Young's work on, 95
time, Galileo's measurement of, 21, 34-35; Newton's definitions of, 49, 316; relativity of, 320-328. *See also* Einstein, Albert; relativity; synchronism
time intervals, and accelerated motion, 20-21
Toepler pump, 230
torque, 64
Torricelli, Evangelista, 36, 41
Torricelli vacuum experiment, 40, 41
torsion, laws of, Coulomb, 61-62, 64
torsion balance, invented by Coulomb, 61; construction of, 61-65; and Cavendish, 77
transformation equations, Einstein, 324, 325, 327
transmutation, induced, Rutherford's experiment on, 253-265
transparency, Roentgen, 201n, 202, 203, 204; of x-rays and cathode rays compared, 204
Trinity College, 42, 43, 218, 251, 252, 284
truss, 78
Tunstall, N., 255
Tyndall, John, 302

ultrared rays, 206

ultraviolet light, and photoelectric effect, 233; and x-rays, 201, 206
unified field theory, Einstein, 234
uniform acceleration, defined by Galileo, 21
uniform motion, Galileo's definition of, 19
unipolar induction, 139
unmodified rays, 356
uranium, fission of, 266, 332
uranium salts, and radiation, 210, 212, 213-215
Urey, Harold C., 266, 278

vacuum pump, invention of, 37, 40
Van de Graaff generator, 266
velocity, instantaneous, 26
velocity of light, 190, 285, 289, 295; absolute, 317
velocity, uniform or constant, Galileo's definition of, 19, 20, 25; in modern notation, 120
Verdet, Émile, 290
Verdet's constant, 290
vernier, 81
Victoria College, 254
Vinci, Leonardo da, 4-5, 9, 14
violent motions, Aristotle's theory of, 25; Galileo's theory of, 18, 20-35
vis inertia, 47, 48, 56, 57
vis insita, 47, 56, 57
viscosity, coefficient of, Stokes' law, 245
visible rays, 206
Volta, Alessandro, 121
volta electricity, Faraday's experiment on, 149-155. *See also* electrolysis
volta-electrometer, 156
voltage-multiplying generator, 266
voltaic cell or pile, 121, 122; Faraday's, 130
voltaic electricity, 136, 149, 150, 153

water, composition of, 76; density of, 91; friction of, 182; Joule's experiment on friction of, 178-182; and light, 105
Walton, E., 266
Washington University, St. Louis, 349-350
water-clock, Galileo's use of, 14, 34-35
Watt, James, 76
wave-particle theory of light, 93-94, 97, 100, 101-102, 108, 109, 110-112, 348, 349. *See also* Compton effect; Fresnel, Augustin; Huygens, Christian; Young, Thomas
wavelength theory of light, 103, 110-120. *See also* Fresnel, Augustin; light
Weber, Wilhelm, 287, 295
Webster, 268
Wheatstone, Sir Charles, 136, 171
Wheatstone bridge, 156

Wheeler, John, 332
white light, 111n
"white-light spectrum," 310
Wien, Wilhelm, 302, 306, 307
Wien's displacement law, 302, 306, 307; and
    Planck, 310-312
Wilson, C. T. R., 241, 349
Wilson, H. A., 238, 241
Wollaston, Rev. Francis John Hyde, 77-78
Wollaston, William Hyde, 129
Wooster, College of, 349
Wren, Christopher, 55
Würzburg, University of, 199

x-rays, 201-209, 334, 348-349; absorption of,
    202; and Becquerel's work on radiation, 214,
    215; Compton's work on, 350-358; fluo-
rescence of, 201-203; heating effect of, 203-
204; interference phenomena of, 206; and
positive and negative electric charges,
208-209; transparency of, compared to
cathode rays, 204. *See also* radiation;
Roentgen, Wilhelm K.

Young, Thomas, 108, 110-111, 112; experiment
    on interference of light, 96-107; life and
    work of, 94-95
Young's modulus, 95

Zeeman, Pieter, 130, 301
Zeeman effect, 301-302
zinc sulphide, and radiation, 212
Zurich, University of, 199
Zurich Polytechnic, 234